PARAMETER ESTIMATION AND INVERSE PROBLEMS

PARAMETER ESTIMATION AND INVERSE PROBLEMS

Third Edition

RICHARD C. ASTER
Colorado State University
Fort Collins, CO, USA

BRIAN BORCHERS
New Mexico Institute of Mining and Technology
Socorro, NM, USA

CLIFFORD H. THURBER
University of Wisconsin-Madison
Madison, WI, USA

ELSEVIER

Elsevier
Radarweg 29, PO Box 211, 1000 AE Amsterdam, Netherlands
The Boulevard, Langford Lane, Kidlington, Oxford OX5 1GB, United Kingdom
50 Hampshire Street, 5th Floor, Cambridge, MA 02139, United States

Notices

Knowledge and best practice in this field are constantly changing. As new research and experience broaden our understanding, changes in research methods, professional practices, or medical treatment may become necessary.

Practitioners and researchers must always rely on their own experience and knowledge in evaluating and using any information, methods, compounds, or experiments described herein. In using such information or methods they should be mindful of their own safety and the safety of others, including parties for whom they have a professional responsibility.

To the fullest extent of the law, neither the Publisher nor the authors, contributors, or editors, assume any liability for any injury and/or damage to persons or property as a matter of products liability, negligence or otherwise, or from any use or operation of any methods, products, instructions, or ideas contained in the material herein.

Library of Congress Cataloging-in-Publication Data
A catalog record for this book is available from the Library of Congress

British Library Cataloguing-in-Publication Data
A catalogue record for this book is available from the British Library

ISBN: 978-0-12-804651-7

For information on all Elsevier publications
visit our website at https://www.elsevier.com/books-and-journals

Working together
to grow libraries in
developing countries

www.elsevier.com • www.bookaid.org

Publisher: Candice Janco
Acquisition Editor: Marisa LaFleur
Editorial Project Manager: Katerina Zaliva
Production Project Manager: Vijayaraj Purushothaman
Designer: Greg Harris

Typeset by VTeX

CONTENTS

PREFACE TO THE THIRD EDITION

This textbook evolved from courses in geophysical inverse methods taught by the authors at New Mexico Tech (R.A., B.B.), the University of Wisconsin–Madison (C.T.), and Colorado State University (R.A.). Reflecting the wide applicability of the subject matter, the audience for these courses has encompassed a broad range of first- or second-year graduate students (and occasionally advanced undergraduates) from geophysics, hydrology, mathematics, astrophysics, engineering, and other disciplines.

Our principal goal for this text continues to be introductory to intermediate level philosophical and methodological understanding of parameter estimation and inverse problems, specifically regarding such key issues as uncertainty, ill-posedness, regularization, bias, and resolution. The accompanying exercises include a mix of applied and theoretical problems.

We emphasize key theoretical points and applications with illustrative examples. MATLAB codes and associated data that implement these examples are available in a GitHub repository at https://github.com/brianborchers/PEIP. We welcome questions, comments, and suggested improvements to the code. The margin icon shown here also indicates where associated code and/or data associated with exercises are available in the repository.

This book has necessarily had to distill a tremendous body of mathematics going back to (at least) Newton and Gauss. We hope that it will continue to find a broad audience of students and professionals interested in the general problem of estimating physical models from data. Because this is an introductory text surveying a very broad field, we have not been able to go into great depth. However, each chapter has a "notes and further reading" section to help guide the reader to further exploration of specific topics.

We expect that readers of this book will have prior familiarity with calculus, differential equations, linear algebra, probability, and statistics at the undergraduate level. In our experience, many students benefit from a review of these topics, and we commonly spend the first two to three weeks of the course reviewing material from Appendices A, B, and C.

Some advanced topics have been left out of this book because of space limitations and/or because we expect that many readers would not be sufficiently familiar with the required mathematics. For example, some readers with a strong mathematical background may be surprised that we primarily consider inverse problems with discrete data and discretized models. By doing this we avoid much of the technical complexity of functional analysis. Some of the important advanced applications and topics that we have necessarily excluded are inverse scattering problems, seismic diffraction tomography, data assimilation, simulated annealing, and expectation maximization (EM) methods.

Chapters 1 through 4 form the heart of the book, and should be covered in sequence. Chapters 5 through 8 are somewhat independent of each other. These may be covered in any order, but draw heavily on Chapters 1 through 4. Chapters 9 and 10 are independent of Chapters 5 through 8, but are most appropriately covered in sequence. Chapter 11 is substantially independent of the material in Chapters 5 through 10, and provides an introduction to the Bayesian perspective on inverse problems and Bayesian solution methods.

If appreciable time is allotted for review of linear algebra, vector calculus, probability, and statistics topics in the appendices, there will probably not be time to cover the entire book in one (e.g., 16-week) university semester. Regardless, it should be possible for instructors to cover selected material beyond Chapter 4 in most cases.

We especially wish to acknowledge our own professors and mentors in this field, including Kei Aki, Robert Parker, Robert Meyer, and Peter Shearer. We also thank our colleagues, including a great many students from our courses, who have provided sustained encouragement and feedback from the initial drafting through to the subsequent revisions of the book. These individuals have included Kent Anderson, James Beck, Elena Resmerita, Charlotte Rowe, Tyson Strand, Aaron Master, Jason Mattax, and Suzan van der Lee. Stuart Anderson, Greg Beroza, Ken Creager, Don Clewett, Ken Dueker, Eliza Michalopoulou, Paul Segall, Anne Sheehan, Carl Tape, and Kristy Tiampo deserve special mention for their classroom testing of early and subsequent versions of this text and their associated helpful suggestions for improvement. Robert Nowack, Gary Pavlis, Randall Richardson, and Steve Roecker provided thorough and helpful reviews during the scoping of this project. We also offer special thanks to Per Christian Hansen of the Technical University of Denmark for the authorship and dissemination of his Regularization Tools, which were an inspiration in writing this book.

Additional feedback that improved subsequent editions has been provided by Carl Tape, Ken Dueker, Anne Sheehan, Pamela Moyer, John Townend, Frederik Tilmann, and Kurt Feigl. Ronni Grapenthin, who co-taught the course with Brian Borchers, and Oleg Makhnin, who co-taught with Rick Aster, in both instances at New Mexico Tech, provided additional valuable contributions. We also thank the editorial staff at Elsevier over the years, especially Frank Cynar, Kyle Sarofeen, Jennifer Helé, John Fedor, Marisa LaFleur, and Katerina Zaliva. Suzanne Borchers and Susan Delap provided valuable proofreading and graphics expertise during the initial drafting of the book. The work of Brian Borchers on the third edition was supported in part by the National Science Foundation under Grant DMS-1439786 while he was in residence at the Institute for

Computational and Experimental Research in Mathematics in Providence, RI, during the Fall 2017 program on Mathematical and Computational Challenges in Radar and Seismic Reconstruction.

Rick Aster
Brian Borchers
Cliff Thurber

March 2018

CHAPTER ONE

Introduction

Synopsis

General issues associated with parameter estimation and inverse problems are introduced through the concepts of the forward problem and its inverse solution. Scaling and superposition properties that characterize linear systems are given, and common situations leading to linear and nonlinear mathematical models are discussed. Examples of discrete and continuous linear and nonlinear parameter estimation problems to be revisited in later chapters are shown. Mathematical demonstrations highlighting the key issues of solution existence, uniqueness, and instability are presented and discussed.

1.1. CLASSIFICATION OF PARAMETER ESTIMATION AND INVERSE PROBLEMS

Scientists and engineers frequently wish to relate physical parameters characterizing a **model**, m, to observations making up some set of **data**, d. We will commonly assume that the fundamental physics are adequately understood, and that a function, G, may be specified relating m and d such that

$$G(m) = d \, . \tag{1.1}$$

In practice, d may be a function or a collection of discrete observations.

Importantly, real-world data always contain noise. Two common ways that noise may arise are unmodeled influences on instrument readings and numerical roundoff. We can thus envision data as generally consisting of noiseless observations from a "perfect" experiment, d_{true}, plus a noise component η,

$$d = G(m_{\text{true}}) + \eta \tag{1.2}$$

$$= d_{\text{true}} + \eta \tag{1.3}$$

where d_{true} exactly satisfies (1.1) for m equal to the true model, m_{true}, and we assume that the forward modeling is exact. We will see that it is commonly mathematically possible, although practically undesirable, to also fit all or part of η by (1.1). It is, remarkably, often the case that a solution for m that is influenced by even a proportionally very small amount of noise can have little or no correspondence to m_{true}. Another key issue that may seem surprising is that there are commonly an infinite number of models aside from m_{true} that fit the perfect data, d_{true}.

When m and d are functions, we typically refer to G as an **operator**. G will be called a function when m and d are vectors. G can take on many forms. In some cases,

Parameter Estimation and Inverse Problems
https://doi.org/10.1016/B978-0-12-804651-7.00006-7

G is an ordinary differential equation (ODE) or partial differential equation (PDE). In other cases, G is a linear or nonlinear system of algebraic equations.

Note that there is some inconsistency between mathematicians and other scientists in the terminology of modeling. Applied mathematicians usually refer to $G(m) = d$ as the "mathematical model" and to m as the "parameters." On the other hand, scientists often refer to G as the "forward operator" and to m as the "model." We will adopt the scientific parlance and refer to m as the "model" while referring to the equation $G(m) = d$ as the "mathematical model."

The **forward problem** is to find d given m. Computing $G(m)$ might involve solving an ODE or PDE, evaluating an integral equation, or applying an algorithm to m for which there is no explicit analytical formulation. Our focus in this textbook is the **inverse problem** of finding m given d. A third problem, not addressed here, is the **model identification problem** of determining G given examples of m and d.

In many cases, we will want to determine a finite number of parameters, n, that define a model. The parameters may define a physical entity directly (e.g., density, voltage, seismic velocity), or may be coefficients or other constants in a functional relationship that describes a physical process. In this case, we can express the model parameters as an n element vector $\mathbf{m}$. Similarly, if there are a finite number of data points then we can express the data as an m element vector $\mathbf{d}$ (note that the frequent use of the integer m for the number of data points is easily distinguishable from a model m by its context). Such problems are called **discrete inverse problems** or **parameter estimation problems**. A general parameter estimation problem can be written as a system of equations

$$G(\mathbf{m}) = \mathbf{d} . \qquad (1.4)$$

Alternatively, where the model and data are functions of continuous variables, such as time or space, the task of estimating m from d is called a **continuous inverse problem**. A central theme of this book is that continuous inverse problems can often be well-approximated, and usefully solved, as discrete inverse problems.

We will generally refer to problems with small numbers of parameters as "parameter estimation problems." Problems with a larger number of parameters, and which will often require the application of additional stabilizing constraints, will be referred to as "inverse problems." The need for stabilizing solution constraints arises because the associated systems of equations or operations are ill-conditioned, in a sense that will be discussed later in this chapter. In both parameter estimation and inverse problems we solve for a set of parameters that characterize a model, and a key point of this textbook is that the treatment of all such problems can be generalized. In practice, what is most important is the distinction between ill-conditioned and well-conditioned parameter estimation problems.

A class of mathematical models for which many useful results exist is **linear systems.** Linear systems obey superposition

$$G(m_1 + m_2) = G(m_1) + G(m_2) \tag{1.5}$$

and scaling

$$G(\alpha m) = \alpha G(m) . \tag{1.6}$$

In the case of a discrete linear inverse problem, (1.4) can always be written in the form of a linear system of algebraic equations

$$G(\mathbf{m}) = \mathbf{Gm} = \mathbf{d} \tag{1.7}$$

where $\mathbf{G}$ is a matrix (see Exercise 1.1). In a continuous linear inverse problem, G can often be expressed as a linear integral operator, where (1.1) has the form

$$\int_a^b g(x, \xi) \, m(\xi) \, d\xi = d(x) \tag{1.8}$$

and the function $g(x, \xi)$ is called the **kernel**. The linearity of (1.8) is easily demonstrated because

$$\int_a^b g(x, \xi)(m_1(\xi) + m_2(\xi)) \, d\xi = \int_a^b g(x, \xi) \, m_1(\xi) \, d\xi + \int_a^b g(x, \xi) m_2(x) \, d\xi \tag{1.9}$$

and

$$\int_a^b g(x, \xi) \, \alpha m(\xi) \, d\xi = \alpha \int_a^b g(x, \xi) \, m(\xi) \, d\xi . \tag{1.10}$$

Equations in the form of (1.8), where $m(x)$ is the unknown, are called **Fredholm integral equations of the first kind (IFK)**. IFK's arise in a surprisingly large number of inverse problems. An important point is that these equations have mathematical properties that make it difficult to obtain useful solutions by straightforward methods.

In many cases the kernel in (1.8) can be written to depend explicitly on $x - \xi$, producing a **convolution equation**

$$\int_{-\infty}^{\infty} g(x - \xi) \, m(\xi) \, d\xi = d(x) . \tag{1.11}$$

Here we have written the interval of integration as extending from minus infinity to plus infinity, but other intervals can easily be accommodated by having $g(x - \xi) = 0$ outside of the interval of interest. When a forward problem has the form of (1.11), determining $d(x)$ from $m(x)$ is called **convolution**, and the inverse problem of determining $m(x)$ from $d(x)$ is called **deconvolution**.

Another IFK arises in the problem of inverting a **Fourier transform**

$$\Phi(f) = \int_{-\infty}^{\infty} e^{-i2\pi fx} \phi(x) \ dx \qquad (1.12)$$

to obtain $\phi(x)$. Although there exist tables and analytic methods of obtaining Fourier transforms and their inverses, numerical estimates of $\phi(x)$ may be of interest, such as when there is no analytic inverse or when we wish to estimate $\phi(x)$ from spectral data collected at discrete frequencies.

Why linearity arises in many interesting physical problems is an intriguing question. One reason is that the physics of many interesting problems is associated with only small departures from equilibrium. An important geophysical example of this kind is seismic wave propagation, where the stresses associated with elastic wavefields are often very small relative to the elastic moduli of rocks. This creates only small strains and a very nearly linear stress-strain relationship. Because of this, seismic wavefield problems in many useful circumstances obey superposition and scaling. Other physical fields, such as gravity and magnetism, also have effectively linear physics at the field strengths typically encountered in geophysics.

Because many important inverse problems are linear, and because linear theory is a key component of nonlinear problem solving methods, Chapters 2 through 8 of this book extensively cover theory and methods for the solution of linear inverse problems. Nonlinear mathematical models arise when model parameters have an inherently nonlinear relationship with respect to data. This situation occurs, for example, in electromagnetic field problems where we wish to relate geometric model parameters such as layer thicknesses to observed field properties. We discuss methods for nonlinear parameter estimation and inverse problems in Chapters 9 and 10, respectively.

1.2. EXAMPLES OF PARAMETER ESTIMATION PROBLEMS

Example 1.1

A canonical parameter estimation problem is the fitting of a function, defined by a collection of parameters, to a data set. In cases where this function fitting procedure can be cast as a linear inverse problem, the procedure is referred to as **linear regression**. An ancient example of linear regression is the characterization of a ballistic trajectory. In a basic take on this problem, the data $\mathbf{y}$ are altitude observations of a ballistic body at a set of times $\mathbf{t}$ (Fig. 1.1). We wish to solve for a model, $\mathbf{m}$, that contains the initial altitude (m_1), initial vertical velocity (m_2), and effective gravitational acceleration (m_3) experienced by the body during its trajectory. This and related problems are naturally of practical interest in aeronautics and weaponry, but are also of fundamental geophysical

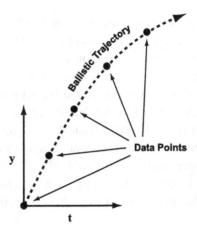

Figure 1.1 The parabolic trajectory problem.

interest, for example, in absolute gravity meters capable of estimating g from the acceleration of a falling object in a vacuum to accuracies on the order of one part in 10^9 or better [115].

The mathematical model is a quadratic function in the $(t,\ y)$ plane

$$y(t) = m_1 + m_2 t - (1/2)m_3 t^2 \tag{1.13}$$

that we expect to apply at all times along the trajectory (and not just at the times t_i when we happen to have made observations). The data consist of m observations of the height of the body y_i at corresponding times t_i. Assuming that the t_i are measured precisely, and applying (1.13) to each observation, we obtain a system of equations with m rows and $n = 3$ columns that relates the data y_i to the model parameters, m_j

$$\begin{bmatrix} 1 & t_1 & -\frac{1}{2}t_1^2 \\ 1 & t_2 & -\frac{1}{2}t_2^2 \\ 1 & t_3 & -\frac{1}{2}t_3^2 \\ \vdots & \vdots & \vdots \\ 1 & t_m & -\frac{1}{2}t_m^2 \end{bmatrix} \begin{bmatrix} m_1 \\ m_2 \\ m_3 \end{bmatrix} = \begin{bmatrix} y_1 \\ y_2 \\ y_3 \\ \vdots \\ y_m \end{bmatrix}. \tag{1.14}$$

Although the mathematical model relating y and t (1.13) is quadratic, the equations for the three parameters m_i in (1.14) are linear, so solving for $\mathbf{m} = [m_1,\ m_2,\ m_3]^T$ is a linear parameter estimation problem.

If there are more data points than model parameters in (1.14) ($m > n$), then the m constraint equations in (1.14) may be inconsistent, in which case it will be impossible to find a model $\mathbf{m}$ that satisfies every equation exactly. The nonexistence of a model that exactly satisfies the observations in such a case can be seen geometrically in that no parabola will exist that can go through all the (t_i, y_i) points (Exercise 1.2). Such a situation could arise because of noise in the determinations of the y_i and/or t_i, and/or because the forward model (1.13) is approximate (for example, because (1.13) neglects atmospheric drag). In elementary linear algebra, where an exact solution is typically expected for a system of linear equations, we might throw up our hands at this point and simply state that no solution exists. However, practically useful solutions to such systems may be found by solving for model parameters that satisfy the data in an approximate "best fit" sense.

A reasonable approach to finding the "best" approximate solution to an inconsistent system of linear equations is to find an $\mathbf{m}$ that minimizes some misfit measure, calculated from the differences between the observations and the theoretical predictions of the forward problem, commonly called **residuals**. A traditional and very widely applied strategy is to find the model that minimizes the **2-norm** (Euclidean length) of the residual vector

$$\|\mathbf{y} - \mathbf{Gm}\|_2 = \sqrt{\sum_{i=1}^{m}(y_i - (\mathbf{Gm})_i)^2} \ . \tag{1.15}$$

However, (1.15) is not the only, or necessarily the best, misfit measure that can be applied to approximately solve systems of equations. An alternative misfit measure that is superior in many situations is the **1-norm**

$$\|\mathbf{y} - \mathbf{Gm}\|_1 = \sum_{i=1}^{m}|y_i - (\mathbf{Gm})_i| \ . \tag{1.16}$$

We shall see in Chapter 2 that a solution that minimizes (1.16) is less sensitive to data points that are wildly discordant with the mathematical model than one that minimizes (1.15). Solution techniques that are resistant to such data **outliers** are called **robust estimation procedures**.

●───

Example 1.2

A classic nonlinear parameter estimation problem in geophysics is determining the space and time coordinates of an earthquake nucleation, the hypocenter, which is specified by the 4-vector

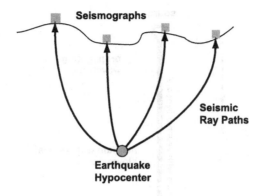

Figure 1.2 The earthquake location problem.

$$\mathbf{m} = \begin{bmatrix} \mathbf{x} \\ \tau \end{bmatrix} \tag{1.17}$$

where $\mathbf{x}$ is the three-dimensional earthquake location and the fourth element, τ, is the earthquake origin time (Fig. 1.2). The hypocenter model we seek best fits a vector of seismic wave **arrival times**, $\mathbf{t}$, observed at an m-station seismic network. The mathematical model is

$$G(\mathbf{m}) = \mathbf{t} \tag{1.18}$$

where G models the physics of seismic wave propagation to map a hypocenter into a vector of predicted arrival times at m stations. G depends on the appropriate seismic velocity structure, $\nu(\mathbf{x})$ for the observed wave types, which we assume here to be known.

The earthquake location problem is nonlinear even if $\nu(\mathbf{x})$ is a constant, c. In this case, all the ray paths in Fig. 1.2 would be straight, and the arrival time of the signal at station i would be

$$t_i = \frac{\|\mathbf{S}_{\cdot,i} - \mathbf{x}\|_2}{c} + \tau \tag{1.19}$$

where the ith column of the matrix $\mathbf{S}$, $\mathbf{S}_{\cdot,i}$, specifies the coordinates for station i. The arrival times cannot be expressed as a linear system of equation because the 2-norm of the ray path length in the numerator of (1.19) is nonlinear with respect to the spatial parameters x_i.

In special cases, a change of variables can convert a nonlinear problem into a linear one. More generally, nonlinear parameter estimation problems can often be solved by choosing a starting model and then iteratively improving it until a good solution is obtained. General methods for solving nonlinear parameter estimation problems are discussed in Chapter 9.

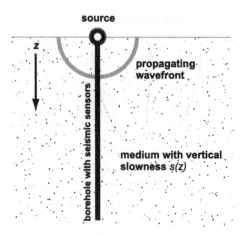

Figure 1.3 The vertical seismic profiling problem.

1.3. EXAMPLES OF INVERSE PROBLEMS

Example 1.3

In **vertical seismic profiling** we wish to know the vertical seismic velocity of the material surrounding a borehole. A downward–propagating seismic wavefront is generated at the surface by a source, and seismic waves are sensed by a string of seismometers in the borehole (Fig. 1.3).

The arrival times of the seismic wavefront at each instrument, which reflects the seismic velocity for vertically traveling waves as a function of depth, is measured from the recorded seismograms. The problem is nonlinear if it is expressed in terms of seismic velocities. However, we can linearize this problem via a simple change of variables by reparameterizing the seismic structure in terms of the reciprocal of the velocity $v(z)$, or **slowness**, $s(z)$. The observed travel time at depth z can then be expressed as the definite integral of the vertical slowness, s, from the surface to z

$$t(z) = \int_0^z s(\xi) \, d\xi \tag{1.20}$$

$$= \int_0^\infty s(\xi) H(z - \xi) \, d\xi \tag{1.21}$$

where the kernel function H is the **Heaviside step function**, which is equal to one when its argument is nonnegative and zero when its argument is negative. The explicit dependence of the kernel on $z - \xi$ shows that (1.21) is a convolution.

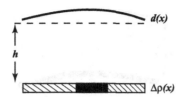

Figure 1.4 A linear inverse problem; determine a buried line mass density deviation, $\Delta\rho(x)$, relative to a background model, from gravity anomaly observations $d(x)$.

In theory, we can easily solve (1.21) because, by the fundamental theorem of calculus,

$$s(z) = \frac{dt(z)}{dz} . \tag{1.22}$$

However, differentiating the observations to obtain a solution may not be a practically useful approach. This is because there will generally be noise present in the observed times $t(z)$, and applying (1.22) may result in a solution that includes unphysical values of $s(z)$ or has other unrealistic model features.

Example 1.4

A further instructive linear inverse problem is the inversion of a vertical gravity anomaly, $d(x)$, observed at some height h, to estimate an unknown buried line mass density distribution deviation from a background model, $m(x) = \Delta\rho(x)$ (Fig. 1.4). The mathematical model for this problem can be expressed as an IFK, because the data are a superposition of the vertical gravity contributions from the differential elements comprising the line mass, i.e., [186]

$$d(x) = \Gamma \int_{-\infty}^{\infty} \frac{h}{\left((\xi - x)^2 + h^2\right)^{3/2}} \, m(\xi) \, d\xi \tag{1.23}$$

$$= \int_{-\infty}^{\infty} g(\xi - x) m(\xi) \, d\xi \tag{1.24}$$

where Γ is Newton's gravitational constant. Note that the kernel has the form $g(\xi - x)$, and (1.24) is thus a convolution. Because this kernel is a smooth function, $d(x)$ will be a smoothed and scaled transformation of $m(x)$. Conversely, solutions for $m(x)$ will be a roughened transformation of $d(x)$. For this reason we again need to be wary of possibly severe deleterious effects arising from data noise.

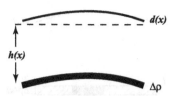

Figure 1.5 A nonlinear inverse problem; determine the depth to a buried line mass density anomaly $h(x)$ from observed gravity anomaly observations $d(x)$.

Example 1.5

Consider a variation on Example 1.4, where the depth of the line density perturbation varies, rather than the density contrast. The gravity anomaly is now attributable to a variation in the burial depth, $m(x) = h(x)$, of an assumed known fixed line density perturbation, $\Delta\rho$ (Fig. 1.5). The physics is the same as in Example 1.4, so the data are still given by the superposition of density perturbation contributions to the gravitational anomaly field, but the mathematical model now takes the form

$$d(x) = \Gamma \int_{-\infty}^{\infty} \frac{m(\xi)}{\left((\xi - x)^2 + m^2(\xi)\right)^{3/2}} \, \Delta\rho \, d\xi \ . \tag{1.25}$$

This problem is nonlinear in $m(x)$ because (1.25) does not follow the superposition and scaling rules (1.5) and (1.6).

Nonlinear inverse problems are generally significantly more difficult to solve than linear ones. In special cases, they may be solvable by coordinate transformations that globally linearize the problem or by other special-case methods. In other cases the problem cannot be globally linearized, so nonlinear optimization techniques must be applied. The essential differences in the treatment of linear and nonlinear problems arise because, as we shall see, all linear problems can be generalized to be the "same" in an important sense, so that a single set of solution methods can be applied to all. In contrast, nonlinear problems tend to be nonlinear in mathematically different ways and either require specific strategies or, more commonly, can be solved by iterative methods employing local linearization.

Example 1.6

A classic pedagogical inverse problem is an experiment in which an angular distribution of illumination passes through a thin slit and produces a diffraction pattern, for which the intensity is observed (Fig. 1.6; [180]).

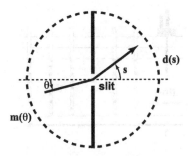

Figure 1.6 The Shaw diffraction intensity problem (1.26), with model and data represented as functions of angles θ and s.

The data $d(s)$ are measurements of diffracted light intensity as a function of outgoing angle $-\pi/2 \le s \le \pi/2$. Our goal is to find the intensity of the incident light on the slit, $m(\theta)$, as a function of the incoming angle $-\pi/2 \le \theta \le \pi/2$.

The forward problem relating d and m can be expressed as the linear mathematical model

$$d(s) = \int_{-\pi/2}^{\pi/2} (\cos(s) + \cos(\theta))^2 \left(\frac{\sin(\pi(\sin(s) + \sin(\theta)))}{\pi(\sin(s) + \sin(\theta))} \right)^2 m(\theta)\, d\theta \ . \tag{1.26}$$

Example 1.7

Consider the problem of recovering the history of groundwater pollution at a source site from later measurements of the contamination, C, at downstream wells to which the contaminant plume has been transported by advection and diffusion (Fig. 1.7). This "source history reconstruction problem" has been considered by a number of authors [148,184,185].

The mathematical model for contaminant transport is an advection–diffusion equation

$$\frac{\partial C}{\partial t} = D\frac{\partial^2 C}{\partial x^2} - v\frac{\partial C}{\partial x} \tag{1.27}$$

$$C(0,\ t) = C_{in}(t)$$

$$C(x,\ t) \to 0 \text{ as } x \to \infty$$

where D is the diffusion coefficient and v is the velocity of groundwater flow. The solution to (1.27) at time T is the convolution

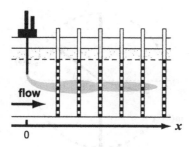

Figure 1.7 The contaminant plume source history reconstruction problem.

$$C(x, \ T) = \int_0^T C_{in}(t) f(x, \ T - t) \, dt \,, \tag{1.28}$$

where $C_{in}(t)$ is the time history of contaminant injection at $x = 0$, and the kernel is

$$f(x, T - t) = \frac{x}{2\sqrt{\pi D (T - t)^3}} \ e^{-\frac{[x - v(T - t)]^2}{4D(T - t)}} \,. \tag{1.29}$$

Example 1.8

An important and instructive inverse problem is **tomography**, from the Greek roots *tomos*, "to section" or "to cut" (the ancient concept of an *atom* was that of an irreducible, uncuttable object) and *graphein*, "to write." Tomography is the general technique of determining models that are consistent with path-integrated properties such as attenuation (e.g., X-ray, radar, muon, seismic), travel time (e.g., electromagnetic, seismic, or acoustic), or source intensity (e.g., positron emission). Although tomography problems originally involved determining models that were two-dimensional slices of three-dimensional objects, the term is now commonly used in situations where the model is two- three- or four- (i.e., time-lapse tomography) dimensional. Tomography has many applications in medicine, engineering, acoustics, and Earth science. One important geophysical example is cross-well tomography, where seismic sources are installed in a borehole, and the signals are received by sensors in another borehole. Another important example is joint earthquake location/velocity structure inversion carried out on scales ranging from a fraction of a cubic kilometer to global [203–205].

The physical model for tomography in its most basic form (Fig. 1.8) assumes that geometric ray theory (essentially the high-frequency limiting case of the wave equation) is valid, so that wave energy traveling between a source and receiver can be considered

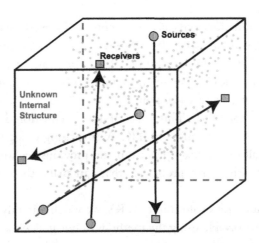

Figure 1.8 Conceptual depiction of ray path tomography. Sources and receivers may, in general, be either at the edges or within the volume, and paths may be either straight, as depicted, or bent by refraction and/or reflection.

to be propagating along infinitesimally narrow ray paths. Importantly, the density of ray paths in a tomographic problem may vary significantly throughout a section or volume, and may thus provide much better constraints on physical properties in densely-sampled regions than in sparsely-sampled ones.

In seismic tomography, if the slowness at a point $\mathbf{x}$ is $s(\mathbf{x})$, and the ray path ℓ is known, then the travel time for seismic energy propagating along that ray path is given by the line integral along ℓ

$$t = \int_{\ell} s(\mathbf{x}(l)) \, dl \, . \tag{1.30}$$

Note that (1.30) is just a higher-dimensional generalization of (1.21), the forward problem for the vertical seismic profiling example. In general, seismic ray paths will be bent due to refraction and/or reflection. In cases where such effects are negligible, ray paths can be usefully approximated as straight lines (e.g., as depicted in Fig. 1.8), and the forward and inverse problems can be cast in a linear form. However, if the ray paths are significantly bent by heterogeneous structure, the resulting inverse problem will be nonlinear.

1.4. DISCRETIZING INTEGRAL EQUATIONS

Consider problems of the form

$$d(x) = \int_{a}^{b} g(x, \, \xi) m(\xi) \, d\xi \, . \tag{1.31}$$

Here $d(x)$ is a known function, typically representing observed data. The kernel $g(x, \xi)$ is considered to be given, and encodes the physics relating an unknown model $m(\xi)$ to corresponding data $d(x)$. The interval $[a, b]$ may be finite or infinite. The function $d(x)$ might in theory be known over an entire interval but in practice we will only have measurements of $d(x)$ at a finite set of points.

We wish to solve for $m(x)$. This type of linear equation is called a Fredholm integral equation of the first kind, or IFK. A surprisingly large number of inverse problems, including all of the examples from the previous section, can be written as IFK's. Unfortunately, IFK's have properties that can make them very challenging to solve.

To obtain useful numerical solutions to IFK's, we will frequently discretize them into forms that are tractably solvable using the methods of linear algebra. We first assume that $d(x)$ is known at a finite number of points $x_1, x_2, \ldots, x_m$. We can then write the forward problem as

$$d_i = d(x_i) = \int_a^b g(x_i, \xi) m(\xi) \, d\xi \quad i = 1, 2, \ldots, m \tag{1.32}$$

or as

$$d_i = \int_a^b g_i(\xi) m(\xi) \, d\xi \quad i = 1, 2, \ldots, m \tag{1.33}$$

where $g_i(\xi) = g(x_i, \xi)$. The functions g_i are referred to as **representers** or **data kernels**.

Here, we will apply a **quadrature rule** to approximate (1.33) numerically. Note that, although quadrature methods are applied in this section to linear integral equations, they also have utility in the discretization of nonlinear problems. The simplest quadrature approach is the **midpoint rule**, where we divide the interval $[a, b]$ into n subintervals, and pick points $x_1, x_2, \ldots, x_n$ in the middle of each interval. The points are given by

$$x_j = a + \frac{\Delta x}{2} + (j - 1)\Delta x \tag{1.34}$$

where

$$\Delta x = \frac{b - a}{n} . \tag{1.35}$$

The integral (1.33) is then approximated by (Fig. 1.9)

$$d_i = \int_a^b g_i(\xi) m(\xi) \, d\xi \approx \sum_{j=1}^n g_i(x_j) m(x_j) \, \Delta x , \quad i = 1, 2, \ldots, m . \tag{1.36}$$

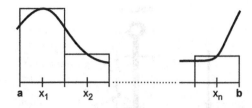

Figure 1.9 Grid for the midpoint rule.

If we let

$$G_{i,j} = g_i(x_j)\Delta x \quad \left(\begin{array}{l} i = 1,\ 2,\ \ldots,\ m \\ j = 1,\ 2,\ \ldots,\ n \end{array} \right) \tag{1.37}$$

and

$$m_j = m(x_j) \quad j = 1,\ 2,\ \ldots,\ n \tag{1.38}$$

then we obtain a linear system of equations $\mathbf{Gm} = \mathbf{d}$.

The approach of using the midpoint rule to approximate the integral is known as **simple collocation**. Of course, there are more sophisticated quadrature rules for numerically approximating integrals (e.g., the trapezoidal rule or Simpson's rule). In each case, we end up with an m by n linear system of equations but the formulas for evaluating the elements of $\mathbf{G}$ will be different.

Example 1.9

Consider vertical seismic profiling (Example 1.3), where we wish to estimate vertical seismic slowness using travel time measurements of downward propagating seismic waves (Fig. 1.10). We discretize the forward problem (1.21) for m travel time data points, t_i, and n model depths, z_j, spaced at constant intervals of Δz. This discretization is depicted in Fig. 1.10.

The discretized problem has

$$t_i = \sum_{j=1}^{n} H(y_i - z_j)s_j\,\Delta z \tag{1.39}$$

where $n/m = \Delta y/\Delta z$ is an integer. The rows, $\mathbf{G}_{i,\cdot}$, of the matrix each consist of $i \cdot n/m$ elements equal to Δz on the left and $n - (i \cdot n/m)$ zeros on the right. For $n = m$, $\mathbf{G}$ is a lower triangular matrix with each nonzero entry equal to Δz.

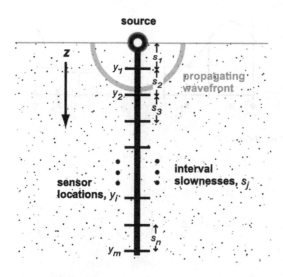

Figure 1.10 Discretization of the vertical seismic profiling problem ($n = m$) into uniform intervals.

Example 1.10

To discretize the Shaw problem (1.26) we apply the method of simple collocation with m and n intervals for the data and model functions, respectively. We additionally define the data and model points at m and n equally spaced respective angles given by

$$s_i = \frac{(i - 0.5)\pi}{m} - \frac{\pi}{2} \quad i = 1,\ 2,\ \ldots,\ m \tag{1.40}$$

and

$$\theta_j = \frac{(j - 0.5)\pi}{n} - \frac{\pi}{2} \quad j = 1,\ 2,\ \ldots,\ n\,. \tag{1.41}$$

Correspondingly discretizing the data and model into m and n-length vectors

$$d_i = d(s_i) \quad i = 1,\ 2,\ \ldots,\ m \tag{1.42}$$

and

$$m_j = m(\theta_j) \quad j = 1,\ 2,\ \ldots,\ n \tag{1.43}$$

leads to a discrete linear system $\mathbf{Gm} = \mathbf{d}$, where

$$G_{i,j} = (\cos(s_i) + \cos(\theta_j))^2 \left(\frac{\sin(\pi(\sin(s_i) + \sin(\theta_j)))}{\pi(\sin(s_i) + \sin(\theta_j))} \right)^2 \Delta\theta \tag{1.44}$$

and

$$\Delta\theta = \frac{\pi}{n} \, . \tag{1.45}$$

Example 1.11

We discretize the advection diffusion problem (1.27), assuming that the parameters D and v in (1.29) are known. We wish to estimate $C_{in}(t)$ from simultaneous concentration observations at the locations x_i at some later time T. The convolution (1.28) for $C(x, T)$ is discretized as

$$\mathbf{Gm} = \mathbf{d} \tag{1.46}$$

where $\mathbf{d}$ is a vector of sampled concentrations at different well locations, $\mathbf{x}$, at a time T, $\mathbf{m}$ is a vector of C_{in} values to be estimated, and

$$G_{i,j} = f(x_i, \ T - t_j)\Delta t \tag{1.47}$$

$$= \frac{x_i}{2\sqrt{\pi D(T - t_j)^3}} \, e^{-\frac{[x_i - v(T - t_j)]^2}{4D(T - t_j)}} \, \Delta t \, . \tag{1.48}$$

Example 1.12

A common way of discretizing the model in a tomographic problem is as uniform blocks. This approach is equivalent to applying the midpoint rule to the travel time forward problem (1.30).

In this parameterization, the elements of $\mathbf{G}$ are just the ray path lengths within corresponding blocks. Consider the example of Fig. 1.11, where 9 homogeneous blocks with sides of unit length and unknown slowness are crossed by 8 ray paths. For straight ray paths, we map the two-dimensional slowness grid to a model vector using a row-by-row indexing convention. The constraint equations in the mathematical model are then

$$\mathbf{Gm} = \begin{bmatrix} 1 & 0 & 0 & 1 & 0 & 0 & 1 & 0 & 0 \\ 0 & 1 & 0 & 0 & 1 & 0 & 0 & 1 & 0 \\ 0 & 0 & 1 & 0 & 0 & 1 & 0 & 0 & 1 \\ 1 & 1 & 1 & 0 & 0 & 0 & 0 & 0 & 0 \\ 0 & 0 & 0 & 1 & 1 & 1 & 0 & 0 & 0 \\ 0 & 0 & 0 & 0 & 0 & 0 & 1 & 1 & 1 \\ \sqrt{2} & 0 & 0 & 0 & \sqrt{2} & 0 & 0 & 0 & \sqrt{2} \\ 0 & 0 & 0 & 0 & 0 & 0 & 0 & 0 & \sqrt{2} \end{bmatrix} \begin{bmatrix} s_{1,1} \\ s_{1,2} \\ s_{1,3} \\ s_{2,1} \\ s_{2,2} \\ s_{2,3} \\ s_{3,1} \\ s_{3,2} \\ s_{3,3} \end{bmatrix} = \begin{bmatrix} t_1 \\ t_2 \\ t_3 \\ t_4 \\ t_5 \\ t_6 \\ t_7 \\ t_8 \end{bmatrix} \, . \tag{1.49}$$

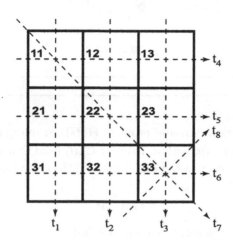

Figure 1.11 Discretization of a tomography problem into uniform blocks. Ray paths correspond to the constraint equations in (1.49).

Because there are 9 unknown parameters $s_{i,j}$ in the model, but only 8 constraints, the **G** matrix is clearly rank deficient. In fact, rank(**G**) is only 7. In addition, there is clearly redundant information in (1.49), in that the slowness $s_{3,3}$ is completely determined by t_8, yet $s_{3,3}$ also influences the observations t_3, t_6, and t_7.

1.5. WHY INVERSE PROBLEMS ARE HARD

Scientists and engineers need to be concerned with far more than simply finding mathematically acceptable answers to parameter estimation and inverse problems. One fundamental reason is that there may be many models that adequately fit the data. It is thus essential to characterize just what solution has been obtained, how "good" it is in terms of physical plausibility, its ability to predict the data, and perhaps how consistent it is with other constraints. Essential issues that must be considered include **solution existence**, **solution uniqueness**, and **instability** of the solution process.

1. *Existence.* There may be no model that exactly fits the data. This can occur in practice because the mathematical model of the system's physics is approximate (or perhaps simply incorrect) or because the data contain noise.

2. *Uniqueness.* If exact solutions do exist, they may not be unique, even for an infinite number of exact data points. That is, there may be other solutions besides m_{true} that exactly satisfy $G(m) = d_{\text{true}}$. This situation commonly occurs in potential field problems. A classic example is the external gravitational field from a spherically-

symmetric mass distribution, which depends only on the total mass, and not on the radial density distribution.

Nonuniqueness is a characteristic of rank deficient discrete linear inverse problems because the matrix $\mathbf{G}$ in this case has a nontrivial **null space**. In linear inverse problems, models, $\mathbf{m}_0$, that lie in the null space of $\mathbf{G}$ are solutions to $\mathbf{Gm}_0 = \mathbf{0}$. By superposition, any linear combination of these **null space models** can be added to a particular model that satisfies (1.7) and not change the fit to the data. There are thus an infinite number of mathematically acceptable models in such situations. In practical terms, suppose that there exists a nonzero model projection $\mathbf{m}_0$ that results in an instrument reading of zero. We cannot discriminate this situation from the situation where $\mathbf{m}_0$ is truly zero.

An important and thorny issue with problems that have nonunique solutions is that an estimated model may be significantly smoothed or otherwise biased relative to the true situation. Characterizing such bias is essential to interpreting models in terms of their possible correspondence to reality. This issue falls under the general topic of **model resolution analysis**.

3. *Instability*. The process of computing an inverse solution can be, and often is, extremely unstable in that a small change in measurement (e.g., a small η in (1.3)) can lead to an enormous change in the estimated model. Inverse problems where this situation arises are referred to as **ill-posed** in the case of continuous systems, or **ill-conditioned** in the case of discrete linear systems. A key point is that it is commonly possible to stabilize the inversion process by imposing additional constraints that bias the solution, a process that is generally referred to as **regularization**. Regularization is frequently essential to producing a usable solution to an otherwise intractable ill-posed or ill-conditioned inverse problem.

To explore existence, uniqueness, and instability issues, let us consider some simple mathematical examples using an IFK

$$\int_0^1 g(x, \xi) \, m(\xi) \, d\xi = \gamma(x) \, . \tag{1.50}$$

First, consider the trivial case where the kernel is a constant, e.g.,

$$g(x, \xi) = 1 \tag{1.51}$$

that produces the integral equation

$$\int_0^1 m(\xi) \, d\xi = \gamma(x) \, . \tag{1.52}$$

Because the left hand side of (1.52) is independent of x, this system has no solution unless $y(x)$ is a constant. Thus, there are an infinite number of mathematically conceivable data sets $y(x)$ that are not constant and for which no exact solution exists. This is a simple illustration of a solution existence issue.

When a solution to (1.52) does exist, the solution is nonunique because there are an infinite number of functions that, when integrated over the unit interval, produce the same constant and thus satisfy the IFK exactly. This demonstrates a uniqueness issue.

A more subtle example of nonuniqueness can be seen for

$$g(x, \; \xi) = x \sin(\pi \xi) \tag{1.53}$$

in (1.50), so that the IFK becomes

$$\int_0^1 x \sin(\pi \xi) \; m(\xi) \; d\xi = y(x) \; . \tag{1.54}$$

The functions $\sin(k \pi x)$ for integer values of k are orthogonal in the sense that

$$\int_0^1 \sin(k \pi x) \sin(l \pi x) \; dx = -\frac{1}{2} \int_0^1 \cos(\pi (k + l)x) - \cos(\pi (k - l)x) \; dx$$

$$= -\frac{1}{2\pi} \left(\frac{\sin(\pi (k + l))}{k + l} - \frac{\sin(\pi (k - l))}{k - l} \right) = 0 \tag{1.55}$$

$$(k \neq \pm l; \; k, \; l \neq 0) \; .$$

Thus, in (1.54), for models of the form $m(x) = \sin(k \pi x)$, for $k = \pm 2, \; \pm 3, \; \ldots,$ we have

$$\int_0^1 g(x, \; \xi) m(\xi) \; d\xi = \int_0^1 g(x, \; \xi) \sin(k \pi \xi) \; d\xi = 0 \; . \tag{1.56}$$

Furthermore, because (1.54) is a linear system, we can add any function of the form

$$m_0(x) = \sum_{k=2}^{\infty} \alpha_k \sin(k \pi x) \tag{1.57}$$

to a solution, $m(x)$, and obtain a new model that fits the data equally well.

$$\int_0^1 x \sin(\pi \xi)(m(\xi) + m_0(\xi)) \; d\xi = \int_0^1 x \sin(\pi \xi) \; m(\xi) \; d\xi + \int_0^1 x \sin(\pi \xi) \; m_0(\xi) \; d\xi$$

$$= \int_0^1 x \sin(\pi \xi) \; m(\xi) \; d\xi + 0 \; . \tag{1.58}$$

There are thus an infinite number of very different solutions that fit the data equally well.

Finally, even if we do not encounter existence or uniqueness issues, instability is a fundamental feature of IFK's. It can be shown that

$$\lim_{k \to \infty} \int_{-\infty}^{\infty} g(x, \, \xi) \sin k\pi\xi \; d\xi = 0 \qquad (1.59)$$

for *all* square-integrable functions $g(x, \, \xi)$. This result is known as the **Riemann–Lebesgue lemma** [173]. Examining (1.59) in more detail, we can intuit why this occurs. The oscillatory sine function is smoothed by integration with the kernel $g(\xi, \, x)$. For sufficiently large values of the sine frequency, k, the integrated positive and negative excursions of the kernel-weighted sine function will approach zero. The inverse problem has this situation reversed, so that an inferred model can be extremely sensitive to small changes in the data. If such small data changes are created by random noise that has nothing to do with the physical system that we are studying, then a model that fits this noise can diverge wildly from the true model.

The unstable character of IFK solutions has similarities to the situation encountered in solving linear systems of equations where the condition number of the matrix is very large, or equivalently, where the matrix is nearly singular. In both cases, the difficulty lies in the character of the mathematical model itself and not in the particular algorithm used to solve the problem.

1.6. EXERCISES

1. Consider a mathematical model of the form $G(\mathbf{m}) = \mathbf{d}$, where $\mathbf{m}$ is a vector of length n, and $\mathbf{d}$ is a vector of length m. Suppose that the model obeys the superposition and scaling laws and is thus linear. Show that $G(\mathbf{m})$ can be written in the form

$$G(\mathbf{m}) = \mathbf{\Gamma m} \qquad (1.60)$$

   where $\mathbf{\Gamma}$ is an m by n matrix. What are the elements of $\mathbf{\Gamma}$? Hint: Write $\mathbf{m}$ as a linear combination of the standard basis vectors (Appendix A). Apply the superposition and scaling laws. Finally, recall the operations of matrix–vector multiplication.

2. Can (1.14) be inconsistent, even with only $m = 3$ data points? How about just $m = 2$ data points? If the system can be inconsistent, give an example. If not, explain why not.

3. Consider the borehole vertical seismic profile problem of Examples 1.3 and 1.9 for $n = 100$ equally spaced seismic sensors located at depths of $z = 0.2, 0.4, \ldots, 20$ m, and for a model $\mathbf{m}$ describing n corresponding equal length seismic slowness values for 0.2 m intervals having midpoints at $z - 0.1$ m.

   a. Calculate the appropriate system matrix $\mathbf{G}$ for discretizing the integral equation (1.21) using the midpoint rule.

   b. For a seismic velocity model having a linear depth gradient specified by

$$v = v_0 + kz \tag{1.61}$$

   where the velocity at $z = 0$ is $v_0 = 1$ km/s and the gradient is $k = 40$ m/s per m, calculate the true slowness values at the midpoints of the n intervals, $\mathbf{m}_{true}$. Additionally, integrate the corresponding slowness function for (1.61) using (1.21) to calculate a noiseless synthetic data vector, $\mathbf{d}$, of predicted seismic travel times at the sensor depths.

   c. Solve for the slowness, $\mathbf{m}$, as a function of depth using your $\mathbf{G}$ matrix and analytically calculated noiseless travel times using the MATLAB backslash operator. Compare your result graphically with $\mathbf{m}_{true}$.

   d. Generate a noisy travel time vector where independent normally distributed noise with a standard deviation of 0.05 ms is added to the elements of $\mathbf{d}$. Re-solve the system for $\mathbf{m}$ and again compare your result graphically with $\mathbf{m}_{true}$. How has the model changed?

   e. Repeat the problem, but for just $n = 4$ sensor depths and corresponding equal length slowness intervals. Is the recovery of the true model improved? Explain in terms of the condition numbers of your $\mathbf{G}$ matrices.

4. Find a journal article that discusses the solution of an inverse problem in a discipline of special interest to you. What are the data? Are the data discrete or continuous? Have the authors discussed possible sources of noise in the data? What is the model? Is the model continuous or discrete? What physical laws determine the forward operator G? Is G linear or nonlinear? Do the authors discuss any issues associated with existence, uniqueness, or instability of solutions?

1.7. NOTES AND FURTHER READING

Ill-posed and ill-conditioned inverse problems arise frequently in practical and theoretical geophysics, e.g., [20,186,201]. Some important references that focus on inverse problems in geophysics and remote sensing include [37,78,125,159,211]. Instructive examples of ill-posed problems and their solution can be found in the book edited by

Tikhonov and Goncharsky [207]. More mathematically oriented references on inverse problems include [8,57,77,84,86,113,120,137,141,199,206]. Tomography, particularly in medical imaging and seismology, is a very large field. Some general references on tomography are [89,99,106,124,126,143,152].

CHAPTER TWO

Linear Regression

Synopsis

Linear regression is introduced as a parameter estimation problem, and least squares solutions are derived. Maximum likelihood is defined, and its association with least squares solutions under normally distributed data errors is demonstrated. Statistical tests based on χ^2 that provide insight into least squares solutions are discussed. The mapping of data errors into model errors in the context of least squares is described. The determination of confidence intervals using the model covariance matrix and the meaning of model parameter correlations is discussed. The problems of estimating unknown data standard deviations and recognizing proportional data errors are addressed. The issue of data outliers and the concept of robust estimation are introduced, and 1-norm minimization is introduced as a robust estimation technique. General propagation of errors between data and model using Monte Carlo methods is discussed in the context of the iteratively reweighted least squares 1-norm minimization algorithm.

2.1. INTRODUCTION TO LINEAR REGRESSION

The problem of finding a parameterized curve that approximately fits a set of data is referred to as **regression**. When the regression model is linear in the fitted parameters, then we have a **linear regression** problem. In this chapter linear regression problems are analyzed as discrete linear inverse problems.

Consider a discrete linear inverse problem. We begin with a data vector **d** of m observations, and a vector **m** of n model parameters that we wish to determine. The forward problem can be expressed as a linear system of equations using an m by n matrix **G**

$$\mathbf{Gm} = \mathbf{d} \ . \tag{2.1}$$

Recall that if rank$(\mathbf{G}) = n$, then the matrix has full column rank. In this chapter we will assume that the matrix **G** has full column rank. In Chapter 3 we will consider rank deficient problems.

For a full column rank matrix with more rows than columns, it is frequently the case that there is no solution **m** that satisfies (2.1) exactly. This happens because the dimension of the range of **G** is smaller than m, and a noisy data vector **d** will generally lie outside of the range of **G** (**d** will lie in R^m).

A useful approximate solution may still be found by finding a particular model **m** that minimizes some measure of the misfit between the actual data and **Gm**. The **residual vector** is the vector of differences between observed data and corresponding model predictions

$$\mathbf{r} = \mathbf{d} - \mathbf{Gm} \,, \tag{2.2}$$

and the elements of $\mathbf{r}$ are frequently referred to simply as residuals. One commonly used measure of the misfit is the 2-norm of the residual vector, and a model that minimizes this 2-norm is called a **least squares solution**. The least squares or 2-norm solution is of special interest both because it is readily amenable to analysis and geometric intuition, and because it turns out to be statistically the most likely solution if data errors are normally distributed.

The least squares solution is, from the normal equations (A.73),

$$\mathbf{m}_{L_2} = (\mathbf{G}^T\mathbf{G})^{-1}\mathbf{G}^T\mathbf{d} \,. \tag{2.3}$$

It can be shown that if $\mathbf{G}$ is of full column rank then $(\mathbf{G}^T\mathbf{G})^{-1}$ exists (Exercise A.13e).

A classic linear regression problem is finding parameters m_1 and m_2 for a line

$$y = m_1 + m_2 x \tag{2.4}$$

that best fits a set of $m > 2$ data points. The system of equations in this case is

$$\mathbf{Gm} = \begin{bmatrix} 1 & x_1 \\ 1 & x_2 \\ . & . \\ . & . \\ . & . \\ 1 & x_m \end{bmatrix} \begin{bmatrix} m_1 \\ m_2 \end{bmatrix} = \begin{bmatrix} d_1 \\ d_2 \\ . \\ . \\ . \\ d_m \end{bmatrix} = \mathbf{d} \tag{2.5}$$

where the d_i are observations of y at each corresponding position x_i. Applying (2.3) to find a least squares solution gives

$$\mathbf{m}_{L_2} = (\mathbf{G}^T\mathbf{G})^{-1}\mathbf{G}^T\mathbf{d} \tag{2.6}$$

$$= \left(\begin{bmatrix} 1 & \cdots & 1 \\ x_1 & \cdots & x_m \end{bmatrix} \begin{bmatrix} 1 & x_1 \\ \vdots & \vdots \\ 1 & x_m \end{bmatrix} \right)^{-1} \begin{bmatrix} 1 & \cdots & 1 \\ x_1 & \cdots & x_m \end{bmatrix} \begin{bmatrix} d_1 \\ d_2 \\ \vdots \\ d_m \end{bmatrix}$$

$$= \begin{bmatrix} m & \sum_{i=1}^m x_i \\ \sum_{i=1}^m x_i & \sum_{i=1}^m x_i^2 \end{bmatrix}^{-1} \begin{bmatrix} \sum_{i=1}^m d_i \\ \sum_{i=1}^m x_i d_i \end{bmatrix}$$

$$= \frac{1}{m\sum_{i=1}^m x_i^2 - (\sum_{i=1}^m x_i)^2} \begin{bmatrix} \sum_{i=1}^m x_i^2 & -\sum_{i=1}^m x_i \\ -\sum_{i=1}^m x_i & m \end{bmatrix} \begin{bmatrix} \sum_{i=1}^m d_i \\ \sum_{i=1}^m x_i d_i \end{bmatrix} \,.$$

2.2. STATISTICAL ASPECTS OF LEAST SQUARES

If we consider data points to be imperfect measurements that include random errors, then we are faced with the problem of finding the solution that is best from a statistical point of view. One approach, **maximum likelihood estimation**, considers the question from the following perspective. Given that we observed a particular data set, that we know the statistical characteristics of these observations, and that we have a mathematical model for the forward problem, what is the model from which these observations would most likely arise?

Maximum likelihood estimation is a general method that can be applied to any estimation problem where a joint probability density function (B.26) can be assigned to the observations. The essential problem is to find the most likely model, as characterized by the elements of the parameter vector $\mathbf{m}$, for the set of observations contained in the vector $\mathbf{d}$. We will assume that the observations are independent so that we can use the product form of the joint probability density function (B.28).

Given a model $\mathbf{m}$, we have a probability density function $f_i(d_i|\mathbf{m})$ for each of the observations. In general, these probability density functions will vary depending on $\mathbf{m}$, so probability densities are conditional on $\mathbf{m}$. The joint probability density for a vector of independent observations $\mathbf{d}$ will be

$$f(\mathbf{d}|\mathbf{m}) = f_1(d_1|\mathbf{m})f_2(d_2|\mathbf{m})\cdots f_m(d_m|\mathbf{m}) \ . \tag{2.7}$$

Note that the $f(d_i|\mathbf{m})$ are probability densities, not probabilities. We can only compute the probability of observing data in some range for a given model $\mathbf{m}$ by integrating $f(\mathbf{d}|\mathbf{m})$ over that range. In fact, the probability of getting any particular set of data exactly is precisely zero! This conceptual conundrum can be resolved by considering the probability of getting a data set that lies within a small m-dimensional box around a particular data set $\mathbf{d}$. This probability will be nearly proportional to the probability density $f(\mathbf{d}|\mathbf{m})$.

In practice, we measure a particular data vector and wish to find the "best" model to match it in the maximum likelihood sense. That is, $\mathbf{d}$ will be a fixed set of observations, and $\mathbf{m}$ will be a vector of parameters to be estimated. The **likelihood function** L of $\mathbf{m}$ given $\mathbf{d}$ is given by the joint probability density function of $\mathbf{d}$ given $\mathbf{m}$

$$L(\mathbf{m}|\mathbf{d}) = f(\mathbf{d}|\mathbf{m}) \ . \tag{2.8}$$

For many possible models (2.8) will be extremely close to zero because such models would be extremely unlikely to produce the observed data set $\mathbf{d}$. The likelihood would be much larger for any models that, conversely, would be relatively likely to produce the observed data. According to the **maximum likelihood principle** we should select the model $\mathbf{m}$ that maximizes the likelihood function (2.8). Model estimates obtained in this manner have many desirable statistical properties [38,52].

It is particularly insightful that when we have a discrete linear inverse problem and the data errors are independent and normally distributed, then the maximum likelihood principle solution is the least squares solution. To show this, assume that the data have independent random errors that are normally distributed with expected value zero, and where the standard deviation of the ith observation d_i is σ_i. The probability density for d_i then takes the form of (B.6)

$$f_i(d_i|\mathbf{m}) = \frac{1}{\sigma_i\sqrt{2\pi}}e^{-\frac{1}{2}(d_i-(\mathbf{Gm})_i)^2/\sigma_i^2} .$$

(2.9)

The likelihood function for the complete data set is the product of the individual likelihoods

$$L(\mathbf{m}|\mathbf{d}) = \frac{1}{(2\pi)^{m/2}\prod_{i=1}^m \sigma_i} \prod_{i=1}^m e^{-\frac{1}{2}(d_i-(\mathbf{Gm})_i)^2/\sigma_i^2} .$$

(2.10)

The constant factor does not affect the maximization of L, so we can solve

$$\max \prod_{i=1}^m e^{-\frac{1}{2}(d_i-(\mathbf{Gm})_i)^2/\sigma_i^2} .$$

(2.11)

The natural logarithm is a monotonically increasing function, so we can equivalently solve

$$\max \log \prod_{i=1}^m e^{-\frac{1}{2}(d_i-(\mathbf{Gm})_i)^2/\sigma_i^2} = \max \left[-\frac{1}{2}\sum_{i=1}^m \frac{(d_i-(\mathbf{Gm})_i)^2}{\sigma_i^2} \right] .$$

(2.12)

Finally, if we turn the maximization into a minimization by changing sign and ignore the constant factor of $1/2$, the problem becomes

$$\min \sum_{i=1}^m \frac{(d_i-(\mathbf{Gm})_i)^2}{\sigma_i^2} .$$

(2.13)

Aside from the distinct $1/\sigma_i^2$ factors in each term of the sum, this is identical to the least squares problem for $\mathbf{Gm} = \mathbf{d}$. (2.13) is commonly referred to as the **weighted least squares problem**.

To incorporate the data standard deviations into the weighted least squares solution, we scale the system of equations to obtain a weighted system of equations. Let a diagonal weighting matrix be

$$\mathbf{W} = \text{diag}(1/\sigma_1, 1/\sigma_2, \ldots, 1/\sigma_m) .$$

(2.14)

Then let

$$\mathbf{G}_w = \mathbf{W}\mathbf{G} \tag{2.15}$$

and

$$\mathbf{d}_w = \mathbf{W}\mathbf{d} \ . \tag{2.16}$$

The weighted system of equations is then

$$\mathbf{G}_w\mathbf{m} = \mathbf{d}_w \ . \tag{2.17}$$

The normal equations (A.73) solution to (2.17) is

$$\mathbf{m}_{L_2} = (\mathbf{G}_w^T\mathbf{G}_w)^{-1}\mathbf{G}_w^T\mathbf{d}_w \ . \tag{2.18}$$

Now,

$$\|\mathbf{d}_w - \mathbf{G}_w\mathbf{m}_{L_2}\|_2^2 = \sum_{i=1}^{m}(d_i - (\mathbf{G}\mathbf{m}_{L_2})_i)^2/\sigma_i^2 \ . \tag{2.19}$$

Thus, the least squares solution to $\mathbf{G}_w\mathbf{m} = \mathbf{d}_w$ is the maximum likelihood solution.

The sum of the squares of the residuals also provides useful statistical information about the quality of the model estimates obtained with least squares. The **chi-square statistic** is

$$\chi_{\text{obs}}^2 = \sum_{i=1}^{m}(d_i - (\mathbf{G}\mathbf{m}_{L_2})_i)^2/\sigma_i^2 \ . \tag{2.20}$$

Since χ_{obs}^2 depends on the random measurement errors in $\mathbf{d}$, it is itself a random variable. It can be shown that under our assumptions χ_{obs}^2 has a χ^2 distribution with $\nu = m - n$ degrees of freedom [38,52].

The probability density function for the χ^2 distribution is

$$f_{\chi^2}(x) = \frac{1}{2^{\nu/2}\Gamma(\nu/2)}x^{\frac{1}{2}\nu - 1}e^{-x/2} \tag{2.21}$$

(Fig. B.4). The χ^2 **test** provides a statistical assessment of the assumptions that we used in finding the least squares solution. In this test, we compute χ_{obs}^2 and compare it to the theoretical χ^2 distribution with $\nu = m - n$ degrees of freedom.

The probability of obtaining a χ^2 value as large or larger than the observed value (and hence a worse misfit between data and model data predictions than that obtained) is called the p-**value** of the test, and is given by

$$p = \int_{\chi_{\text{obs}}^2}^{\infty} f_{\chi^2}(x) \ dx \ . \tag{2.22}$$

When data errors are independent and normally distributed, and the mathematical model is correct, it can be shown that the p-value will be uniformly distributed between zero and one (Exercise 2.4). In practice, particular p-values that are very close to either extreme indicate that one or more of these assumptions are incorrect.

There are three general cases.

1. The p-value is not too small and not too large. Our least squares solution produces an acceptable data fit and our statistical assumptions of data errors are consistent. Practically, p does not actually have to be very large to be deemed marginally "acceptable" in many cases (e.g. $p \approx 10^{-2}$), as truly "wrong" models will typically produce extraordinarily small p-values (e.g. 10^{-12}) because of the short-tailed nature of the normal distribution.

    Because the p-value will be uniformly distributed when we have a correct mathematical model and our statistical data assumptions are valid, it is inappropriate to conclude anything based on the differences between p-values in this range. For example, one should not conclude that a p-value of 0.7 is "better" than a p-value of 0.2.

2. The p-value is very small. We are faced with three non-exclusive possibilities, but something is clearly wrong.

    a. The data truly represent an extremely unlikely realization. This is easy to rule out for p-values very close to zero. For example, suppose an experiment produced a data realization where the probability of a worse fit was 10^{-9}. If the model was correct, then we would have to perform on the order of a billion experiments to get a comparably poor fit to the data. It is far more likely that something else is wrong.

    b. The mathematical model $\mathbf{Gm} = \mathbf{d}$ is incorrect. Most often this happens because we have left some important aspect of the physics out of the mathematical model.

    c. The data errors are underestimated or not normally distributed. In particular, we may have underestimated the σ_i.

3. The p-value is very close to one. The fit of the model predictions to the data is almost exact. We should investigate the possibility that we have overestimated the data errors. A more sinister possibility is that a very high p-value is indicative of data fraud, such as might happen if data were cooked-up ahead of time to fit a particular model!

A rule of thumb for problems with a large number of degrees of freedom ν is that the expected value of χ^2 approaches ν. This arises because, by the central limit theorem (Section B.6), the χ^2 random variable, which is itself a sum of random variables, will become normally distributed as the number of terms in the sum becomes large. The mean of the resulting distribution will approach ν and the standard deviation will approach $(2\nu)^{1/2}$.

In addition to examining χ^2_{obs}, it is important to examine the individual weighted residual vector elements corresponding to the model

$$r_{w,i} = (\mathbf{d} - \mathbf{Gm})_i/\sigma_i = (\mathbf{d}_w - \mathbf{G}_w\mathbf{m})_i \ . \tag{2.23}$$

The elements of $\mathbf{r}_w$ should be roughly normally distributed with standard deviation one and should show no systematic patterns. In some cases, when an incorrect model has been fit to the data, the residuals (2.23) will reveal the nature of the modeling error. For example, in linear regression to a line, it might be that all residuals are negative for small and large values of the independent variable x but positive for intermediate values of x. This might suggest that additional complexity (e.g., an additional quadratic term) could be required in the regression model.

Parameter estimates obtained via linear regression are linear combinations of the data (2.18). If the data errors are normally distributed, then the parameter estimates will also be normally distributed because a linear combination of normally distributed random variables is normally distributed [4,38]. To derive the mapping between data and model covariances, consider the covariance of a data vector $\mathbf{d}$ of normally distributed, independent random variables, operated on by a general linear transformation specified by a matrix $\mathbf{A}$. The appropriate covariance mapping is (B.64)

$$\text{Cov}(\mathbf{Ad}) = \mathbf{A}\text{Cov}(\mathbf{d})\mathbf{A}^T \ . \tag{2.24}$$

The least squares solution has $\mathbf{A} = (\mathbf{G}_w^T\mathbf{G}_w)^{-1}\mathbf{G}_w^T$. The general covariance matrix of the model parameters for a least squares solution is thus

$$\text{Cov}(\mathbf{m}_{L_2}) = (\mathbf{G}_w^T\mathbf{G}_w)^{-1}\mathbf{G}_w^T\text{Cov}(\mathbf{d}_w)\mathbf{G}_w(\mathbf{G}_w^T\mathbf{G}_w)^{-1} \ . \tag{2.25}$$

If the weighted data are independent, and thus have an identity covariance matrix, this simplifies to

$$\text{Cov}(\mathbf{m}_{L_2}) = (\mathbf{G}_w^T\mathbf{G}_w)^{-1}\mathbf{G}_w^T\mathbf{I}_m\mathbf{G}_w(\mathbf{G}_w^T\mathbf{G}_w)^{-1} = (\mathbf{G}_w^T\mathbf{G}_w)^{-1} \ . \tag{2.26}$$

In the case of independent and identically distributed normal data errors, so that the data covariance matrix $\text{Cov}(\mathbf{d})$ is simply the variance σ^2 times the m by m identity matrix $\mathbf{I}_m$ (2.26) can be written in terms of the unweighted system matrix as

$$\text{Cov}(\mathbf{m}_{L_2}) = \sigma^2(\mathbf{G}^T\mathbf{G})^{-1} \ . \tag{2.27}$$

Note that even for a diagonal data covariance matrix, the model covariance matrix is typically not diagonal, and the model parameters are thus correlated. Because elements of least squares models are each constructed from linear combinations of the data vector elements, this statistical dependence between the elements of $\mathbf{m}$ should not be surprising.

The expected value of the least squares solution is

$$E[\mathbf{m}_{L_2}] = (\mathbf{G}_w^T \mathbf{G}_w)^{-1} \mathbf{G}_w^T E[\mathbf{d}_w] \ . \tag{2.28}$$

Because $E[\mathbf{d}_w] = \mathbf{d}_{\text{true},w}$, and $\mathbf{G}_w \mathbf{m}_{\text{true}} = \mathbf{d}_{\text{true},w}$, we have

$$\mathbf{G}_w^T \mathbf{G}_w \mathbf{m}_{\text{true}} = \mathbf{G}_w^T \mathbf{d}_{\text{true},w} \ . \tag{2.29}$$

Thus

$$E[\mathbf{m}_{L_2}] = (\mathbf{G}_w^T \mathbf{G}_w)^{-1} \mathbf{G}_w^T \mathbf{G}_w \mathbf{m}_{\text{true}} \tag{2.30}$$
$$= \mathbf{m}_{\text{true}} \ . \tag{2.31}$$

In statistical terms, the least squares solution is said to be **unbiased**.

We can compute 95% confidence intervals for individual model parameters using the fact that each model parameter m_i has a normal distribution with mean given by the corresponding element of $\mathbf{m}_{\text{true}}$ and variance $\text{Cov}(\mathbf{m}_{L_2})_{i,i}$. The 95% confidence intervals are given by

$$\mathbf{m}_{L_2} \pm 1.96 \ \text{diag}(\text{Cov}(\mathbf{m}_{L_2}))^{1/2} \tag{2.32}$$

where the 1.96 factor arises from

$$\frac{1}{\sigma\sqrt{2\pi}} \int_{-1.96\sigma}^{1.96\sigma} e^{-\frac{x^2}{2\sigma^2}} \ dx \approx 0.95 \ . \tag{2.33}$$

Example 2.1

Let us recall Example 1.1 of linear regression of ballistic observations to a quadratic model, where the regression model is

$$y(t) = m_1 + m_2 t - (1/2)m_3 t^2 \ . \tag{2.34}$$

Here y is measured in the upward direction, and the minus sign is applied to the third term because gravitational acceleration is downward. Consider a synthetic data set with $m = 10$ observations and independent normal data errors ($\sigma = 8$ m), generated using

$$\mathbf{m}_{\text{true}} = [10 \text{ m}, \ 100 \text{ m/s}, \ 9.8 \text{ m/s}^2]^T. \tag{2.35}$$

The synthetic data set is in the file **data1.mat**.

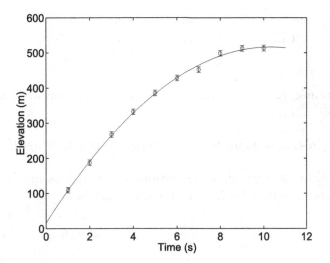

Figure 2.1 Data and model predictions for the ballistics example, with 1 standard deviation data error bounds indicated.

To obtain the least squares solution, we construct the **G** matrix. The ith row of **G** is given by

$$\mathbf{G}_{i,\cdot} = [1, \ t_i, \ -(1/2)t_i^2] \tag{2.36}$$

so that

$$\mathbf{G} = \begin{bmatrix} 1 & 1 & -0.5 \\ 1 & 2 & -2.0 \\ 1 & 3 & -4.5 \\ 1 & 4 & -8.0 \\ 1 & 5 & -12.5 \\ 1 & 6 & -18.0 \\ 1 & 7 & -24.5 \\ 1 & 8 & -32.0 \\ 1 & 9 & -40.5 \\ 1 & 10 & -50.0 \end{bmatrix}. \tag{2.37}$$

We solve for the parameters using the weighted normal equations, (2.18), to obtain a model estimate

$$\mathbf{m}_{L_2} = [16.4 \text{ m}, \ 97.0 \text{ m/s}, \ 9.4 \text{ m/s}^2]^T . \tag{2.38}$$

Fig. 2.1 shows the observed data and the fitted curve. The model covariance matrix associated with $\mathbf{m}_{L_2}$ is

$$\text{Cov}(\mathbf{m}_{L_2}) = \begin{bmatrix} 88.53 & -33.60 & -5.33 \\ -33.60 & 15.44 & 2.67 \\ -5.33 & 2.67 & 0.48 \end{bmatrix}. \tag{2.39}$$

For this example, (2.32) gives the following parameter estimates with 95% confidence intervals:

$$\mathbf{m}_{L_2} = [16.42 \pm 18.44 \text{ m}, \ 96.97 \pm 7.70 \text{ m/s}, \ 9.41 \pm 1.36 \text{ m/s}^2]^T. \tag{2.40}$$

The χ^2 value for this regression is approximately 4.2, and the number of degrees of freedom is $\nu = m - n = 10 - 3 = 7$, so the p-value, (2.22), is

$$p = \int_{4.20}^{\infty} \frac{1}{2^{7/2} \Gamma(7/2)} x^{\frac{5}{2}} e^{-\frac{x}{2}} \ dx \approx 0.76 \tag{2.41}$$

which is in the realm of plausibility. This means that the fitted model is consistent with the modeling and data uncertainty assumptions.

If we consider combinations of model parameters, the interpretation of the uncertainty in the model parameters becomes more complicated. To characterize model uncertainty more effectively, we can examine 95% **confidence regions** for pairs or larger sets of parameters. When joint parameter confidence regions are projected onto the coordinate axes m_i we obtain intervals for parameters that may be significantly larger than we would estimate when considering parameters individually, as in (2.40).

For a vector of estimated model parameters $\mathbf{m}_{L_2}$ characterized by an n-dimensional multivariate normal distribution with mean $\mathbf{m}_{\text{true}}$ and covariance matrix $\mathbf{C} = \text{Cov}(\mathbf{m}_{L_2})$,

$$(\mathbf{m}_{\text{true}} - \mathbf{m}_{L_2})^T \mathbf{C}^{-1} (\mathbf{m}_{\text{true}} - \mathbf{m}_{L_2}) \tag{2.42}$$

can be shown to have a χ^2 distribution with n degrees of freedom [111]. Thus, if Δ^2 is the 95th percentile of the χ^2 distribution with n degrees of freedom (B.9), the 95% confidence region is defined by the inequality

$$(\mathbf{m}_{\text{true}} - \mathbf{m}_{L_2})^T \mathbf{C}^{-1} (\mathbf{m}_{\text{true}} - \mathbf{m}_{L_2}) \leq \Delta^2. \tag{2.43}$$

The confidence region defined by this inequality is an n-dimensional ellipsoid.

If we wish to find an error ellipsoid for a lower dimensional subset of the model parameters, we can project the n-dimensional error ellipsoid onto the lower dimensional subspace by taking only those rows and columns of $\mathbf{C}$ and elements of $\mathbf{m}$ that correspond to the dimensions we want to keep [1]. In this case, the number of degrees of freedom in the associated Δ^2 calculation should be correspondingly reduced from n to match the number of model parameters in the projected error ellipsoid.

Since the covariance matrix and its inverse are symmetric and positive definite, we can diagonalize $\mathbf{C}^{-1}$ using (A.77) as

$$\mathbf{C}^{-1} = \mathbf{Q}\boldsymbol{\Lambda}\mathbf{Q}^{T} \tag{2.44}$$

where $\boldsymbol{\Lambda}$ is a diagonal matrix of positive eigenvalues and the columns of $\mathbf{Q}$ are orthonormal eigenvectors. The semiaxes defined by the columns of $\mathbf{Q}$ and associated eigenvalues are referred to as the **principal axes** of the error ellipsoid. The ith semimajor error ellipsoid axis direction is defined by $\mathbf{Q}_{\cdot,i}$ and its corresponding length is $\Delta/\sqrt{\Lambda_{i,i}}$. For example, if the vector $\mathbf{m}_{\text{true}} - \mathbf{m}_{L_2}$ is parallel to $\mathbf{Q}_{\cdot,i}$ and has length $\frac{\Delta}{\sqrt{\Lambda_{i,i}}}$, i.e.,

$$\mathbf{m}_{\text{true}} - \mathbf{m}_{L_2} = \frac{\Delta}{\sqrt{\Lambda_{i,i}}}\mathbf{Q}_{\cdot,i} , \tag{2.45}$$

then

$$(\mathbf{m}_{\text{true}} - \mathbf{m}_{L_2})^{T}\mathbf{C}^{-1}(\mathbf{m}_{\text{true}} - \mathbf{m}_{L_2}) = \frac{\Delta}{\sqrt{\Lambda_{i,i}}}\mathbf{Q}_{\cdot,i}^{T}\mathbf{Q}\boldsymbol{\Lambda}\mathbf{Q}^{T}\mathbf{Q}_{\cdot,i}\frac{\Delta}{\sqrt{\Lambda_{i,i}}} \tag{2.46}$$

$$= \frac{\Delta}{\sqrt{\Lambda_{i,i}}}\mathbf{e}_{i}^{T}\boldsymbol{\Lambda}\mathbf{e}_{i}\frac{\Delta}{\sqrt{\Lambda_{i,i}}} \tag{2.47}$$

$$= \Delta^{2} . \tag{2.48}$$

Because the model covariance matrix is typically not diagonal, the principal axes are typically not aligned in the m_i axis directions. However, we can project the appropriate confidence ellipsoid onto the m_i axes to obtain an n-dimensional "box" that includes the entire 95% error ellipsoid, along with some additional external volume. Such a box provides a conservative confidence interval for a joint collection of model parameters.

Correlations for parameter pairs (m_i, m_j) are measures of the inclination of the error ellipsoid principal axes with respect to the parameter axes. A correlation approaching $+1$ means the projection is needle-like with its long principal axis having a positive slope, a near-zero correlation means that the projection has principal axes that are nearly aligned with the axes of the (m_i, m_j) plane, and a correlation approaching -1 means that the projection is needle-like with its long principal axis having a negative slope.

Example 2.2

The parameter correlations for Example 2.1 are

$$\rho_{m_i, m_j} = \frac{\text{Cov}(m_i, m_j)}{\sqrt{\text{Var}(m_i)\,\text{Var}(m_j)}} \tag{2.49}$$

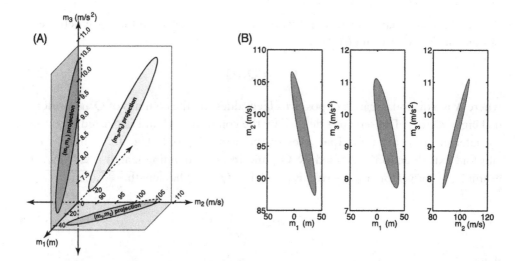

Figure 2.2 Projections of the 95% error ellipsoid onto model axes. (A) Projections in perspective; (B) projections onto the parameter axis planes.

which give

$$\rho_{m_1,m_2} = -0.91 \tag{2.50}$$

$$\rho_{m_1,m_3} = -0.81 \tag{2.51}$$

$$\rho_{m_2,m_3} = 0.97 \;. \tag{2.52}$$

The three model parameters are highly statistically dependent, and the error ellipsoid is thus inclined in model space. Fig. 2.2 shows the 95% confidence ellipsoid.

Diagonalization of $\mathbf{C}^{-1}$ (2.44) shows that the directions of the semiaxes for the error ellipsoid are

$$\mathbf{Q} = [\mathbf{Q}_{\cdot,1}, \; \mathbf{Q}_{\cdot,2}, \; \mathbf{Q}_{\cdot,3}] \approx \begin{bmatrix} 0.93 & 0.36 & -0.03 \\ -0.36 & 0.90 & -0.23 \\ -0.06 & 0.23 & 0.97 \end{bmatrix} \tag{2.53}$$

with corresponding eigenvalues

$$[\lambda_1, \; \lambda_2, \; \lambda_3] \approx [0.0098, \; 0.4046, \; 104.7] \;. \tag{2.54}$$

The corresponding 95% confidence ellipsoid semiaxis lengths are ((2.42) and [111])

$$\sqrt{F_{\chi^2_{\cdot,3}}^{-1}(0.95)}[1/\sqrt{\lambda_1}, \; 1/\sqrt{\lambda_2}, \; 1/\sqrt{\lambda_3}] \approx [28.2, \; 4.4, \; 0.27] \tag{2.55}$$

where $F_{\chi^2,3}^{-1}(0.95) \approx 7.80$ is the 95th percentile of the χ^2 distribution with three degrees of freedom.

Projecting the 95% confidence ellipsoid defined by (2.53) and (2.55) into the (m_1, m_2, m_3) coordinate system, and selecting maximum absolute values in the m_i directions to define an ellipsoid-bounding box, we obtain 95% confidence intervals for the parameters considered jointly

$$[m_1, m_2, m_3] = [16.42 \pm 26.25 \text{ m}, \ 96.97 \pm 10.24 \text{ m/s}, \ 9.41 \pm 1.65 \text{ m/s}^2] \qquad (2.56)$$

that are appreciably broader than the single parameter confidence estimates obtained using only the diagonal covariance matrix terms in (2.40). Note that there is actually a greater than 95% probability that the box defined by (2.56) will include the true values of the parameters. The reason is that these intervals, considered together as a rectangular prism-shaped region, include a significant parameter space volume that lies outside of the 95% confidence ellipsoid.

It is insightful to note that the model covariance matrix (2.25) does not depend on the estimated model, but depends solely on the system matrix and data covariance. Model covariance is thus exclusively a characteristic of experimental design that reflects how much influence the noise in a *general* data set will have on a model estimate, not upon *particular* data values from an individual experiment. In assessing a model, it is essential to evaluate the *p*-value, or another "goodness-of-fit" measure, because an examination of the solution parameters and the covariance matrix alone does *not* reveal whether we are actually fitting the data adequately.

2.3. AN ALTERNATIVE VIEW OF THE 95% CONFIDENCE ELLIPSOID

Recall (2.31) that in linear regression, the least squares solution $\mathbf{m}_{L_2}$ for zero-mean, multivariate-distributed normal data errors itself has a multivariate normal distribution with

$$E[\mathbf{m}_{L_2}] = \mathbf{m}_{\text{true}} \ . \qquad (2.57)$$

By (2.26), the model covariance matrix is

$$\mathbf{C} = \text{Cov}(\mathbf{m}_{L_2}) = (\mathbf{G}_w^T \mathbf{G}_w)^{-1} \ , \qquad (2.58)$$

where the rows of $\mathbf{G}_w$ are those of $\mathbf{G}$ that have been weighted by respective reciprocal data standard deviations (2.15), and we assume that the data errors are independent and that $(\mathbf{G}_w^T \mathbf{G}_w)$ is nonsingular. By Theorem B.6,

$$(\mathbf{m}_{\text{true}} - \mathbf{m}_{L_2})^T \mathbf{C}^{-1} (\mathbf{m}_{\text{true}} - \mathbf{m}_{L_2}) \qquad (2.59)$$

has a χ^2 distribution with degrees of freedom equal to the number of model parameters n. Let Δ^2 be the 95th percentile of the χ^2 distribution with n degrees of freedom. Then the probability

$$P\left((\mathbf{m}_{\text{true}} - \mathbf{m}_{L_2})^T \mathbf{C}^{-1}(\mathbf{m}_{\text{true}} - \mathbf{m}_{L_2}) \le \Delta^2\right) \tag{2.60}$$

will be 0.95.

Although (2.60) describes an ellipsoid centered at $\mathbf{m}_{\text{true}}$, the inequality is symmetric in $\mathbf{m}_{\text{true}}$ and $\mathbf{m}_{L_2}$, and can also therefore be thought of as defining an ellipsoid centered around $\mathbf{m}_{L_2}$. Thus, there is a 95% probability that when we gather our data and compute $\mathbf{m}_{L_2}$, the true model $\mathbf{m}_{\text{true}}$ will lie within the model space ellipsoid defined by

$$(\mathbf{m} - \mathbf{m}_{L_2})^T \mathbf{C}^{-1}(\mathbf{m} - \mathbf{m}_{L_2}) \le \Delta^2 . \tag{2.61}$$

Since $\mathbf{C} = (\mathbf{G}_w^T \mathbf{G}_w)^{-1}$, $\mathbf{C}^{-1} = \mathbf{G}_w^T \mathbf{G}_w$, and the 95th percentile confidence ellipsoid can also be written as

$$(\mathbf{m} - \mathbf{m}_{L_2})^T \mathbf{G}_w^T \mathbf{G}_w (\mathbf{m} - \mathbf{m}_{L_2}) \le \Delta^2 . \tag{2.62}$$

If we let

$$\chi^2(\mathbf{m}) = \|\mathbf{G}_w \mathbf{m} - \mathbf{d}_w\|_2^2 = (\mathbf{G}_w \mathbf{m} - \mathbf{d}_w)^T (\mathbf{G}_w \mathbf{m} - \mathbf{d}_w) \tag{2.63}$$

then

$$\begin{aligned}
\chi^2(\mathbf{m}) - \chi^2(\mathbf{m}_{L_2}) &= (\mathbf{G}_w \mathbf{m} - \mathbf{d}_w)^T (\mathbf{G}_w \mathbf{m} - \mathbf{d}_w) - (\mathbf{G}_w \mathbf{m}_{L_2} - \mathbf{d}_w)^T (\mathbf{G}_w \mathbf{m}_{L_2} - \mathbf{d}_w) \\
&= \mathbf{m}^T \mathbf{G}_w^T \mathbf{G}_w \mathbf{m} - \mathbf{d}_w^T \mathbf{G}_w \mathbf{m} - \mathbf{m}^T \mathbf{G}_w^T \mathbf{d}_w + \mathbf{d}_w^T \mathbf{d}_w \\
&\quad - \mathbf{m}_{L_2}^T \mathbf{G}_w^T \mathbf{G}_w \mathbf{m}_{L_2} + \mathbf{d}_w^T \mathbf{G}_w \mathbf{m}_{L_2} + \mathbf{m}_{L_2}^T \mathbf{G}_w^T \mathbf{d}_w - \mathbf{d}_w^T \mathbf{d}_w \\
&= \mathbf{m}^T \mathbf{G}_w^T \mathbf{G}_w \mathbf{m} - \mathbf{d}_w^T \mathbf{G}_w \mathbf{m} - \mathbf{m}^T \mathbf{G}_w^T \mathbf{d}_w \\
&\quad - \mathbf{m}_{L_2}^T \mathbf{G}_w^T \mathbf{G}_w \mathbf{m}_{L_2} + \mathbf{d}_w^T \mathbf{G}_w \mathbf{m}_{L_2} + \mathbf{m}_{L_2}^T \mathbf{G}_w^T \mathbf{d}_w .
\end{aligned} \tag{2.64}$$

Since $\mathbf{m}_{L_2}$ is a least squares solution to the weighted system of equations, it satisfies the corresponding normal equations. We can thus replace all occurrences of $\mathbf{G}_w^T \mathbf{d}_w$ with $\mathbf{G}_w^T \mathbf{G}_w \mathbf{m}_{L_2}$ using (2.3) to obtain

$$\begin{aligned}
\chi^2(\mathbf{m}) - \chi^2(\mathbf{m}_{L_2}) &= \mathbf{m}^T \mathbf{G}_w^T \mathbf{G}_w \mathbf{m} - \mathbf{m}_{L_2}^T \mathbf{G}_w^T \mathbf{G}_w \mathbf{m} - \mathbf{m}^T \mathbf{G}_w^T \mathbf{G}_w \mathbf{m}_{L_2} \\
&\quad - \mathbf{m}_{L_2}^T \mathbf{G}_w^T \mathbf{G}_w \mathbf{m}_{L_2} + \mathbf{m}_{L_2}^T \mathbf{G}_w^T \mathbf{G}_w \mathbf{m}_{L_2} + \mathbf{m}_{L_2}^T \mathbf{G}_w^T \mathbf{G}_w \mathbf{m}_{L_2} \\
&= \mathbf{m}^T \mathbf{G}_w^T \mathbf{G}_w \mathbf{m} - \mathbf{m}_{L_2}^T \mathbf{G}_w^T \mathbf{G}_w \mathbf{m} \\
&\quad - \mathbf{m}^T \mathbf{G}_w^T \mathbf{G}_w \mathbf{m}_{L_2} + \mathbf{m}_{L_2}^T \mathbf{G}_w^T \mathbf{G}_w \mathbf{m}_{L_2}
\end{aligned} \tag{2.65}$$

and, finally,

$$\chi^2(\mathbf{m}) - \chi^2(\mathbf{m}_{L_2}) = (\mathbf{m} - \mathbf{m}_{L_2})^T \mathbf{G}_w^T \mathbf{G}_w (\mathbf{m} - \mathbf{m}_{L_2}) . \tag{2.66}$$

Thus our 95% confidence ellipsoid can also be written as

$$\chi^2(\mathbf{m}) - \chi^2(\mathbf{m}_{L_2}) \le \Delta^2 \qquad (2.67)$$

and the contour of the $\chi^2(\mathbf{m})$ function at $\chi^2(\mathbf{m}_{L_2}) + \Delta^2$ gives the boundary of the 95th percentile confidence ellipsoid.

2.4. UNKNOWN MEASUREMENT STANDARD DEVIATIONS

Suppose that we do not know the standard deviations of the measurement errors *a priori*. In this case, if we assume that the measurement errors are independent and normally distributed with expected value of zero and standard deviation σ, then we can perform the linear regression and estimate σ from the residuals.

First, we find the least squares solution to the unweighted problem $\mathbf{Gm} = \mathbf{d}$, and let

$$\mathbf{r} = \mathbf{d} - \mathbf{Gm}_{L_2} . \qquad (2.68)$$

To estimate a constant data element standard deviation from the residuals, let

$$s = \frac{\|\mathbf{r}\|_2}{\sqrt{\nu}} \qquad (2.69)$$

where $\nu = m - n$ is the number of degrees of freedom [52].

As you might expect, there is a statistical cost associated with not knowing the true standard deviation. If the data standard deviations are known ahead of time, then each

$$\tilde{m}_i = \frac{m_i - m_{\text{true}_i}}{\sqrt{C_{i,i}}} \qquad (2.70)$$

where $\mathbf{C}$ is the covariance matrix (2.27), has a standard normal distribution. However, if instead of a known standard deviation we have an estimate s obtained using (2.69), then, if $\tilde{\mathbf{C}}$ is given by the covariance matrix formula (2.27), but with $\sigma = s$, each

$$\tilde{m}_i = \frac{m_i - m_{\text{true}_i}}{\sqrt{\tilde{C}_{i,i}}} \qquad (2.71)$$

has a Student's t distribution (B.7) with $\nu = m - n$ degrees of freedom. For smaller numbers of degrees of freedom this produces appreciably broader confidence intervals than the standard normal distribution. As ν becomes large, (2.69) becomes an increasingly better estimate of σ as the two distributions converge. Confidence ellipsoids corresponding to this case can also be computed, but the formula is somewhat more complicated than in the case of known standard deviations [52].

In assessing goodness-of-fit in this case, a problem arises in that we can no longer apply the χ^2 test. Recall that the χ^2 test was based on the assumption that the data errors

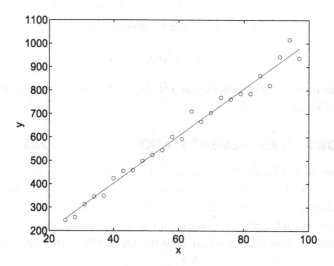

Figure 2.3 Data for Example 2.3, and corresponding linear regression line.

were normally distributed with known standard deviations σ_i. If the actual residuals were too large relative to the σ_i, then χ^2 would be large, and we would reject the linear regression fit based on a very small p-value. However, if we substitute (2.69) into (2.20), we find that $\chi^2_{\text{obs}} = \nu$, so such a model will always pass the χ^2 test.

Example 2.3

Consider the analysis of a linear regression problem in which the measurement errors are assumed to be independent and normally distributed, with equal but unknown standard deviations σ. We are given data y_i collected at points x_i (Fig. 2.3) that appear to follow a linear relationship.

In this case, the system matrix for the forward problem is

$$\mathbf{G} = \begin{bmatrix} 1 & x_1 \\ 1 & x_2 \\ \cdot & \cdot \\ \cdot & \cdot \\ \cdot & \cdot \\ 1 & x_m \end{bmatrix}. \tag{2.72}$$

The least squares solution to

$$\mathbf{Gm} = \mathbf{y} \tag{2.73}$$

has

$$y_i = -1.03 + 10.09x_i \, . \tag{2.74}$$

Fig. 2.3 shows the data and the linear regression line. Our estimate of the standard deviation of the measurement errors from (2.69) is $s = 30.74$. The estimated covariance matrix for the fitted parameters is

$$\tilde{\mathbf{C}} = s^2(\mathbf{G}^T\mathbf{G})^{-1} = \begin{bmatrix} 338.24 & -4.93 \\ -4.93 & 0.08 \end{bmatrix} . \tag{2.75}$$

The parameter confidence intervals, evaluated for each parameter separately, are

$$m_1 = -1.03 \pm \sqrt{338.24} \; t_{m-2,0.975} = -1.03 \pm 38.05 \tag{2.76}$$

and

$$m_2 = 10.09 \pm \sqrt{0.08} \; t_{m-2,0.975} = 10.09 \pm 0.59 \, . \tag{2.77}$$

Since the actual standard deviation of the measurement errors is unknown, we cannot perform a χ^2 test of goodness-of-fit. However, we can still examine the residuals. Fig. 2.4 shows the residuals. It is clear that although they appear to be random, the standard deviation seems to increase as x and y increase. This is a common phenomenon in linear regression, called a **proportional effect**. One possible way that such an effect might occur is if the measurement errors were proportional to the measurement magnitude. This could occur, for example, due to characteristics of the instrumentation.

For independent errors where the standard deviations of the data points are proportional to the observation, we can rescale the system of equations (2.73) by dividing each equation by y_i, to obtain

$$\mathbf{G}_w\mathbf{m} = \mathbf{y}_w \, . \tag{2.78}$$

If statistical assumptions are correct, (2.78) has a least squares solution that approximates (2.18). We obtain a revised least squares estimate of

$$y_i = -12.24 + 10.25x_i \tag{2.79}$$

with 95% parameter confidence intervals, evaluated as in (2.76) and (2.77), of

$$m_1 = -12.24 \pm 22.39 \tag{2.80}$$

and

$$m_2 = 10.25 \pm 0.47 \, . \tag{2.81}$$

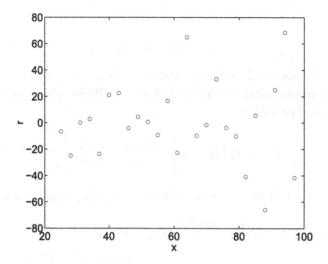

Figure 2.4 Unweighted residuals for Example 2.3.

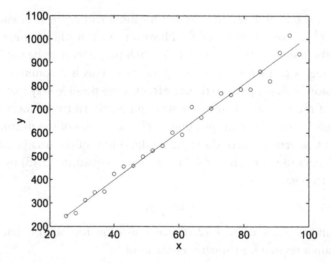

Figure 2.5 Data for Example 2.3, and corresponding linear regression line, weighted system.

Fig. 2.5 shows the data and least squares fit. Fig. 2.6 shows the scaled residuals. Note that there is now no obvious trend in the magnitude of the residuals as x and y increase, as there was in Fig. 2.4. The estimated standard deviation is 0.045, or 4.5% of the y value. In fact, these data were generated according to the true model $y_i = 10x_i + 0$, using standard deviations for the measurement errors that were 5% of the y value.

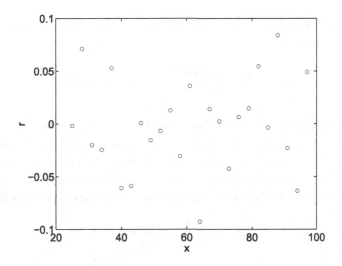

Figure 2.6 Weighted residuals for Example 2.3.

2.5. L_1 REGRESSION

Least squares solutions are highly susceptible to even small numbers of discordant observations, or **outliers**. Outliers are data points that are highly inconsistent with the other data. Outliers may arise from procedural measurement error, for example from incorrectly recording the position of a decimal point in an observation, or from instrumental glitches. Outliers should be assessed carefully, since the data might actually be showing us that the form of the mathematical model that we are trying to fit is incorrect. However, if we conclude that there are only a small number of outliers in the data due to incorrect measurements, we need to analyze the data in a way that minimizes their effect on the estimated model.

We can readily appreciate the strong effect of outliers on least squares solutions from a maximum likelihood perspective by noting the very rapid fall-off of the tails of the normal distribution. For example, the probability of a single data point drawn from a normal distribution being more than five standard deviations away from its expected value is less than one in one million

$$P(|X - E[X]| \geq 5\sigma) = \frac{2}{\sqrt{2\pi}} \int_5^\infty e^{-\frac{1}{2}x^2} \, dx \approx 6 \times 10^{-7} \, . \qquad (2.82)$$

If an outlier occurs in the data set due to a non-normal error process, the least squares solution will go to great lengths to accommodate it, and thus prevent its corresponding factor in the total likelihood product (2.10) from being vanishingly small.

As an alternative to least squares, consider the solution that minimizes the 1-norm of the residual vector,

$$\mu^{(1)} = \sum_{i=1}^{m} \frac{|d_i - (\mathbf{Gm})_i|}{\sigma_i} = \|\mathbf{d}_w - \mathbf{G}_w\mathbf{m}\|_1 . \qquad (2.83)$$

This **1-norm solution** $\mathbf{m}_{L_1}$ will be more outlier resistant, or **robust**, than the least squares solution $\mathbf{m}_{L_2}$, because (2.83) does not square each of the terms in the misfit measure, as (2.13) does. The 1-norm solution $\mathbf{m}_{L_1}$ also has a maximum likelihood interpretation, in that it is the maximum likelihood estimator for data with errors distributed according to a double-sided exponential distribution (Appendix B)

$$f(x) = \frac{1}{\sigma\sqrt{2}} e^{-\sqrt{2}|x-\mu|/\sigma} . \qquad (2.84)$$

Data sets with errors distributed as (2.84) are unusual. Nevertheless, it is often worthwhile to consider a solution in which (2.83) is minimized rather than (2.13), even if most of the measurement errors are normally distributed, should there be reason to suspect the presence of outliers. This solution strategy may be useful if the data outliers occur for reasons that do not undercut our belief that the mathematical model is otherwise correct.

Example 2.4

We can demonstrate the advantages of 1-norm minimization using the quadratic ballistics regression example discussed earlier. Fig. 2.7 shows the original sequence of independent data points with unit standard deviations, except one of the points (d_4) is now an outlier with respect to a mathematical model of the form (2.34). It is the original data point with 200 m subtracted from it. The least squares model for this data set is

$$\mathbf{m}_{L_2} = [26.4 \text{ m}, \ 75.6 \text{ m/s}, \ 4.86 \text{ m/s}^2]^T . \qquad (2.85)$$

The least squares solution is skewed away from the majority of data points in trying to accommodate the outlier and is a poor estimate of the true model. We can also see that (2.85) fails to fit these data acceptably because of its huge χ^2 value (≈ 489). This is clearly astronomically out of bounds for a problem with 7 degrees of freedom, where the χ^2 value should not be far from 7. The corresponding p-value for $\chi^2 = 489$ is effectively zero.

The upper curve in Fig. 2.7

$$\mathbf{m}_{L_1} = [17.6 \text{ m}, \ 96.4 \text{ m/s}, \ 9.31 \text{ m/s}^2]^T \qquad (2.86)$$

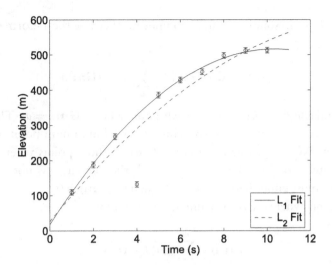

Figure 2.7 L_1 (solid) and L_2 (dashed) solutions for a parabolic data set with an outlier at $t = 4$ s.

is obtained using the 1-norm solution that minimizes (2.83). The data prediction from (2.86) faithfully fits the quadratic trend for the majority of the data points and is only slightly influenced by the outlier at $t = 4$. It is also much closer than (2.85) to the true model (2.35), and to the least squares model for the data set without the outlier (2.38).

In examining the differences between 2- and 1-norm models, it is instructive to consider the almost trivial regression problem of estimating the value of a single parameter from repeated measurements. The system of equations $\mathbf{Gm} = \mathbf{d}$ is

$$\begin{bmatrix} 1 \\ 1 \\ 1 \\ . \\ . \\ . \\ 1 \end{bmatrix} \mathbf{m} = \begin{bmatrix} d_1 \\ d_2 \\ d_3 \\ . \\ . \\ . \\ d_m \end{bmatrix} \tag{2.87}$$

where $\mathbf{m}$ is the 1 by 1 vector containing the parameter that we wish to estimate.

The least squares solution to (2.87) can be seen from the normal equations (A.73) to be simply the observational average

$$\mathbf{m}_{L_2} = (\mathbf{G}^T\mathbf{G})^{-1}\mathbf{G}^T\mathbf{d} = m^{-1} \sum_{i=1}^{m} d_i . \tag{2.88}$$

Finding the 1-norm solution is more complicated because the 1-norm of the residual vector

$$f(\mathbf{m}) = \|\mathbf{d} - \mathbf{Gm}\|_1 = \sum_{i=1}^{m} |d_i - (\mathbf{Gm})_i| \qquad (2.89)$$

is a nondifferentiable function of $\mathbf{m}$ at each point where $(\mathbf{Gm})_i = d_i$. The good news is that $f(\mathbf{m})$ is a convex function of $\mathbf{m}$. Thus any local minimum point is also a global minimum point. We can proceed by finding $f'(\mathbf{m})$ at those points where it is defined, and then separately consider the points at which the derivative is not defined. Every minimum point must either have $f'(\mathbf{m})$ undefined or $f'(\mathbf{m}) = 0$.

At those points where $f'(\mathbf{m})$ is defined, it is given by

$$f'(\mathbf{m}) = \sum_{i=1}^{m} \mathrm{sgn}(d_i - \mathbf{m}_1) \,, \qquad (2.90)$$

where the **signum function** $\mathrm{sgn}(x)$ is -1 if its argument is negative, 1 if its argument is positive, and 0 if its argument is zero. The derivative (2.90) is zero when exactly half of the data are less than $\mathbf{m}$ and half of the data are greater than $\mathbf{m}$. Of course, this can only happen when the number of observations m is even. In this case, any value of $\mathbf{m}$ lying between the two middle observations is a 1-norm solution. When there are an odd number of data, the median data point is the unique 1-norm solution. Even an extreme outlier will not have a large effect on the median of an otherwise clustered set of observations. This illuminates the robustness of the 1-norm solution.

The general problem of finding solutions that minimize $\|\mathbf{d} - \mathbf{Gm}\|_1$ is a bit complicated. One practical method is **iteratively reweighted least squares**, or **IRLS** [177]. The IRLS algorithm solves a sequence of weighted least squares problems whose solutions converge to a 1-norm minimizing solution. Beginning with the residual vector

$$\mathbf{r} = \mathbf{d} - \mathbf{Gm} \qquad (2.91)$$

we want to minimize

$$f(\mathbf{m}) = \|\mathbf{r}\|_1 = \sum_{i=1}^{m} |r_i| \,. \qquad (2.92)$$

The function in (2.92), like the function in (2.89), is nondifferentiable at any point where one of the elements of $\mathbf{r}$ is zero. Ignoring this issue for a moment, we can go ahead and compute the derivatives of f at other points.

$$\frac{\partial f(\mathbf{m})}{\partial m_k} = \sum_{i=1}^{m} \frac{\partial |r_i|}{\partial m_k} = -\sum_{i=1}^{m} G_{i,k} \, \mathrm{sgn}(r_i) \,. \qquad (2.93)$$

Writing $\text{sgn}(r_i)$ as $r_i/|r_i|$ gives

$$\frac{\partial f(\mathbf{m})}{\partial m_k} = -\sum_{i=1}^{m} G_{i,k} \frac{1}{|r_i|} r_i \ . \tag{2.94}$$

The gradient of f is

$$\nabla f(\mathbf{m}) = -\mathbf{G}^T \mathbf{R} \mathbf{r} = -\mathbf{G}^T \mathbf{R}(\mathbf{d} - \mathbf{G}\mathbf{m}) \tag{2.95}$$

where $\mathbf{R}$ is a diagonal weighting matrix with diagonal elements that are the absolute values of the reciprocals of the residuals, so that

$$R_{i,i} = 1/|r_i| \ . \tag{2.96}$$

To find the 1-norm minimizing solution, we solve $\nabla f(\mathbf{m}) = \mathbf{0}$, which gives

$$\mathbf{G}^T \mathbf{R}(\mathbf{d} - \mathbf{G}\mathbf{m}) = \mathbf{0} \tag{2.97}$$

or

$$\mathbf{G}^T \mathbf{R} \mathbf{G} \mathbf{m} = \mathbf{G}^T \mathbf{R} \mathbf{d} \ . \tag{2.98}$$

Because $\mathbf{R}$ is a nonlinear function of $\mathbf{m}$, (2.98) is a nonlinear system of equations that we cannot solve directly. IRLS is an iterative algorithm to find the appropriate weights and solve the system. The algorithm begins with the least squares solution $\mathbf{m}^{(0)} = \mathbf{m}_{L_2}$. We calculate the corresponding residual vector $\mathbf{r}^{(0)} = \mathbf{d} - \mathbf{G}\mathbf{m}^{(0)}$ to construct the weighting matrix $\mathbf{R}$ using (2.96). We then solve (2.98) to obtain a new model $\mathbf{m}^{(1)}$ and associated residual vector $\mathbf{r}^{(1)}$. The process is repeated until the model and residual vectors converge. A typical rule is to stop the iteration when

$$\frac{\|\mathbf{m}^{(k+1)} - \mathbf{m}^{(k)}\|_2}{1 + \|\mathbf{m}^{(k+1)}\|_2} < \tau \tag{2.99}$$

for a specified tolerance τ.

If any element of the residual vector becomes zero, then (2.96) becomes undefined. However, this problem can be easily addressed by selecting a tolerance ϵ below which we consider the residuals to be effectively zero. If $|r_i| < \epsilon$, then we set $R_{i,i} = 1/\epsilon$. With this modification it can be shown that this procedure will always converge to an approximate 1-norm minimizing solution. We have implemented the IRIS algorithm in the library function **irls**.

As with the χ^2 misfit measure for least squares solutions, there is a corresponding p-value that can be evaluated under the assumption of normal data errors for the assessment of 1-norm solutions. Let

$$\mu_{\text{obs}}^{(1)} = \|\mathbf{G}\mathbf{m}_{L_1} - \mathbf{d}\|_1 \ . \tag{2.100}$$

For an observed 1-norm misfit measure (2.100) the probability that a worse misfit could have occurred given independent and normally distributed data and v degrees of freedom is approximately given by [160]

$$p^{(1)}(x,\ v) = P(\mu^{(1)} > \mu^{(1)}_{\text{obs}}) = 1 - S(x) + \frac{\gamma Z^{(2)}(x)}{6} \tag{2.101}$$

where

$$S(x) = \frac{1}{\sigma_1\sqrt{2\pi}} \int_{-\infty}^{x} e^{-\frac{\xi^2}{2\sigma_1^2}}\, d\xi \tag{2.102}$$

$$\sigma_1 = \sqrt{(1 - 2/\pi)v} \tag{2.103}$$

$$\gamma = \frac{2 - \pi/2}{(\pi/2 - 1)^{3/2}v^{\frac{1}{2}}} \tag{2.104}$$

$$Z^{(2)}(x) = \frac{x^2 - 1}{\sqrt{2\pi}} e^{-\frac{x^2}{2}} \tag{2.105}$$

$$x = \frac{\mu^{(1)} - \sqrt{2/\pi}\ v}{\sigma_1}. \tag{2.106}$$

2.6. MONTE CARLO ERROR PROPAGATION

For solution techniques that are nonlinear and/or algorithmic, such as IRLS, there is typically no analytical way to propagate uncertainties in the data to uncertainties in the estimated model parameters. In such cases, however, one can apply **Monte Carlo error propagation** techniques, in which we simulate a collection of noisy data vectors and then examine the statistics of the resulting ensemble of models.

For L_1 minimizing solutions, we can obtain an approximate covariance matrix by first forward-propagating the solution into an assumed noise-free baseline data vector

$$\mathbf{Gm}_{L_1} = \mathbf{d}_b. \tag{2.107}$$

We next re-solve the IRLS problem many times for independent noise realizations, obtaining a suite of q 1-norm solutions to

$$\mathbf{Gm}_{L_1,i} = \mathbf{d}_b + \boldsymbol{\eta}_i \tag{2.108}$$

where $\boldsymbol{\eta}_i$ is the ith noise vector realization. If $\mathbf{A}$ is the q by n matrix where the ith row contains the difference between the ith model estimate and the average model

$$\mathbf{A}_{i,\cdot} = \mathbf{m}^T_{L_1,i} - \bar{\mathbf{m}}^T_{L_1} \tag{2.109}$$

then an empirical estimate of the covariance matrix is

$$\text{Cov}(\mathbf{m}_{L_1}) = \frac{\mathbf{A}^T \mathbf{A}}{q} . \tag{2.110}$$

Example 2.5

Recall Example 2.4. An estimate of $\text{Cov}(\mathbf{m}_{L_1})$ using 10,000 iterations of the Monte Carlo procedure is

$$\text{Cov}(\mathbf{m}_{L_1}) = \begin{bmatrix} 128.9 & -47.15 & -7.312 \\ -47.15 & 21.28 & 3.630 \\ -7.31 & 3.630 & 0.6554 \end{bmatrix} \tag{2.111}$$

which contains elements that are about 1.4 times as large as those of the least squares solution (2.39). Unlike least squares solutions, model parameters obtained with the IRLS algorithm will not generally be normally distributed. However, we can compute approximate confidence intervals for the parameters from the covariance matrix diagonal, provided that analysis of the obtained Monte Carlo solution parameter distributions reveals that they are approximately normally distributed. Such an analysis can be performed by examining the parameter distributions with a Q–Q plot (Appendix B) and/or generating an ellipsoidal confidence boundary under normal assumptions and counting the proportion of points within the ellipsoid to check for consistency. In this example, Q–Q plots reveal the estimates to be approximately normally distributed, and calculating corresponding 95% confidence intervals from (2.111) using (2.32) gives

$$\mathbf{m}_{L_1} = [17.1 \pm 22.2 \text{ m}, \ 97.0 \pm 9.04 \text{ m/s}, \ 9.48 \pm 1.59 \text{ m/s}^2]^T . \tag{2.112}$$

2.7. EXERCISES

1. A seismic profiling experiment is performed where the first arrival times of seismic energy from a mid-crustal refractor are observed at a number of distances (in kilometers) from the source, at various times (in seconds after the source origin time). The data can be found in the MATLAB data file **profile.mat**. A two-layer flat Earth structure gives the mathematical model

$$t_i = t_0 + s_2 x_i \tag{2.113}$$

where the intercept time t_0 depends on the thickness and slowness of the upper layer, and s_2 is the slowness of the lower layer. The estimated noise in the first arrival time measurements is believed to be independent and normally distributed with expected value 0 and standard deviation $\sigma = 0.1$ s.

a. Find the least squares solution for the model parameters t_0 and s_2. Plot the data, the fitted model, and the residuals.

b. Calculate and comment on the model parameter correlation matrix (e.g., 2.49).

c. Plot the error ellipsoid in the (t_0, s_2) plane. How are the correlations manifested in the general appearance of the error ellipsoid in (t_0, s_2) space? Calculate conservative 95% confidence intervals for t_0 and s_2 for the appropriate value of Δ^2. Hint: The following MATLAB function will plot a two-dimensional covariance ellipse about the model parameters, where C is the covariance matrix, DELTA2 is Δ^2, and m is the 2-vector of model parameters.

```
%set the number of points on the ellipse to generate and plot
function plot_ellipse(DELTA2,C,m)
n=100;
%construct a vector of n equally-spaced angles from (0,2*pi)
theta=linspace(0,2*pi,n)';
%corresponding unit vector
xhat=[cos(theta),sin(theta)];
Cinv=inv(C);
%preallocate output array
r=zeros(n,2);
for i=1:n
  %store each (x,y) pair on the confidence ellipse
  %in the corresponding row of r
  r(i,:)=sqrt(DELTA2/(xhat(i,:)*Cinv*xhat(i,:)'))*xhat(i,:);
end
%
% Plot the ellipse and set the axes.
%
plot(m(1)+r(:,1), m(2)+r(:,2));
axis equal
```

d. Evaluate the p-value for this model. You may find the library function **chi2cdf** to be useful here.

e. Evaluate the value of χ^2 for 1000 Monte Carlo simulations using the data prediction from your model perturbed by noise that is consistent with the data assumptions. Compare a histogram of these χ^2 values with the theoretical χ^2 distribution for the correct number of degrees of freedom. You may find the library function **chi2pdf** to be useful here.

f. Are your p-value and Monte Carlo χ^2 distribution consistent with the theoretical modeling and the data set? If not, explain what is wrong.

g. Use IRLS to find 1-norm estimates for t_0 and s_2. Plot the data predictions from your model relative to the true data and compare with (a).

    h. Use Monte Carlo error propagation and IRLS to estimate symmetric 95% confidence intervals on the 1-norm solution for t_0 and s_2.

    i. Examining the contributions from each of the data points to the 1-norm misfit measure, can you make a case that any of the data points are statistical outliers?

2. In this chapter we have largely assumed that the data errors are independent. Suppose instead that the data errors have an MVN distribution with expected value **0** and a covariance matrix $\mathbf{C}_D$. It can be shown that the likelihood function is then (B.61)

$$L(\mathbf{m}|\mathbf{d}) = \frac{1}{(2\pi)^{m/2}} \frac{1}{\sqrt{\det(\mathbf{C}_D)}} e^{-(\mathbf{Gm}-\mathbf{d})^T \mathbf{C}_D^{-1}(\mathbf{Gm}-\mathbf{d})/2} . \tag{2.114}$$

    a. Show that the maximum likelihood estimate can be obtained by solving the minimization problem

$$\min (\mathbf{Gm} - \mathbf{d})^T \mathbf{C}_D^{-1}(\mathbf{Gm} - \mathbf{d}) . \tag{2.115}$$

    b. Show that (2.115) can be solved using the system of equations

$$\mathbf{G}^T \mathbf{C}_D^{-1} \mathbf{Gm} = \mathbf{G}^T \mathbf{C}_D^{-1} \mathbf{d} . \tag{2.116}$$

    c. Show that (2.115) is equivalent to the linear least squares problem

$$\min \|\mathbf{C}_D^{-1/2}\mathbf{Gm} - \mathbf{C}_D^{-1/2}\mathbf{d}\|_2 \tag{2.117}$$

    where $\mathbf{C}_D^{-1/2}$ is the matrix square root of $\mathbf{C}_D^{-1}$.

    d. The Cholesky factorization of $\mathbf{C}_D^{-1}$ can also be used instead of the matrix square root. Show that (2.115) is equivalent to the linear least squares problem

$$\min \|\mathbf{RGm} - \mathbf{Rd}\|_2 \tag{2.118}$$

    where $\mathbf{R}$ is the Cholesky factor of $\mathbf{C}_D^{-1}$.

3. Use MATLAB to generate 10,000 realizations of a data set of $m = 5$ points $\mathbf{d} = a + b\mathbf{x} + \boldsymbol{\eta}$, where $\mathbf{x} = [1, 2, 3, 4, 5]^T$, the $n = 2$ true model parameters are $a = b = 1$, and $\boldsymbol{\eta}$ is an m-element vector of independent $N(0, 1)$ noise.

    a. Assuming that the noise standard deviation is known *a priori* to be 1, solve for the parameters a and b using least squares for each realization and histogram them in 100 bins.

    b. Calculate the parameter covariance matrix, $\mathbf{C} = \sigma^2(\mathbf{G}^T\mathbf{G})^{-1}$, assuming independent $N(0, 1)$ data errors, and give standard deviations, σ_a and σ_b, for your estimates of a and b estimated from $\mathbf{C}$,

c. Calculate standardized parameter estimates

$$\tilde{a} = \frac{a-1}{\sqrt{C_{1,1}}} \qquad (2.119)$$

and

$$\tilde{b} = \frac{b-1}{\sqrt{C_{2,2}}} \qquad (2.120)$$

for your solutions for a and b. Demonstrate using a Q–Q plot (Appendix B) that your estimates for $\tilde{a}$ and $\tilde{b}$ are distributed as $N(0, 1)$.

d. Show using a Q–Q plot that the squared residual lengths

$$\|\mathbf{r}\|_2^2 = \|\mathbf{d} - \mathbf{Gm}\|_2^2 \qquad (2.121)$$

for your solutions in (a) are distributed as χ^2 with $m - n = \nu = 3$ degrees of freedom.

e. Assume that the noise standard deviation for the synthetic data set is not known, and instead estimate it for each realization k as

$$s_k = \sqrt{\frac{1}{m-n}\sum_{i=1}^{m} r_i^2} \, . \qquad (2.122)$$

Histogram your standardized solutions

$$\tilde{a} = \frac{a - \bar{a}}{\sqrt{\tilde{C}_{1,1}}} \qquad (2.123)$$

and

$$\tilde{b} = \frac{b - \bar{b}}{\sqrt{\tilde{C}_{2,2}}} \qquad (2.124)$$

where $\tilde{C} = s_k^2 (\mathbf{G}^T\mathbf{G})^{-1}$ is the covariance matrix estimation for the kth realization.

f. Demonstrate using a Q–Q plot that your estimates for $\tilde{a}$ and $\tilde{b}$ are distributed as the Student's t distribution with $\nu = 3$ degrees of freedom.

4. Suppose that we analyze a large number of data sets $\mathbf{d}$ in a linear regression problem and compute p-values for each data set. The χ^2_{obs} values should be distributed according to a χ^2 distribution with $m - n$ degrees of freedom. Show that the corresponding p-values will be uniformly distributed between 0 and 1.

5. Setup the normal equations to obtain the 20 coefficients a_i in the polynomial function

$$y_i = a_0 + a_1 x_i + a_2 x_i^2 + \cdots + a_{19} x_i^{19} \qquad (2.125)$$

in a least squares fit to the 21 noise-free data points

$$(x_i, \ y_i) = (-1, \ 1), \ (-0.9, \ -0.9), \ \ldots, (1, \ 1). \qquad (2.126)$$

Plot the fitted polynomial for $x = -1, -0.999, -0.998, \ldots, 1$, and show the data points. List the parameters a_i obtained in your regression. Clearly, the correct solution has $a_1 = 1$, and all other $a_i = 0$. Explain why your answer differs.

2.8. NOTES AND FURTHER READING

Linear regression is a major subfield within statistics, and there are literally hundreds of associated textbooks. Many of these references focus on applications of linear regression in the social sciences. In such applications, the primary focus is often on determining which variables have an effect on response variables of interest (rather than on estimating parameter values for a predetermined model). In this context it is important to compare alternative regression models and to test the hypothesis that a predictor variable has a nonzero coefficient in the regression model. Since we normally know which predictor variables are important in the physical sciences, the approach commonly differs. Useful linear regression references from the standpoint of estimating parameters in the context considered here include [52,142].

Robust statistical methods are an important topic. Huber discusses a variety of robust statistical procedures [97]. The problem of computing a 1-norm solution has been extensively researched. Techniques for 1-norm minimization include methods based on the simplex method for linear programming, interior point methods, and iteratively reweighted least squares [6,43,167,177]. The IRLS method is the simplest to implement, but interior point methods can be the most efficient approaches for very large problems. Watson reviews the history of methods for finding p-norm solutions including the 1-norm case [219].

We have assumed that **G** is known exactly. However, in some cases, elements of **G** might be subject to measurement error. This problem has been studied as the **total least squares problem** [98]. An alternative approach to least squares problems with uncertainties in **G** that has received considerable attention is called **robust least squares** [14,55].

CHAPTER THREE

Rank Deficiency and Ill-Conditioning

Synopsis

The characteristics of rank deficient and ill-conditioned linear systems of equations are explored using the singular value decomposition. The connection between model and data null spaces and solution uniqueness and ability to fit data is examined. Model and data resolution matrices are defined. The relationship between singular value size and singular vector roughness and its connection to the effect of noise on solutions are discussed in the context of the fundamental tradeoff between model resolution and instability. Specific manifestations of these issues in rank deficient and ill-conditioned discrete problems are shown in several examples.

3.1. THE SVD AND THE GENERALIZED INVERSE

A method of analyzing and solving least squares problems that is of particular interest in ill-conditioned and/or rank deficient systems is the **singular value decomposition**, or **SVD**. In the SVD [72,117,194], an m by n matrix $\mathbf{G}$ is factored into

$$\mathbf{G} = \mathbf{U}\mathbf{S}\mathbf{V}^T \tag{3.1}$$

where

- $\mathbf{U}$ is an m by m orthogonal matrix with columns that are unit basis vectors spanning the **data space**, R^m.
- $\mathbf{V}$ is an n by n orthogonal matrix with columns that are unit basis vectors spanning the **model space**, R^n.
- $\mathbf{S}$ is an m by n diagonal matrix with diagonal elements called **singular values**.

The SVD matrices can be computed in MATLAB with the **svd** command. It can be shown that every matrix has a singular value decomposition [72].

The singular values along the diagonal of $\mathbf{S}$ are customarily arranged in decreasing size, $s_1 \geq s_2 \geq \cdots \geq s_{\min(m,\,n)} \geq 0$. Note that some of the singular values may be zero. If only the first p singular values are nonzero, we can partition $\mathbf{S}$ as

$$\mathbf{S} = \begin{bmatrix} \mathbf{S}_p & \mathbf{0} \\ \mathbf{0} & \mathbf{0} \end{bmatrix} \tag{3.2}$$

where $\mathbf{S}_p$ is a p by p diagonal matrix composed of the positive singular values. Expanding the SVD representation of $\mathbf{G}$ in terms of the columns of $\mathbf{U}$ and $\mathbf{V}$ gives

$$\mathbf{G} = \begin{bmatrix} \mathbf{U}_{\cdot,1}, & \mathbf{U}_{\cdot,2}, & \ldots, & \mathbf{U}_{\cdot,m} \end{bmatrix} \begin{bmatrix} \mathbf{S}_p & \mathbf{0} \\ \mathbf{0} & \mathbf{0} \end{bmatrix} \begin{bmatrix} \mathbf{V}_{\cdot,1}, & \mathbf{V}_{\cdot,2}, & \ldots, & \mathbf{V}_{\cdot,n} \end{bmatrix}^T \tag{3.3}$$

Parameter Estimation and Inverse Problems
https://doi.org/10.1016/B978-0-12-804651-7.00008-0

$$= [\mathbf{U}_p, \ \mathbf{U}_0] \begin{bmatrix} \mathbf{S}_p & 0 \\ 0 & 0 \end{bmatrix} [\mathbf{V}_p, \ \mathbf{V}_0]^T \tag{3.4}$$

where $\mathbf{U}_p$ denotes the first p columns of $\mathbf{U}$, $\mathbf{U}_0$ denotes the last $m - p$ columns of $\mathbf{U}$, $\mathbf{V}_p$ denotes the first p columns of $\mathbf{V}$, and $\mathbf{V}_0$ denotes the last $n - p$ columns of $\mathbf{V}$. Because the last $m - p$ columns of $\mathbf{U}$ and the last $n - p$ columns of $\mathbf{V}$ in (3.4) are multiplied by zeros in $\mathbf{S}$, we can simplify the SVD of $\mathbf{G}$ into its **compact form**

$$\mathbf{G} = \mathbf{U}_p \mathbf{S}_p \mathbf{V}_p^T. \tag{3.5}$$

For any vector $\mathbf{y}$ in the range of $\mathbf{G}$, applying (3.5) gives

$$\mathbf{y} = \mathbf{G}\mathbf{x} \tag{3.6}$$

$$= \mathbf{U}_p \left(\mathbf{S}_p \mathbf{V}_p^T \mathbf{x} \right) . \tag{3.7}$$

Thus every vector in $R(\mathbf{G})$ can be written as $\mathbf{y} = \mathbf{U}_p \mathbf{z}$ where $\mathbf{z} = \mathbf{S}_p \mathbf{V}_p^T \mathbf{x}$. Writing out this matrix–vector multiplication, we see that any vector $\mathbf{y}$ in $R(\mathbf{G})$ can be written as a linear combination of the columns of $\mathbf{U}_p$

$$\mathbf{y} = \sum_{i=1}^{p} z_i \mathbf{U}_{\cdot,i} . \tag{3.8}$$

The columns of $\mathbf{U}_p$ span $R(\mathbf{G})$, are linearly independent, and form an orthonormal basis for $R(\mathbf{G})$. Because this orthonormal basis has p vectors, $\mathrm{rank}(\mathbf{G}) = p$.

Since $\mathbf{U}$ is an orthogonal matrix, the columns of $\mathbf{U}$ form an orthonormal basis for R^m. By Theorem A.5, $N(\mathbf{G}^T) + R(\mathbf{G}) = R^m$, so the remaining $m - p$ columns of $\mathbf{U}_0$ form an orthonormal basis for the null space of $\mathbf{G}^T$. Note that, because the null space basis is nonunique, and because basis vectors contain an inherent sign ambiguity, basis vectors calculated and illustrated in this chapter and elsewhere may not match ones calculated locally using the provided MATLAB code. We will sometimes refer to $N(\mathbf{G}^T)$ as the **data null space**. Similarly, because $\mathbf{G}^T = \mathbf{V}_p \mathbf{S}_p \mathbf{U}_p^T$, the columns of $\mathbf{V}_p$ form an orthonormal basis for $R(\mathbf{G}^T)$ and the columns of $\mathbf{V}_0$ form an orthonormal basis for $N(\mathbf{G})$. We will sometimes refer to $N(\mathbf{G})$ as the **model null space**.

Two other important SVD properties are similar to properties of eigenvalues and eigenvectors (Appendix A). Because the columns of $\mathbf{V}$ are orthonormal,

$$\mathbf{V}^T \mathbf{V}_{\cdot,i} = \mathbf{e}_i . \tag{3.9}$$

Thus

$$\mathbf{G}\mathbf{V}_{\cdot,i} = \mathbf{U}\mathbf{S}\mathbf{V}^T \mathbf{V}_{\cdot,i} \tag{3.10}$$

$$= \mathbf{U}\mathbf{S}\mathbf{e}_i \qquad (3.11)$$

$$= s_i \mathbf{U}_{\cdot,i} \qquad (3.12)$$

and, similarly,

$$\mathbf{G}^T\mathbf{U}_{\cdot,i} = \mathbf{V}\mathbf{S}^T\mathbf{U}^T\mathbf{U}_{\cdot,i} \qquad (3.13)$$

$$= \mathbf{V}\mathbf{S}^T\mathbf{e}_i \qquad (3.14)$$

$$= s_i \mathbf{V}_{\cdot,i} . \qquad (3.15)$$

There is a connection between the singular values of $\mathbf{G}$ and the eigenvalues of $\mathbf{G}\mathbf{G}^T$ and $\mathbf{G}^T\mathbf{G}$.

$$\mathbf{G}\mathbf{G}^T\mathbf{U}_{\cdot,i} = \mathbf{G}s_i\mathbf{V}_{\cdot,i} \qquad (3.16)$$

$$= s_i\mathbf{G}\mathbf{V}_{\cdot,i} \qquad (3.17)$$

$$= s_i^2\mathbf{U}_{\cdot,i} . \qquad (3.18)$$

Similarly,

$$\mathbf{G}^T\mathbf{G}\mathbf{V}_{\cdot,i} = s_i^2\mathbf{V}_{\cdot,i} . \qquad (3.19)$$

These relations show that we could, in theory, compute the SVD by finding the eigenvalues and eigenvectors of $\mathbf{G}^T\mathbf{G}$ and $\mathbf{G}\mathbf{G}^T$. In practice, more efficient specialized algorithms are used [50,72,209].

The SVD can be used to compute a generalized inverse of $\mathbf{G}$, called the **Moore–Penrose pseudoinverse**, because it has desirable inverse properties originally identified by Moore and Penrose [139,162]. The generalized inverse is

$$\mathbf{G}^\dagger = \mathbf{V}_p\mathbf{S}_p^{-1}\mathbf{U}_p^T . \qquad (3.20)$$

MATLAB has a **pinv** command that generates $\mathbf{G}^\dagger$. This command allows the user to select a tolerance such that singular values smaller than the tolerance are not included in the computation.

Using (3.20), we define the pseudoinverse solution to be

$$\mathbf{m}_\dagger = \mathbf{G}^\dagger\mathbf{d} \qquad (3.21)$$

$$= \mathbf{V}_p\mathbf{S}_p^{-1}\mathbf{U}_p^T\mathbf{d} . \qquad (3.22)$$

Among the desirable properties of (3.22) is that $\mathbf{G}^\dagger$, and hence $\mathbf{m}_\dagger$, always exist. In contrast, the inverse of $\mathbf{G}^T\mathbf{G}$ that appears in the normal equations (2.3) does not exist when $\mathbf{G}$ is not of full column rank. We will shortly show that $\mathbf{m}_\dagger$ is a least squares solution.

To encapsulate what the SVD tells us about our linear system, $\mathbf{G}$, and the corresponding generalized inverse system $\mathbf{G}^\dagger$, consider four cases:

1. $m = n = p$. Both the model and data null spaces, $N(\mathbf{G})$ and $N(\mathbf{G}^T)$, respectively, are trivial. $\mathbf{U}_p = \mathbf{U}$ and $\mathbf{V}_p = \mathbf{V}$ are square orthogonal matrices, so that $\mathbf{U}_p^T = \mathbf{U}_p^{-1}$, and $\mathbf{V}_p^T = \mathbf{V}_p^{-1}$. Eq. (3.22) gives

$$\mathbf{G}^\dagger = \mathbf{V}_p \mathbf{S}_p^{-1} \mathbf{U}_p^T \tag{3.23}$$

$$= (\mathbf{U}_p \mathbf{S}_p \mathbf{V}_p^T)^{-1} \tag{3.24}$$

$$= \mathbf{G}^{-1} \tag{3.25}$$

which is the matrix inverse for a square full rank matrix. The solution is unique, and the data are fit exactly.

2. $m = p$ and $p < n$. $N(\mathbf{G})$ is nontrivial because $p < n$, but $N(\mathbf{G}^T)$ is trivial. $\mathbf{U}_p^T = \mathbf{U}_p^{-1}$ and $\mathbf{V}_p^T \mathbf{V}_p = \mathbf{I}_p$. $\mathbf{G}$ applied to the generalized inverse solution gives

$$\mathbf{G}\mathbf{m}_\dagger = \mathbf{G}\mathbf{G}^\dagger \mathbf{d} \tag{3.26}$$

$$= \mathbf{U}_p \mathbf{S}_p \mathbf{V}_p^T \mathbf{V}_p \mathbf{S}_p^{-1} \mathbf{U}_p^T \mathbf{d} \tag{3.27}$$

$$= \mathbf{U}_p \mathbf{S}_p \mathbf{I}_p \mathbf{S}_p^{-1} \mathbf{U}_p^T \mathbf{d} \tag{3.28}$$

$$= \mathbf{d} . \tag{3.29}$$

The data are fit exactly but the solution is nonunique because of the existence of the nontrivial model null space $N(\mathbf{G})$.

If $\mathbf{m}$ is any least squares solution, then it satisfies the normal equations. This is shown in Exercise C.5.

$$(\mathbf{G}^T\mathbf{G})\mathbf{m} = \mathbf{G}^T\mathbf{d} . \tag{3.30}$$

Since $\mathbf{m}_\dagger$ is a least squares solution, it also satisfies the normal equations.

$$(\mathbf{G}^T\mathbf{G})\mathbf{m}_\dagger = \mathbf{G}^T\mathbf{d} . \tag{3.31}$$

Subtracting (3.30) from (3.31), we find that

$$(\mathbf{G}^T\mathbf{G})(\mathbf{m}_\dagger - \mathbf{m}) = \mathbf{0} . \tag{3.32}$$

Thus $\mathbf{m}_\dagger - \mathbf{m}$ lies in $N(\mathbf{G}^T\mathbf{G})$. It can be shown (Exercise A.13f) that $N(\mathbf{G}^T\mathbf{G}) = N(\mathbf{G})$. This implies that $\mathbf{m}_\dagger - \mathbf{m}$ lies in $N(\mathbf{G})$.

The general solution is thus the sum of $\mathbf{m}_\dagger$ and an arbitrary model null space vector, $\mathbf{m}_0$, that can be written as a linear combination of a set of basis vectors for $N(\mathbf{G})$. In terms of the columns of $\mathbf{V}$, we can thus write

$$\mathbf{m} = \mathbf{m}_\dagger + \mathbf{m}_0$$

$$= \mathbf{m}_\dagger + \sum_{i=p+1}^{n} \alpha_i \mathbf{V}_{\cdot,i} \tag{3.33}$$

for any coefficients, α_i. Because the columns of $\mathbf{V}$ are orthonormal, the square of the 2-norm of a general solution always equals or exceeds that of $\mathbf{m}_\dagger$

$$\|\mathbf{m}\|_2^2 = \|\mathbf{m}_\dagger\|_2^2 + \sum_{i=p+1}^{n} \alpha_i^2 \geq \|\mathbf{m}_\dagger\|_2^2 \tag{3.34}$$

where we have equality only if all α_i are zero. The generalized inverse solution is thus a **minimum length least squares solution**.

We can also write this solution in terms of $\mathbf{G}$ and $\mathbf{G}^T$.

$$\mathbf{m}_\dagger = \mathbf{V}_p \mathbf{S}_p^{-1} \mathbf{U}_p^T \mathbf{d} \tag{3.35}$$

$$= \mathbf{V}_p \mathbf{S}_p \mathbf{U}_p^T \mathbf{U}_p \mathbf{S}_p^{-2} \mathbf{U}_p^T \mathbf{d} \tag{3.36}$$

$$= \mathbf{G}^T (\mathbf{U}_p \mathbf{S}_p^{-2} \mathbf{U}_p^T) \mathbf{d} \tag{3.37}$$

$$= \mathbf{G}^T (\mathbf{G}\mathbf{G}^T)^{-1} \mathbf{d} . \tag{3.38}$$

In practice it is better to compute a solution using the SVD than to use (3.38) because of numerical accuracy issues.

3. $n = p$ and $p < m$. $N(\mathbf{G})$ is trivial but $N(\mathbf{G}^T)$ is nontrivial. Because $p < m$, $R(\mathbf{G})$ is a subspace of R^m. Here

$$\mathbf{G}\mathbf{m}_\dagger = \mathbf{U}_p \mathbf{S}_p \mathbf{V}_p^T \mathbf{V}_p \mathbf{S}_p^{-1} \mathbf{U}_p^T \mathbf{d} \tag{3.39}$$

$$= \mathbf{U}_p \mathbf{U}_p^T \mathbf{d} . \tag{3.40}$$

The product $\mathbf{U}_p \mathbf{U}_p^T \mathbf{d}$ gives the projection of $\mathbf{d}$ onto $R(\mathbf{G})$. Thus $\mathbf{G}\mathbf{m}_\dagger$ is the point in $R(\mathbf{G})$ that is closest to $\mathbf{d}$, and $\mathbf{m}_\dagger$ is a least squares solution to $\mathbf{G}\mathbf{m} = \mathbf{d}$. Only if $\mathbf{d}$ is actually in $R(\mathbf{G})$ will $\mathbf{m}_\dagger$ be an exact solution to $\mathbf{G}\mathbf{m} = \mathbf{d}$.

We can see that this solution is exactly that obtained from the normal equations because

$$(\mathbf{G}^T \mathbf{G})^{-1} = (\mathbf{V}_p \mathbf{S}_p \mathbf{U}_p^T \mathbf{U}_p \mathbf{S}_p \mathbf{V}_p^T)^{-1} \tag{3.41}$$

$$= (\mathbf{V}_p \mathbf{S}_p^2 \mathbf{V}_p^T)^{-1} \tag{3.42}$$

$$= \mathbf{V}_p \mathbf{S}_p^{-2} \mathbf{V}_p^T \tag{3.43}$$

and

$$\mathbf{m}_\dagger = \mathbf{G}^\dagger \mathbf{d} \tag{3.44}$$

$$= \mathbf{V}_p \mathbf{S}_p^{-1} \mathbf{U}_p^T \mathbf{d} \tag{3.45}$$

$$= \mathbf{V}_p \mathbf{S}_p^{-2} \mathbf{V}_p^T \mathbf{V}_p \mathbf{S}_p \mathbf{U}_p^T \mathbf{d} \tag{3.46}$$

$$= (\mathbf{G}^T \mathbf{G})^{-1} \mathbf{G}^T \mathbf{d} . \tag{3.47}$$

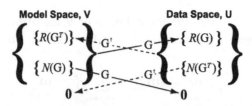

Figure 3.1 SVD model and data space mappings, where $\mathbf{G}^\dagger$ is the generalized inverse. $N(\mathbf{G}^T)$ and $N(\mathbf{G})$ are the data and model null spaces, respectively.

This solution is unique, but cannot fit general data exactly. As with (3.38), it is better in practice to use the generalized inverse solution than to use (3.47) because of numerical accuracy issues.

4. $p < m$ and $p < n$. Both $N(\mathbf{G}^T)$ and $N(\mathbf{G})$ are nontrivial. In this case, the generalized inverse solution encapsulates the behavior of both of the two previous cases, minimizing both $\|\mathbf{Gm} - \mathbf{d}\|_2$ and $\|\mathbf{m}\|_2$.

As in case 3,

$$\mathbf{Gm}_\dagger = \mathbf{U}_p \mathbf{S}_p \mathbf{V}_p^T \mathbf{V}_p \mathbf{S}_p^{-1} \mathbf{U}_p^T \mathbf{d} \tag{3.48}$$

$$= \mathbf{U}_p \mathbf{U}_p^T \mathbf{d} \tag{3.49}$$

$$= \text{proj}_{R(\mathbf{G})} \mathbf{d} . \tag{3.50}$$

Thus $\mathbf{m}_\dagger$ is a least squares solution to $\mathbf{Gm} = \mathbf{d}$.

As in case 2 we can write the model and its norm using (3.33) and (3.34). Thus $\mathbf{m}_\dagger$ is the least squares solution of minimum length.

We have shown that the generalized inverse provides an inverse solution (3.22) that always exists, is both least squares and minimum length, and properly accommodates the rank and dimensions of $\mathbf{G}$. Relationships between the subspaces $R(\mathbf{G})$, $N(\mathbf{G}^T)$, $R(\mathbf{G}^T)$, $N(\mathbf{G})$, and the operators $\mathbf{G}$ and $\mathbf{G}^\dagger$, are schematically depicted in Fig. 3.1. Table 3.1 summarizes the SVD and its properties.

The existence of a nontrivial model null space (one that includes more than just the zero vector) is at the heart of solution nonuniqueness. There are an infinite number of solutions that will fit the data equally well, because model components in $N(\mathbf{G})$ have no effect on data fit. To select a particular preferred solution from this infinite set thus requires more constraints (such as minimum length or smoothing constraints) than are encoded in the matrix $\mathbf{G}$.

To see the significance of the $N(\mathbf{G}^T)$ subspace, consider an arbitrary data vector, $\mathbf{d}_0$, that lies in $N(\mathbf{G}^T)$

$$\mathbf{d}_0 = \sum_{i=p+1}^{m} \beta_i \mathbf{U}_{\cdot,i} . \tag{3.51}$$

Table 3.1 Summary of the SVD and its associated scalars and matrices.

Object	Size	Properties
p	scalar	rank$(\mathbf{G}) = p$; Number of nonzero singular values
m	scalar	Dimension of the data space
n	scalar	Dimension of the model space
$\mathbf{G}$	m by n	Forward problem matrix; $\mathbf{G} = \mathbf{U}\mathbf{S}\mathbf{V}^T = \mathbf{U}_p\mathbf{S}_p\mathbf{V}_p^T$
$\mathbf{U}$	m by m	Orthogonal matrix; $\mathbf{U} = [\mathbf{U}_p, \ \mathbf{U}_0]$
s_i	scalar	ith singular value
$\mathbf{S}$	m by n	Diagonal matrix of singular values; $\mathbf{S}_{i,i} = s_i$
$\mathbf{V}$	n by n	Orthogonal matrix. $\mathbf{V} = [\mathbf{V}_p, \ \mathbf{V}_0]$
$\mathbf{U}_p$	m by p	Columns form an orthonormal basis for $R(\mathbf{G})$
$\mathbf{S}_p$	p by p	Diagonal matrix of nonzero singular values
$\mathbf{V}_p$	n by p	Columns form an orthonormal basis for $R(\mathbf{G}^T)$
$\mathbf{U}_0$	m by $m - p$	Columns form an orthonormal basis for $N(\mathbf{G}^T)$
$\mathbf{V}_0$	n by $n - p$	Columns form an orthonormal basis for $N(\mathbf{G})$
$\mathbf{U}_{.,i}$	m by 1	Eigenvector of $\mathbf{G}\mathbf{G}^T$ with eigenvalue s_i^2
$\mathbf{V}_{.,i}$	n by 1	Eigenvector of $\mathbf{G}^T\mathbf{G}$ with eigenvalue s_i^2
$\mathbf{G}^\dagger$	n by m	Pseudoinverse of $\mathbf{G}$; $\mathbf{G}^\dagger = \mathbf{V}_p\mathbf{S}_p^{-1}\mathbf{U}_p^T$
$\mathbf{m}_\dagger$	n by 1	Generalized inverse solution; $\mathbf{m}_\dagger = \mathbf{G}^\dagger\mathbf{d}$

The generalized inverse operating on such a data vector gives

$$\mathbf{m}_\dagger = \mathbf{V}_p\mathbf{S}_p^{-1}\mathbf{U}_p^T\mathbf{d}_0 \tag{3.52}$$

$$= \mathbf{V}_p\mathbf{S}_p^{-1}\sum_{i=p+1}^{n}\beta_i\mathbf{U}_p^T\mathbf{U}_{.,i} = \mathbf{0} \tag{3.53}$$

because the $\mathbf{U}_{.,i}$ are orthogonal. $N(\mathbf{G}^T)$ is a subspace of R^m consisting of all vectors $\mathbf{d}_0$ that have no influence on the generalized inverse model, $\mathbf{m}_\dagger$. If $p < n$ there are an infinite number of potential data sets that will produce the same model when (3.22) is applied.

3.2. COVARIANCE AND RESOLUTION OF THE GENERALIZED INVERSE SOLUTION

The generalized inverse always gives us a solution, $\mathbf{m}_\dagger$, with well-determined properties, but it is essential to investigate how faithful a representation any model is likely to be of the true situation.

In Chapter 2, we found that under the assumption of independent and normally distributed measurement errors with constant standard deviation, the least squares solution was an unbiased estimator of the true model, and that the estimated model parameters had a multivariate normal distribution with covariance

$$\text{Cov}(\mathbf{m}_{L_2}) = \sigma^2 (\mathbf{G}^T \mathbf{G})^{-1} . \tag{3.54}$$

We can attempt the same analysis for the generalized inverse solution $\mathbf{m}_\dagger$. The covariance matrix would be given by

$$\text{Cov}(\mathbf{m}_\dagger) = \mathbf{G}^\dagger \text{Cov}(\mathbf{d})(\mathbf{G}^\dagger)^T \tag{3.55}$$

$$= \sigma^2 \mathbf{G}^\dagger (\mathbf{G}^\dagger)^T \tag{3.56}$$

$$= \sigma^2 \mathbf{V}_p \mathbf{S}_p^{-2} \mathbf{V}_p^T \tag{3.57}$$

$$= \sigma^2 \sum_{i=1}^{p} \frac{\mathbf{V}_{\cdot,i} \mathbf{V}_{\cdot,i}^T}{s_i^2} . \tag{3.58}$$

Since the s_i are decreasing, successive terms in this sum make larger and larger contributions to the covariance. If we were to truncate (3.58), we could actually decrease the variance in our model estimate! This is discussed further in Section 3.3.

Unfortunately, unless $p = n$, the generalized inverse solution is *not* an unbiased estimator of the true solution. This occurs because the true solution may have nonzero projections onto those basis vectors in $\mathbf{V}$ that are unused in the generalized inverse solution. In practice, the bias introduced by restricting the solution to the subspace spanned by the columns of $\mathbf{V}_p$ may be far larger than the uncertainty due to measurement error.

The concept of **model resolution** is an important way to characterize the bias of the generalized inverse solution. In this approach we see how closely the generalized inverse solution matches a given model, assuming that there are no errors in the data. We begin with a model $\mathbf{m}_{\text{true}}$. By multiplying $\mathbf{G}$ times $\mathbf{m}_{\text{true}}$, we can find a corresponding data vector $\mathbf{d}_{\text{true}}$. If we then multiply $\mathbf{G}^\dagger$ times $\mathbf{d}_{\text{true}}$, we obtain a generalized inverse solution

$$\mathbf{m}_\dagger = \mathbf{G}^\dagger \mathbf{d}_{\text{true}} = \mathbf{G}^\dagger \mathbf{G} \mathbf{m}_{\text{true}} . \tag{3.59}$$

We would like to recover the true model, so that $\mathbf{m}^\dagger = \mathbf{m}_{\text{true}}$. However, because $\mathbf{m}_{\text{true}}$ may have a nonzero projection onto the model null space $N(\mathbf{G})$, $\mathbf{m}_\dagger$ will not in general be equal to $\mathbf{m}_{\text{true}}$. The **model resolution matrix**

$$\mathbf{R}_m = \mathbf{G}^\dagger \mathbf{G} \tag{3.60}$$

$$= \mathbf{V}_p \mathbf{S}_p^{-1} \mathbf{U}_p^T \mathbf{U}_p \mathbf{S}_p \mathbf{V}_p^T \tag{3.61}$$

$$= \mathbf{V}_p \mathbf{V}_p^T , \tag{3.62}$$

characterizes the linear relationship (3.59) between $\mathbf{m}_\dagger$ and $\mathbf{m}_{\text{true}}$ for a linear inverse problem when solved using the generalized inverse.

If $N(\mathbf{G})$ is trivial, then rank$(\mathbf{G}) = p = n$, and $\mathbf{R}_m$ is the n by n identity matrix. In this case the original model is recovered exactly and we say that the model resolution is perfect. If $N(\mathbf{G})$ is a nontrivial subspace of R^n, then $p = \text{rank}(\mathbf{G}) < n$, so that $\mathbf{R}_m$ is a non-identity symmetric and singular matrix. The trace of $\mathbf{R}_m$ provides a simple quantitative measure of the resolution. If $\text{Tr}(\mathbf{R}_m)$ is close to n, then $\mathbf{R}_m$ is, in this sense, close to the identity matrix.

The model resolution matrix can be used to quantify the bias introduced by the pseudoinverse when $\mathbf{G}$ does not have full column rank. We begin by showing that the expected value of $\mathbf{m}_\dagger$ is $\mathbf{R}_m \mathbf{m}_{\text{true}}$.

$$E[\mathbf{m}_\dagger] = E[\mathbf{G}^\dagger \mathbf{d}] \tag{3.63}$$
$$= \mathbf{G}^\dagger E[\mathbf{d}] \tag{3.64}$$
$$= \mathbf{G}^\dagger \mathbf{G} \mathbf{m}_{\text{true}} \tag{3.65}$$
$$= \mathbf{R}_m \mathbf{m}_{\text{true}} . \tag{3.66}$$

Thus, the generalized inverse solution bias is

$$E[\mathbf{m}_\dagger] - \mathbf{m}_{\text{true}} = \mathbf{R}_m \mathbf{m}_{\text{true}} - \mathbf{m}_{\text{true}} \tag{3.67}$$
$$= (\mathbf{R}_m - \mathbf{I}) \mathbf{m}_{\text{true}} \tag{3.68}$$

where

$$\mathbf{R}_m - \mathbf{I} = \mathbf{V}_p \mathbf{V}_p^T - \mathbf{V}\mathbf{V}^T \tag{3.69}$$
$$= -\mathbf{V}_0 \mathbf{V}_0^T . \tag{3.70}$$

We can formulate a bound on the norm of the bias using (3.68) as

$$\|E[\mathbf{m}_\dagger] - \mathbf{m}_{\text{true}}\| \le \|\mathbf{R}_m - \mathbf{I}\| \|\mathbf{m}_{\text{true}}\| . \tag{3.71}$$

$\|\mathbf{R}_m - \mathbf{I}\|$ thus characterizes the bias introduced by the generalized inverse solution. However, the detailed effects of limited resolution on the recovered model will depend on $\|\mathbf{m}_{\text{true}}\|$, about which we may have quite limited a priori knowledge.

In practice, the model resolution matrix is commonly used in two different ways. First, we can examine diagonal elements of $\mathbf{R}_m$. Diagonal elements that are close to one correspond to parameters for which we can claim good resolution. Conversely, if any of the diagonal elements are small, then the corresponding model parameters will be poorly resolved. Second, we can multiply $\mathbf{R}_m$ times a particular test model $\mathbf{m}$ to see how that model would be resolved by the inverse solution. This strategy is called a **resolution test**. One commonly used test model is a **spike model**, which is a model vector with all zero elements, except for one specified element that is set to one. Multiplying $\mathbf{R}_m$

times a spike model extracts the corresponding column of the resolution matrix. These columns of the resolution matrix are called **resolution kernels**. Such functions are also analogous to the averaging kernels in the method of Backus and Gilbert discussed in Chapter 5.

We can multiply $\mathbf{G}^\dagger$ and $\mathbf{G}$ in the opposite order from (3.62) to obtain the **data space resolution matrix**, $\mathbf{R}_d$

$$\mathbf{d}_\dagger = \mathbf{G}\mathbf{m}_\dagger \tag{3.72}$$

$$= \mathbf{G}\mathbf{G}^\dagger\mathbf{d} \tag{3.73}$$

$$= \mathbf{R}_d\mathbf{d} \tag{3.74}$$

where

$$\mathbf{R}_d = \mathbf{U}_p\mathbf{S}_p\mathbf{V}_p^T\mathbf{V}_p\mathbf{S}_p^{-1}\mathbf{U}_p^T \tag{3.75}$$

$$= \mathbf{U}_p\mathbf{U}_p^T \ . \tag{3.76}$$

If $N(\mathbf{G}^T)$ contains only the zero vector, then $p = m$, and $\mathbf{R}_d = \mathbf{I}$. In this case, $\mathbf{d}_\dagger = \mathbf{d}$, and the generalized inverse solution $\mathbf{m}_\dagger$ fits the data exactly. However, if $N(\mathbf{G}^T)$ is nontrivial, then $p < m$, and $\mathbf{R}_d$ is not the identity matrix. In this case $\mathbf{m}_\dagger$ does not exactly fit the data.

It is important to note that model and data space resolution matrices (3.62) and (3.76) do *not* depend on specific data or models, but are exclusively properties of $\mathbf{G}$. They thus ultimately reflect the physics and geometry of the forward and inverse problem, and can be assessed during the design phase of an experiment.

3.3. INSTABILITY OF THE GENERALIZED INVERSE SOLUTION

The generalized inverse solution $\mathbf{m}_\dagger$ has zero projection onto $N(\mathbf{G})$. However, it may include terms involving column vectors in $\mathbf{V}_p$ with very small nonzero singular values. In analyzing the generalized inverse solution it is useful to examine the **singular value spectrum**, which is simply the range of singular values. Small singular values cause the generalized inverse solution to be extremely sensitive to small amounts of noise in the data. As a practical matter, it can also be difficult to distinguish between zero singular values and extremely small singular values. We can quantify the instabilities created by small singular values by recasting the generalized inverse solution to make the effect of small singular values explicit. We start with the formula for the generalized inverse solution

$$\mathbf{m}_\dagger = \mathbf{V}_p\mathbf{S}_p^{-1}\mathbf{U}_p^T\mathbf{d} \ . \tag{3.77}$$

The elements of the vector $\mathbf{U}_p^T\mathbf{d}$ are the dot products of the first p columns of $\mathbf{U}$ with $\mathbf{d}$

$$\mathbf{U}_p^T\mathbf{d} = \begin{bmatrix} (\mathbf{U}_{\cdot,1})^T\mathbf{d} \\ (\mathbf{U}_{\cdot,2})^T\mathbf{d} \\ \cdot \\ \cdot \\ \cdot \\ (\mathbf{U}_{\cdot,p})^T\mathbf{d} \end{bmatrix}. \tag{3.78}$$

When we left-multiply $\mathbf{S}_p^{-1}$ times (3.78), we obtain

$$\mathbf{S}_p^{-1}\mathbf{U}_p^T\mathbf{d} = \begin{bmatrix} \frac{(\mathbf{U}_{\cdot,1})^T\mathbf{d}}{s_1} \\ \frac{(\mathbf{U}_{\cdot,2})^T\mathbf{d}}{s_2} \\ \cdot \\ \cdot \\ \cdot \\ \frac{(\mathbf{U}_{\cdot,p})^T\mathbf{d}}{s_p} \end{bmatrix}. \tag{3.79}$$

Finally, when we left-multiply $\mathbf{V}_p$ times (3.79), we obtain a linear combination of the columns of $\mathbf{V}_p$ that can be written as

$$\mathbf{m}_\dagger = \mathbf{V}_p\mathbf{S}_p^{-1}\mathbf{U}_p^T\mathbf{d} = \sum_{i=1}^{p} \frac{\mathbf{U}_{\cdot,i}^T\mathbf{d}}{s_i}\mathbf{V}_{\cdot,i}. \tag{3.80}$$

In the presence of random noise, $\mathbf{d}$ will generally have a nonzero projection onto each of the directions specified by the columns of $\mathbf{U}$. The presence of a very small s_i in the denominator of (3.80) can thus give us a very large coefficient for the corresponding model space basis vector $\mathbf{V}_{\cdot,i}$, and these basis vectors can dominate the solution. In the worst case, the generalized inverse solution is just a noise amplifier, and the answer is practically useless.

A measure of the instability of the solution is the **condition number**. Note that the condition number considered here for an m by n matrix is a generalization of the condition number for an n by n matrix (A.106), and the two formulations are equivalent when $m = n$.

Suppose that we have a data vector $\mathbf{d}$ and an associated generalized inverse solution $\mathbf{m}_\dagger = \mathbf{G}^\dagger\mathbf{d}$. If we consider a slightly perturbed data vector $\mathbf{d}'$ and its associated generalized inverse solution $\mathbf{m}'_\dagger = \mathbf{G}^\dagger\mathbf{d}'$, then

$$\mathbf{m}_\dagger - \mathbf{m}'_\dagger = \mathbf{G}^\dagger(\mathbf{d} - \mathbf{d}') \tag{3.81}$$

and

$$\|\mathbf{m}_\dagger - \mathbf{m}'_\dagger\|_2 \le \|\mathbf{G}^\dagger\|_2 \|\mathbf{d} - \mathbf{d}'\|_2 \ . \tag{3.82}$$

From (3.80), it is clear that the largest difference in the inverse models will occur when $\mathbf{d} - \mathbf{d}'$ is in the direction $\mathbf{U}_{\cdot,p}$. If

$$\mathbf{d} - \mathbf{d}' = \alpha \mathbf{U}_{\cdot,p} \tag{3.83}$$

then

$$\|\mathbf{d} - \mathbf{d}'\|_2 = \alpha \ . \tag{3.84}$$

We can then compute the effect on the generalized inverse solution as

$$\mathbf{m}_\dagger - \mathbf{m}'_\dagger = \frac{\alpha}{s_p}\mathbf{V}_{\cdot,p} \tag{3.85}$$

with

$$\|\mathbf{m}_\dagger - \mathbf{m}'_\dagger\|_2 = \frac{\alpha}{s_p} \ . \tag{3.86}$$

Thus, we have a bound on the instability of the generalized inverse solution

$$\|\mathbf{m}_\dagger - \mathbf{m}'_\dagger\|_2 \le \frac{1}{s_p}\|\mathbf{d} - \mathbf{d}'\|_2 \ . \tag{3.87}$$

Similarly, we can see that the generalized inverse model is smallest in norm when $\mathbf{d}$ points in a direction parallel to $\mathbf{U}_{\cdot,1}$. Thus

$$\|\mathbf{m}_\dagger\|_2 \ge \frac{1}{s_1}\|\mathbf{d}\|_2 \ . \tag{3.88}$$

Combining these inequalities, we obtain

$$\frac{\|\mathbf{m}_\dagger - \mathbf{m}'_\dagger\|_2}{\|\mathbf{m}_\dagger\|_2} \le \frac{s_1}{s_p}\frac{\|\mathbf{d} - \mathbf{d}'\|_2}{\|\mathbf{d}\|_2} \ . \tag{3.89}$$

The bound (3.89) is applicable to pseudoinverse solutions, regardless of what value of p we use. If we decrease p and thus eliminate model space vectors associated with small singular values, then the solution becomes more stable. However, this stability comes at the expense of reducing the dimension of the subspace of R^n where the solution lies. As a result, the model resolution matrix for the stabilized solution obtained by decreasing p becomes less like the identity matrix, and the fit to the data worsens.

The condition number of $\mathbf{G}$ is the coefficient in (3.89)

$$\mathrm{cond}(\mathbf{G}) = \frac{s_1}{s_k} \tag{3.90}$$

where $k = \min(m, n)$. The MATLAB command **cond** can be used to compute (3.90). If $\mathbf{G}$ is of full rank, and we use all of the singular values in the pseudoinverse solution

($p = k$), then the condition number is exactly (3.90). If **G** is of less than full rank, then the condition number is effectively infinite. As with the model and data resolution matrices ((3.62) and (3.76)), the condition number is a property of **G** that can be computed in the design phase of an experiment before any data are collected.

A condition that insures solution stability and arises naturally from consideration of (3.80) is the **discrete Picard condition** [86]. The discrete Picard condition is satisfied when the dot products of the columns of **U** and the data vector decay to zero more quickly than the singular values, s_i. Under this condition, we should not see instability due to small singular values. The discrete Picard condition can be assessed by plotting the ratios of $\mathbf{U}_{.,i}^T \mathbf{d}$ to s_i across the singular value spectrum.

If the discrete Picard condition is not satisfied, we may still be able to recover a useful model by truncating the series for $\mathbf{m}_\dagger$ (3.80) at term $p' < p$, to produce a **truncated SVD**, or **TSVD** solution. One way to decide where to truncate the series is to apply the **discrepancy principle**. Under the discrepancy principle, we choose p' so that the model fits the data to a specified tolerance, δ, i.e.,

$$\|\mathbf{G}_w \mathbf{m} - \mathbf{d}_w\|_2 \leq \delta \qquad (3.91)$$

where $\mathbf{G}_w$ and $\mathbf{d}_w$ are the weighted system matrix and data vector, respectively.

How should we select δ? We discussed in Chapter 2 that when we estimate the solution to a full column rank least squares problem, $\|\mathbf{G}_w \mathbf{m}_{L_2} - \mathbf{d}_w\|_2^2$ has a χ^2 distribution with $m - n$ degrees of freedom, so we could set δ equal to $\sqrt{m - n}$ if $m > n$. However, when the number of model parameters n is greater than or equal to the number of data m, this formulation fails because there is no χ^2 distribution with fewer than one degree of freedom. In practice, a common heuristic is to require $\|\mathbf{G}_w \mathbf{m} - \mathbf{d}_w\|_2^2$ to be smaller than m, because the approximate median of a χ^2 distribution with m degrees of freedom is m (Fig. B.4).

A TSVD solution will not fit the data as well as solutions that include the model space basis vectors with small singular values. However, fitting the data vector too precisely in ill-posed problems (sometimes referred to as **over-fitting**) will allow data noise to control major features or even completely dominate the model.

The TSVD solution is but one example of **regularization**, where solutions are selected to sacrifice fit to the data in exchange for solution stability. Understanding the tradeoff between fitting the data and solution stability involved in regularization is of fundamental importance.

3.4. A RANK DEFICIENT TOMOGRAPHY PROBLEM

A linear least squares problem is said to be **rank deficient** if there is a clear distinction between the nonzero and zero singular values and rank(**G**) is less than n. Numerically computed singular values will often include some that are extremely small but not quite

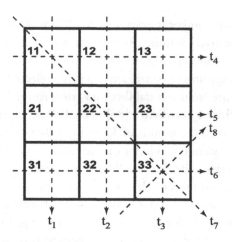

Figure 3.2 A simple tomography example (revisited).

zero, because of roundoff errors. If there is a substantial gap between the largest of these tiny singular values and the first truly nonzero singular value, then it can be easy to distinguish between the two populations. Rank deficient problems can often be solved in a straightforward manner by applying the generalized inverse solution. After truncating the effectively zero singular values, a least squares model of limited resolution will be produced, and stability will seldom be an issue.

Example 3.1

Using the SVD, let us revisit the straight ray path tomography example that we considered earlier in Examples 1.6 and 1.12 (Fig. 3.2). We introduced a rank deficient system in which we were constraining an $n = 9$-parameter slowness model with $m = 8$ travel time observations. We map the two-dimensional grid of slownesses into a model vector by using a row-by-row indexing convention to obtain

$$\mathbf{Gm} = \begin{bmatrix} 1 & 0 & 0 & 1 & 0 & 0 & 1 & 0 & 0 \\ 0 & 1 & 0 & 0 & 1 & 0 & 0 & 1 & 0 \\ 0 & 0 & 1 & 0 & 0 & 1 & 0 & 0 & 1 \\ 1 & 1 & 1 & 0 & 0 & 0 & 0 & 0 & 0 \\ 0 & 0 & 0 & 1 & 1 & 1 & 0 & 0 & 0 \\ 0 & 0 & 0 & 0 & 0 & 0 & 1 & 1 & 1 \\ \sqrt{2} & 0 & 0 & 0 & \sqrt{2} & 0 & 0 & 0 & \sqrt{2} \\ 0 & 0 & 0 & 0 & 0 & 0 & 0 & 0 & \sqrt{2} \end{bmatrix} \begin{bmatrix} s_{11} \\ s_{12} \\ s_{13} \\ s_{21} \\ s_{22} \\ s_{23} \\ s_{31} \\ s_{32} \\ s_{33} \end{bmatrix} = \begin{bmatrix} t_1 \\ t_2 \\ t_3 \\ t_4 \\ t_5 \\ t_6 \\ t_7 \\ t_8 \end{bmatrix}. \tag{3.92}$$

The 8 singular values of **G** are

$$
\text{diag}(\mathbf{S}) = \begin{bmatrix} 3.180 \\ 2.000 \\ 1.732 \\ 1.732 \\ 1.732 \\ 1.607 \\ 0.553 \\ 0 \end{bmatrix}. \tag{3.93}
$$

The ratio of the largest to smallest of the nonzero singular values is about 6, and the generalized inverse solution (3.80) will thus be stable in the presence of noise. Because rank$(\mathbf{G}) = p = 7$ is less than both m and n, the problem is both rank deficient and will in general have no exact solution. The model null space, $N(\mathbf{G})$, is spanned by the two orthonormal vectors that form the 8th and 9th columns of **V**. An orthonormal basis for the null space is

$$
\mathbf{V}_0 = \begin{bmatrix} \mathbf{V}_{.,8}, & \mathbf{V}_{.,9} \end{bmatrix} = \begin{bmatrix} -0.0620 & -0.4035 \\ -0.4035 & 0.0620 \\ 0.4655 & 0.3415 \\ 0.4035 & -0.0620 \\ 0.0620 & 0.4035 \\ -0.4655 & -0.3415 \\ -0.3415 & 0.4655 \\ 0.3415 & -0.4655 \\ 0.0000 & 0.0000 \end{bmatrix}. \tag{3.94}
$$

To obtain a geometric appreciation for the two model null space vectors in this example, we can reshape them into 3 by 3 matrices corresponding to the geometry of the blocks (e.g. by using the MATLAB **reshape** command) to plot their elements in proper physical positions. Here, we have

$$
\text{reshape}(\mathbf{V}_{.,8}, \ 3, \ 3)' = \begin{bmatrix} -0.0620 & -0.4035 & 0.4655 \\ 0.4035 & 0.0620 & -0.4655 \\ -0.3415 & 0.3415 & 0.0000 \end{bmatrix} \tag{3.95}
$$

$$
\text{reshape}(\mathbf{V}_{.,9}, \ 3, \ 3)' = \begin{bmatrix} -0.4035 & 0.0620 & 0.3415 \\ -0.0620 & 0.4035 & -0.3415 \\ 0.4655 & -0.4655 & 0.0000 \end{bmatrix} \tag{3.96}
$$

(Figs. 3.3 and 3.4).

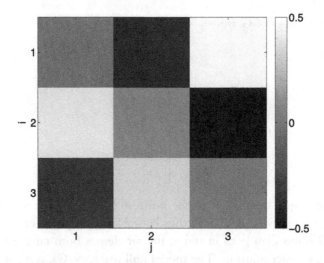

Figure 3.3 Image of the null space model $\mathbf{V}_{.,8}$.

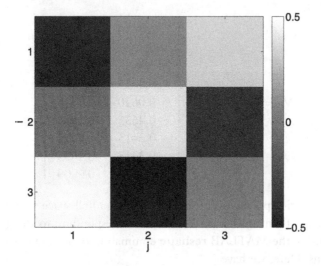

Figure 3.4 Image of the null space model $\mathbf{V}_{.,9}$.

Recall that if $\mathbf{m}_0$ is in the model null space, then (because $\mathbf{Gm}_0 = \mathbf{0}$) we can add such a model to any solution and not change the fit to the data (3.33). When mapped to their physical locations, three common features of the model null space basis vector elements in this example stand out:

1. The sums along all rows and columns are zero
2. The upper left to lower right diagonal sum is zero
3. There is no projection in the $m_9 = s_{33}$ model space direction.

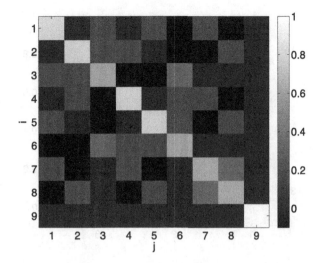

Figure 3.5 Elements of the model resolution matrix, $\mathbf{R}_m$ (3.62), for the generalized inverse solution.

The zero sum conditions (1) and (2) arise because paths passing through any three horizontal or vertical sets of blocks can only constrain the sum of those block values. The condition (3) that $m_9 = 0$ occurs because that model element is uniquely constrained by the 8th ray, which passes exclusively through the $s_{3,3}$ block. Thus, any variation in m_9 will clearly affect the predicted data, and any vector in the model null space must have a value of zero in m_9.

The single basis vector spanning the data null space in this example is

$$\mathbf{U}_0 = \mathbf{U}_{.,8} = \begin{bmatrix} -0.408 \\ -0.408 \\ -0.408 \\ 0.408 \\ 0.408 \\ 0.408 \\ 0.000 \\ 0.000 \end{bmatrix}. \tag{3.97}$$

This indicates that increasing the times t_1, t_2, and t_3 and decreasing the times t_4, t_5, and t_6 by equal amounts will result in no change in the pseudoinverse solution.

Recall that, even for noise-free data, we will not recover a general $\mathbf{m}_{\text{true}}$ in a rank deficient problem using (3.22), but will instead recover a "smeared" model $\mathbf{R}_m \mathbf{m}_{\text{true}}$. Because $\mathbf{R}_m$ for a rank deficient problem is itself rank deficient, this smearing is irreversible. The full $\mathbf{R}_m$ matrix dictates precisely how this smearing occurs. The elements of $\mathbf{R}_m$ for this example are shown in Fig. 3.5.

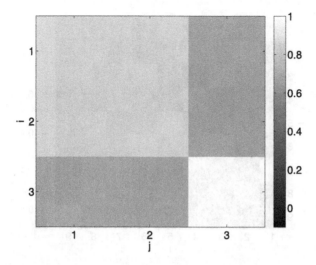

Figure 3.6 Diagonal elements of the resolution matrix plotted in their respective geometric model locations.

Examining the entire n by n model resolution matrix becomes cumbersome in large problems. The n diagonal elements of $\mathbf{R}_m$ can be examined more easily to provide basic information on how well recovered each model parameter will be. The reshaped diagonal of $\mathbf{R}_m$ from Fig. 3.5 is

$$
\text{reshape}(\text{diag}(\mathbf{R}_m),\ 3,\ 3)' = \begin{bmatrix} 0.833 & 0.833 & 0.667 \\ 0.833 & 0.833 & 0.667 \\ 0.667 & 0.667 & 1.000 \end{bmatrix}. \tag{3.98}
$$

These values are plotted in Fig. 3.6.

Fig. 3.6 and (3.98) tell us that m_9 is perfectly resolved, but that we can expect loss of resolution (and hence smearing of the true model into other blocks) for all of the other solution parameters.

We next assess the smoothing effects of limited model resolution by performing a resolution test using synthetic data for a test model of interest, and assessing the recovery of the test model by examining the corresponding inverse solution. Consider a spike model consisting of the vector with its 5th element equal to one and zeros elsewhere (Fig. 3.7). Forward modeling gives the predicted data set for $\mathbf{m}_{\text{test}}$

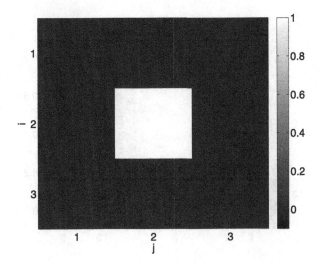

Figure 3.7 A spike test model.

$$\mathbf{d}_{\text{test}} = \mathbf{G}\mathbf{m}_{\text{test}} = \begin{bmatrix} 0 \\ 1 \\ 0 \\ 0 \\ 1 \\ 0 \\ \sqrt{2} \\ 0 \end{bmatrix} \tag{3.99}$$

and the corresponding (reshaped) generalized inverse model is the fifth column of $\mathbf{R}_m$, which is

$$\text{reshape}(\mathbf{m}_\dagger, 3, 3)' = \begin{bmatrix} 0.167 & 0 & -0.167 \\ 0 & 0.833 & 0.167 \\ -0.167 & 0.167 & 0 \end{bmatrix}. \tag{3.100}$$

(Fig. 3.8). The recovered model in this spike test shows that limited resolution causes information about the central block slowness to smear into some, but not all, of the adjacent blocks *even for noise-free data*, with the exact form of the smearing dictated by the model resolution matrix.

It is important to reemphasize that the ability to recover the true model in practice is affected by two separate and important factors. The first factor is the bias caused by limited resolution, which is a characteristic of the matrix **G** and hence applies even

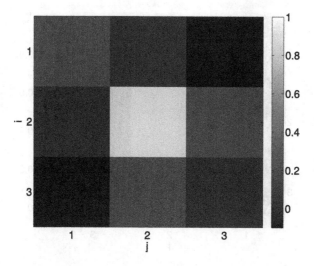

Figure 3.8 The generalized inverse solution for the noise-free spike test.

to noise-free data. The second factor is the mapping of any data noise into the model parameters. In specific cases one effect or the other may dominate.

3.5. DISCRETE ILL-POSED PROBLEMS

In many problems the singular values decay gradually towards zero and do not show an obvious jump between nonzero and zero singular values. This happens frequently when we discretize Fredholm integral equations of the first kind as in Chapter 1. In particular, as we increase the number of points in the discretization, we typically find that $\mathbf{G}$ becomes more and more poorly conditioned. We will refer to these as **discrete ill-posed problems**.

The rate of singular value spectrum decay can be used to characterize a discrete ill-posed problem as mildly, moderately, or severely ill-posed. If $s_j = O(j^{-\alpha})$ for $\alpha \leq 1$ (where O means "on the order of") then we call the problem mildly ill-posed. If $s_j = O(j^{-\alpha})$ for $\alpha > 1$, then the problem is moderately ill-posed. If $s_j = O(e^{-\alpha j})$ for some $\alpha > 0$, then the problem is severely ill-posed.

In discrete ill-posed problems, singular vectors $\mathbf{V}_{.,j}$ associated with large singular values are typically smooth, whereas those corresponding to smaller singular values are highly oscillatory [86]. The influence of rough basis functions becomes increasingly apparent in the character of the generalized inverse solution as more singular values and vectors are included. When we attempt to solve such a problem with the TSVD in the presence of data noise, it is critical to decide where to truncate (3.80). If we truncate the sum too early, then our solution will lack details that require model vectors asso-

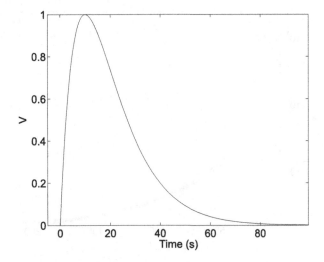

Figure 3.9 Example instrument response; output voltage in response to a unit area ground acceleration impulse (delta function).

ciated with the smaller singular values for their representation. However, if we include too many terms, then the solution becomes dominated by the influence of the data noise.

Example 3.2

Consider an inverse problem where we have a physical process (e.g. seismic ground motion) recorded by a linear instrument of limited bandwidth (e.g. a vertical-component accelerometer; Fig. 3.9). The response of such a device is commonly characterized by an **instrument impulse response**, which is the response of the system to a delta function input. Consider the instrument impulse response

$$g(t) = \begin{cases} g_0 t e^{-t/T_0} & (t \geq 0) \\ 0 & (t < 0) \ . \end{cases} \tag{3.101}$$

Fig. 3.9 shows the displacement response of a critically damped accelerometer with a characteristic time constant T_0 and gain g_0 to a unit area (1 m/s$^2 \cdot$ s) impulsive ground acceleration input.

Assuming that the displacement of the instrument is electronically converted to output volts, we conveniently choose g_0 to be $T_0 e^{-1}$ V/m $\cdot$ s to produce a 1 V maximum output value for the impulse response, and $T_0 = 10$ s.

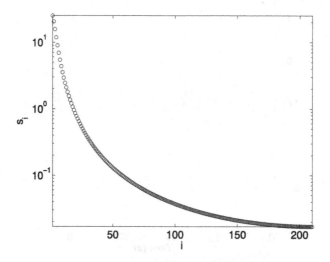

Figure 3.10 Singular values for the discretized convolution matrix.

The instrument output $v(t)$ is a voltage record given by the convolution of the true ground acceleration, $m_{\text{true}}(t)$, with (3.101)

$$v(t) = \int_{-\infty}^{\infty} g(t - \tau) \, m_{\text{true}}(\tau) \, d\tau \ . \tag{3.102}$$

We are interested in the inverse **deconvolution** operation that will remove the smoothing effect of $g(t)$ in (3.102) and allow us to recover the true ground acceleration m_{true}.

Discretizing (3.102) using the midpoint rule with a time interval Δt, we obtain

$$\mathbf{d} = \mathbf{Gm} \tag{3.103}$$

where

$$G_{i,j} = \begin{cases} (t_i - t_j) e^{-(t_i - t_j)/T_0} \Delta t & (t_i \geq t_j) \\ 0 & (t_i < t_j) \ . \end{cases} \tag{3.104}$$

The rows of $\mathbf{G}$ in (3.104) are time-reversed, and the columns of $\mathbf{G}$ are non-time-reversed, sampled representations of the impulse response $g(t)$, lagged by i and j, respectively. Using a time interval of $[-5, \ 100]$ s, outside of which (3.101) and any model, $\mathbf{m}$, of interest are assumed to be very small or zero, and a discretization interval of $\Delta t = 0.5$ s, we obtain a discretized m by n system matrix $\mathbf{G}$ with $m = n = 210$.

The singular values of $\mathbf{G}$ are all nonzero and range from about 25.3 to 0.017, giving a condition number of $s_1/s_m \approx 1488$, and showing that the problem, when discretized in this manner, is moderately ill-posed (Fig. 3.10). However, adding noise at the level of

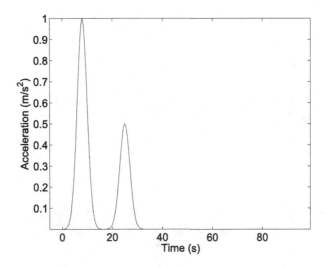

Figure 3.11 The true model.

1 part in 1000 will be sufficient to make the generalized inverse solution unstable. The reason for this can be seen by examining successive rows of **G**, which are nearly but not quite identical, with

$$\frac{\mathbf{G}_{i,\cdot}\mathbf{G}^T_{i+1,\cdot}}{\|\mathbf{G}_{i,\cdot}\|_2\|\mathbf{G}_{i+1,\cdot}\|_2} \approx 0.999 \ . \tag{3.105}$$

This near-identicalness of the rows of **G** makes the system of equations nearly singular, hence resulting in a large condition number.

Now, consider a true ground acceleration signal that consists of two acceleration pulses with widths of $\sigma = 2$ s, centered at $t = 8$ s and $t = 25$ s

$$m_{\text{true}}(t) = e^{-(t-8)^2/(2\sigma^2)} + 0.5e^{-(t-25)^2/(2\sigma^2)} \ . \tag{3.106}$$

We sample $\mathbf{m}_{\text{true}}(t)$ on the time interval $[-5, \ 100]$ s to obtain a 210-element vector $\mathbf{m}_{\text{true}}$, and generate the noise-free data set

$$\mathbf{d}_{\text{true}} = \mathbf{G}\mathbf{m}_{\text{true}} \tag{3.107}$$

and a second data set with independent $N(0, \ (0.05 \text{ V})^2)$ noise added. The true model is shown in Fig. 3.11 and the corresponding data set, with noise added, is shown in Fig. 3.12.

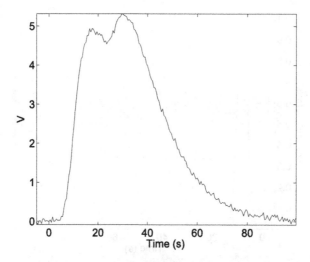

Figure 3.12 Predicted data from the true model plus independent $N(0, (0.05)^2)$ (in units of volts) noise.

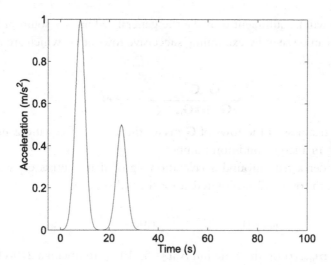

Figure 3.13 Generalized inverse solution using all 210 singular values for the noise-free data.

The recovered least squares model from the full ($p = 210$) generalized inverse solution

$$\mathbf{m} = \mathbf{V}\mathbf{S}^{-1}\mathbf{U}^T\mathbf{d}_{\text{true}} \qquad (3.108)$$

is shown in Fig. 3.13. The model fits its noiseless data vector, $\mathbf{d}_{\text{true}}$, perfectly, and is essentially identical to the true model (Fig. 3.11).

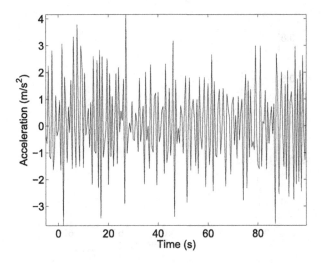

Figure 3.14 Generalized inverse solution using all 210 singular values for the noisy data of Fig. 3.12.

The least squares solution for the noisy data vector, $\mathbf{d}_{\text{true}} + \boldsymbol{\eta}$,

$$\mathbf{m} = \mathbf{VS}^{-1}\mathbf{U}^T(\mathbf{d}_{\text{true}} + \boldsymbol{\eta}) \tag{3.109}$$

is shown in Fig. 3.14.

Although this solution fits its particular data vector, $\mathbf{d}_{\text{true}} + \boldsymbol{\eta}$, exactly, it is worthless in divining information about the true ground motion. Information about $\mathbf{m}_{\text{true}}$ is overwhelmed by the small amount of added noise, amplified enormously by the inversion process.

Can a useful model be recovered by the TSVD? Using the discrepancy principle as our guide and selecting a range of solutions by varying p', we can in fact obtain an appropriate solution when we use just $p' = 26$ columns of $\mathbf{V}$ (Fig. 3.15).

Essential features of the true model are resolved in the solution of Fig. 3.15, but the solution technique introduces oscillations and loss of resolution. Specifically, we see that the widths of the inferred pulses are somewhat wider, and the inferred amplitudes somewhat less, than those of the true ground acceleration. These effects are both hallmarks of limited resolution, as characterized by a non–identity model resolution matrix. An image of the model resolution matrix in Fig. 3.16 shows a finite-width central band and oscillatory side lobes.

A typical (80th) column of the model resolution matrix displays the smearing of the true model into the recovered model for the choice of the $p = 26$ inverse operator (Fig. 3.17). The smoothing is over a characteristic width of about 5 s, which is why our

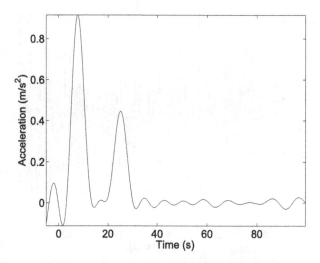

Figure 3.15 TSVD solution using $p' = 26$ singular values for the noisy data shown in Fig. 3.12.

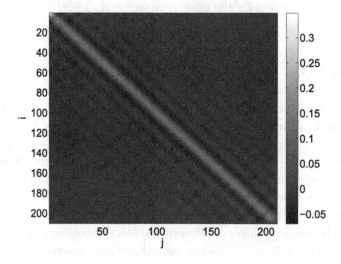

Figure 3.16 The model resolution matrix elements, $\mathbf{R}_{mi,j}$ for the TSVD solution including $p' = 26$ singular values.

recovered model, although it does a decent job of rejecting noise, underestimates the amplitude and narrowness of the pulses in the true model (Fig. 3.11). The oscillatory behavior of the resolution matrix is attributable to our abrupt truncation of the model space.

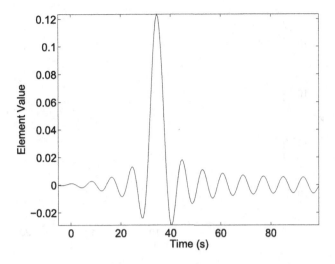

Figure 3.17 The 80th column of the model resolution matrix, $\mathbf{R}_m$ for the TSVD solution including $p' = 26$ singular values.

Each of the n columns of $\mathbf{V}$ is an oscillatory model basis function, with $j - 1$ zero crossings, where j is the column number. In truncating (3.80) at 26 terms to stabilize the inverse solution, we place a limit on the most oscillatory model space basis vectors that we will allow. This truncation gives us a model, and model resolution, that contain oscillatory structure with up to around $p - 1 = 25$ zero crossings. We will examine this perspective further in Chapter 8, where issues associated with oscillatory model basis functions will be revisited in the context of Fourier theory.

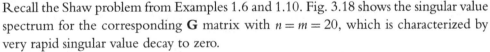

Example 3.3

Recall the Shaw problem from Examples 1.6 and 1.10. Fig. 3.18 shows the singular value spectrum for the corresponding $\mathbf{G}$ matrix with $n = m = 20$, which is characterized by very rapid singular value decay to zero.

The Shaw problem is severely ill-posed, and in its discretized form there is no obvious break point above which the singular values can reasonably be considered to be nonzero and below which the singular values can be considered to be zero. The MATLAB rank command gives $p' = 18$, suggesting that the last two singular values are effectively 0. The condition number of this problem is enormous (larger than 10^{14}).

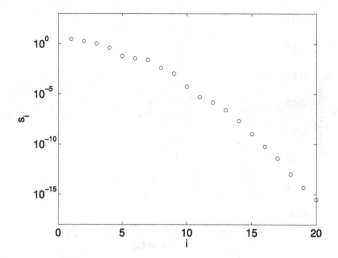

Figure 3.18 Singular values of **G** for the Shaw example ($n = m = 20$).

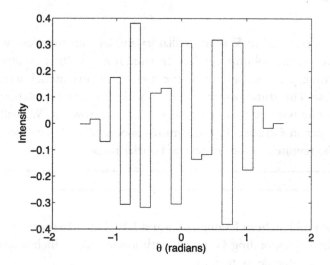

Figure 3.19 $\mathbf{V}_{\cdot,18}$ for the Shaw example problem.

The 18th column of **V**, which corresponds to the smallest nonzero singular value, is shown in Fig. 3.19. In contrast, the first column of **V**, which corresponds to the largest singular value, represents a much smoother model (Fig. 3.20). This behavior is typical of discrete ill-posed problems.

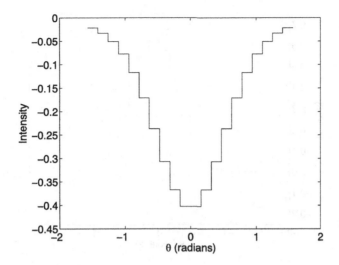

Figure 3.20 $V_{\cdot,1}$ for the Shaw example problem.

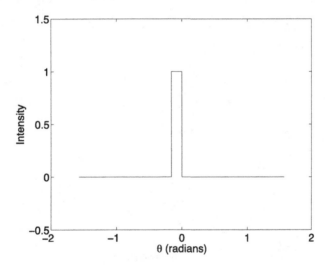

Figure 3.21 The spike model.

Next, we will perform a simple resolution test. Suppose that the input to the system is given by

$$m_i = \begin{cases} 1 & i = 10 \\ 0 & \text{otherwise} . \end{cases} \tag{3.110}$$

(Fig. 3.21). We use the model to obtain noise-free data and then apply the generalized inverse (3.22) with various values of p to obtain TSVD inverse solutions. The corre-

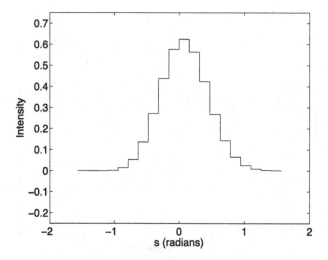

Figure 3.22 Noise-free data predicted for the spike model.

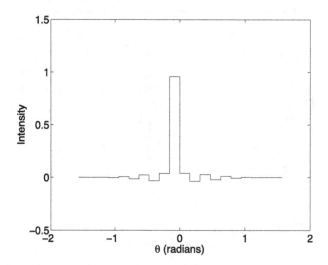

Figure 3.23 The generalized inverse solution for the spike model, no noise.

sponding data are shown in Fig. 3.22. If we compute the generalized inverse from these data using MATLAB's algorithms, we get fairly good recovery of (3.110), although there are still some lower amplitude negative intensity values (Fig. 3.23).

However, if we add a very small amount of noise to the data in Fig. 3.22, things change dramatically. Adding $N(0, (10^{-6})^2)$ noise to the data of Fig. 3.22 and computing

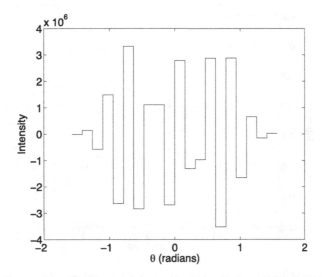

Figure 3.24 Recovery of the spike model with $N(0, (10^{-6})^2)$ noise using the TSVD method ($p' = 18$). Note that the intensity scale ranges from -4×10^6 to 4×10^6.

a generalized inverse solution using $p' = 18$ produces the wild solution of Fig. 3.24, which bears no resemblance to the true model. Note that the vertical scale in Fig. 3.24 is multiplied by 10^6! Furthermore, the solution involves negative intensities, which are not physically possible. This inverse solution is even more sensitive to noise than that of the previous deconvolution example, to the extent that even noise on the order of 1 part in 10^6 will destabilize the solution.

Next, we consider what happens when we use only the 10 largest singular values and their corresponding model space vectors to construct a TSVD solution. Fig. 3.25 shows the solution using 10 singular values with the same noise as Fig. 3.24. Because we have cut off a number of singular values, we have reduced the model resolution. The inverse solution is smeared out, but it is still possible to conclude that there is some significant spike-like feature near $\theta = 0$. In contrast to the situation that we observed in Fig. 3.24, the model recovery is not visibly affected by the noise. The tradeoff is that we must now accept the imperfect resolution of this solution and its attendant bias towards smoother models.

What happens if we discretize the problem with a larger number of intervals? Fig. 3.26 shows the singular values for the **G** matrix with $n = m = 100$ intervals. The first 20 or so singular values are apparently nonzero, whereas the last 80 or so singular values are effectively zero.

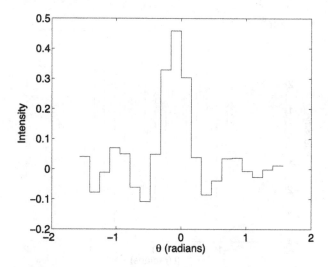

Figure 3.25 Recovery of the spike model with noise using the TSVD method ($p' = 10$).

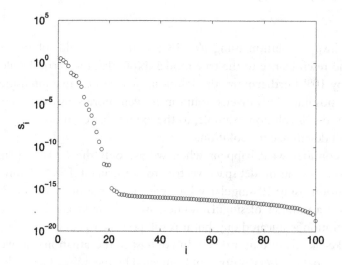

Figure 3.26 Singular values of **G** for the Shaw example ($n = m = 100$).

Fig. 3.27 shows the inverse solution for the spike model with $n = m = 100$ and $p = 10$. This solution is very similar to the solution shown in Fig. 3.24. In general, discretizing over more intervals does not hurt as long as the solution is appropriately regularized and the additional computation time is acceptable.

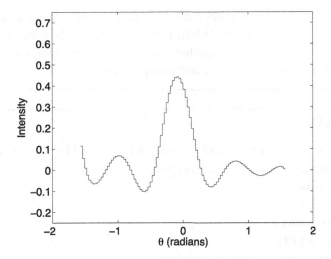

Figure 3.27 Recovery of the spike model with noise using the TSVD method ($n = m = 100, p' = 10$).

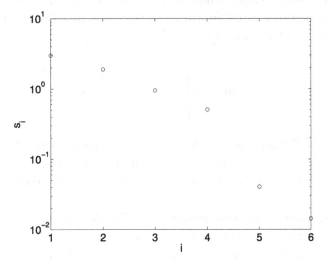

Figure 3.28 Singular values of **G** for the Shaw example ($n = m = 6$).

What about a smaller number of intervals? Fig. 3.28 shows the singular values of the **G** matrix with $n = m = 6$. In this case there are no terribly small singular values. However, with only 6 elements in the model vector, we cannot hope to resolve the details of a source intensity distribution with a complex structure. This is an example of regularization by discretization (see also Exercise 1.3).

This example again demonstrates a fundamental dilemma. If we include small singular values in the series solution (3.80), then our solution becomes unstable in the presence of data noise. If we do not include these terms, our solution is less sensitive to data noise, but we sacrifice resolution and introduce bias.

3.6. EXERCISES

1. The pseudoinverse of a matrix $\mathbf{G}$ was originally defined by Moore and Penrose as the unique matrix $\mathbf{G}^\dagger$ with the properties
   a. $\mathbf{G}\mathbf{G}^\dagger\mathbf{G} = \mathbf{G}$.
   b. $\mathbf{G}^\dagger\mathbf{G}\mathbf{G}^\dagger = \mathbf{G}^\dagger$.
   c. $(\mathbf{G}\mathbf{G}^\dagger)^T = \mathbf{G}\mathbf{G}^\dagger$.
   d. $(\mathbf{G}^\dagger\mathbf{G})^T = \mathbf{G}^\dagger\mathbf{G}$.
   Show that $\mathbf{G}^\dagger$ as given by (3.20) satisfies these four properties.

2. Another resolution test commonly performed in tomography studies is a **checkerboard test**, which consists of using a test model composed of alternating positive and negative perturbations. Perform a checkerboard test on the tomography problem in Example 3.1 using the test model

$$\mathbf{m}_{\text{true}} = \begin{bmatrix} -1 & 1 & -1 \\ 1 & -1 & 1 \\ -1 & 1 & -1 \end{bmatrix}. \tag{3.111}$$

   Evaluate the difference between the true (checkerboard) model and the recovered model in your test, and interpret the pattern of differences. Are any block values recovered exactly? If so, does this imply perfect resolution for these model parameters?

3. Using the parameter estimation problem described in Example 1.1 for determining the three parameters defining a ballistic trajectory, construct synthetic examples that demonstrate the following four cases using the SVD. In each case, display and interpret the SVD components $\mathbf{U}$, $\mathbf{V}$, and $\mathbf{S}$ in terms of the rank, p, of your forward problem $\mathbf{G}$ matrix. Display and interpret any model and data null space vector(s) and calculate model and data space resolution matrices.
   a. Three data points that are exactly fit by a unique model. Plot your data points and the predicted data for your model.
   b. Two data points that are exactly fit by an infinite suite of parabolas. Plot your data points and the predicted data for a suite of these models.
   c. Four data points that are only approximately fit by a parabola. Plot your data points and the predicted data for the least squares model.

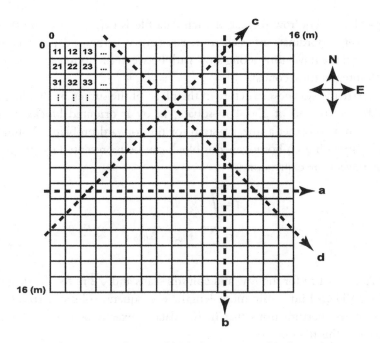

Figure 3.29 Tomography exercise, showing block discretization, block numbering convention, and representative ray paths going east-west (a), north-south (b), southwest-northeast (c), and northwest-southeast (d).

      d. Two data points that are only approximately fit by any parabola, and for which there are an infinite number of least squares solutions. Plot your data points and the predicted data for a suite of least squares models.

4. A large N-S by E-W oriented, nearly square plan view, sandstone quarry block (16 m by 16 m) with a bulk compressional wave seismic velocity of approximately 3000 m/s is suspected of harboring higher-velocity dinosaur remains. An ultrasonic tomography scan is performed in a horizontal plane bisecting the block, producing a data set consisting of 16 E→W, 16 N→S, 31 SW→NE, and 31 NW→SE travel times. See Fig. 3.29. The travel time data (units of s) have statistically independent errors and the travel time contribution for a uniform background model (with a velocity of 3000 m/s) has been subtracted from each travel time measurement.

The MATLAB data files that you will need to load containing the travel time data are: **rowscan.mat**, **colscan.mat**, **diag1scan.mat**, and **diag2scan.mat**. The standard deviations of all data measurements are 1.5×10^{-5} s. Because the travel time contributions for a uniform background model (with a velocity of 3000 m/s) have been subtracted from each travel time measurement, you will be solving for slowness and velocity perturbations relative to a uniform slowness model of $1/3000$ s/m. Use a row-by-row mapping between the slowness grid and the model vector (e.g.,

Example 1.12). The row format of each data file is (x_1, y_1, x_2, y_2, t) where the starting point coordinate of each source is (x_1, y_1), the end point coordinate is (x_2, y_2), and the travel time along a ray path between the source and receiver points is a path integral (in seconds).

Parameterize the slowness structure in the plane of the survey by dividing the block into a 16 by 16 grid of 256 1-m-square, N by E oriented blocks, as shown in Fig. 3.29, and construct a linear system for the forward problem. Assume that the ray paths through each homogeneous block can be represented by straight lines, so that the travel time expression is

$$t = \int_{\ell} s(\mathbf{x}) \, d\ell \qquad (3.112)$$

$$= \sum_{\text{blocks}} s_{\text{block}} \cdot \Delta l_{\text{block}} \qquad (3.113)$$

where Δl_{block} is 1 m for the row and column scans and $\sqrt{2}$ m for the diagonal scans. Use the SVD to find a minimum-length/least squares solution, $\mathbf{m}_\dagger$, for the 256 block slowness perturbations that fit the data as exactly as possible. Perform two inversions in this manner:

A. Using the row and column scans only, and
B. Using the complete data set.

For each inversion:

a. Note the rank of your $\mathbf{G}$ matrix relating the data and model.
b. State and discuss the general solution and data fit significance of the elements and dimensions of the data and model null spaces. Plot and interpret an element of each space and contour or otherwise display a nonzero model that fits the trivial data set $\mathbf{Gm} = \mathbf{d} = \mathbf{0}$ exactly.
   Show the model resolution by contouring or otherwise displaying the 256 diagonal elements of the model resolution matrix, reshaped into an appropriate 16 by 16 grid. Note if there are any model parameters that have perfect resolution.
c. Produce a 16 by 16 element contour or other plot of your slowness perturbation model, displaying the maximum and minimum slowness perturbations in the title of the plot. Interpret any internal structures geometrically and in terms of seismic velocity (in m/s).
d. Describe how one could use nonzero solutions to $\mathbf{Gm} = \mathbf{d} = \mathbf{0}$ to demonstrate that very rough models exist that will fit any data set just as well as a generalized inverse model. Show one such wild model.

5. Consider the data in the file **ifk.mat**.
   The function $d(y)$, $0 \le y \le 1$, is related to an unknown function $m(x)$, $0 \le x \le 1$, by the mathematical model

$$d(\gamma) = \int_0^1 xe^{-xy} m(x) \; dx \; . \tag{3.114}$$

a. Using the data provided, discretize the integral equation using simple collocation to create a square **G** matrix and solve the resulting system of equations.

b. What is the condition number for this system of equations? Given that the
data $d(\gamma)$ are only accurate to about 4 digits, what does this tell you about the
accuracy of your solution?

c. Use the TSVD to compute a solution to this problem. You may find a plot of
the Picard ratios $|\mathbf{U}_{.,i}^T \mathbf{d}/s_i|$ to be especially useful in deciding how many singular
values to include.

3.7. NOTES AND FURTHER READING

The Moore–Penrose generalized inverse was independently discovered by Moore in
1920 and Penrose in 1955 [139,162]. Penrose is generally credited with first showing
that the SVD can be used to compute the generalized inverse [162]. Books that discuss
the linear algebra of the generalized inverse in more detail include [13,31].

There was significant early work on the SVD in the 19th century by Beltrami,
Jordan, Sylvester, Schmidt, and Weyl [193]. However, the singular value decomposition
in matrix form is typically credited to Eckart and Young [53]. Some books that discuss
the properties of the SVD and prove its existence include [72,138,194]. Lanczos presents
an alternative derivation of the SVD [117]. Algorithms for the computation of the SVD
are discussed in [50,72,209]. Books that discuss the use of the SVD and truncated SVD
in solving discrete linear inverse problems include [84,86,137,196].

Resolution tests with spike and checkerboard models as in Example 3.1 are commonly used in practice. However, Leveque, Rivera, and Wittlinger discuss some serious
problems with such resolution tests [123]. Complementary information can be acquired
by examining the diagonal elements of the resolution matrix, which can be efficiently
calculated in isolation from off-diagonal elements even for very large inverse problems
[12,131] (Chapter 6).

Matrices such as those in Example 3.2 in which the elements along diagonals are
constant are called **Toeplitz matrices** [94]. Specialized methods for regularization of
problems involving Toeplitz matrices are available [85].

It is possible to effectively regularize the solution to a discretized version of a continuous inverse problem by selecting a coarse discretization (e.g., Exercise 1.3 and
Example 3.3). This approach is analyzed in [57]. However, in doing so we lose the
ability to analyze the bias introduced by the regularization. In general, we prefer to
use a fine discretization that is consistent with the physics of the forward problem and
explicitly regularize the resulting discretized problem.

CHAPTER FOUR

Tikhonov Regularization

Synopsis

The method of Tikhonov regularization for stabilizing inverse problem solutions is introduced and illustrated with examples. Zeroth-order Tikhonov regularization is explored, including its resolution, bias, and uncertainty properties. The concepts of filter factors (which control the contribution of singular values and their corresponding singular vectors to the solution) and the discrepancy and L-curve criteria (strategies for selecting the regularization parameter) are presented. Higher-order Tikhonov regularization techniques and their computation by the generalized SVD (GSVD) and truncated GSVD are discussed. Generalized cross-validation is introduced as an alternative method for selecting the regularization parameter. Theorems for bounding the error in the regularized solution are discussed. The BVLS method for applying strict upper and lower bounds to least-squares solutions is introduced.

4.1. SELECTING A GOOD SOLUTION

We saw in Chapter 3 that, given the SVD of $\mathbf{G}$ (3.1), we can express a generalized inverse solution by a series (3.80)

$$\mathbf{m}_\dagger = \mathbf{V}_p \mathbf{S}_p^{-1} \mathbf{U}_p^T \mathbf{d} = \sum_{i=1}^{p} \frac{\mathbf{U}_{\cdot,i}^T \mathbf{d}}{s_i} \mathbf{V}_{\cdot,i} \, . \tag{4.1}$$

We also saw that the generalized inverse solution can become extremely unstable when one or more of the singular values, s_i, is small. One approach for dealing with this difficulty, the Truncated Singular Value Decomposition (TSVD), was to truncate the series to remove singular vectors, $\mathbf{V}_{\cdot,i}$, associated with smaller singular values, s_i. This stabilized, or regularized, the solution in the sense that it made the result less sensitive to data noise. The cost of this stabilization approach is that the regularized solution had reduced resolution and was no longer unbiased.

In this chapter we will introduce and discuss Tikhonov regularization, a very widely applied and easily implemented technique for regularizing discrete ill-posed problems. We will show a series formula for the Tikhonov solution that is a modified version of the generalized inverse series (4.1). The Tikhonov series solution has coefficients that are functions of the regularization parameter controlling the degree of regularization and that give greater weight to model elements associated with larger singular values.

For a general linear least squares problem there may be infinitely many least squares solutions. If we consider that the data contain noise, and that there is no point in

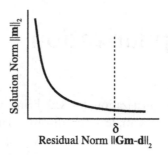

Figure 4.1 A particular misfit norm, δ, and its position on the tradeoff curve between residual misfit, $\|\mathbf{Gm} - \mathbf{d}\|_2$ and model norm, $\|\mathbf{m}\|_2$.

fitting such noise exactly, it becomes evident that there can be many solutions that can adequately fit the data in the sense that $\|\mathbf{Gm} - \mathbf{d}\|_2$ is small enough.

In zeroth-order Tikhonov regularization, we consider all solutions with $\|\mathbf{Gm} - \mathbf{d}\|_2 \leq \delta$, and select the one that minimizes the norm of $\mathbf{m}$

$$\begin{aligned} &\min \|\mathbf{m}\|_2 \\ &\|\mathbf{Gm} - \mathbf{d}\|_2 \leq \delta \ . \end{aligned} \tag{4.2}$$

The fundamental motivation for minimizing the norm of $\mathbf{m}$ is to arrive at a solution that contains just sufficient feature complexity, as quantified by a norm measure, to adequately fit the data. Note that as δ increases, the set of feasible models expands, and the minimum value of $\|\mathbf{m}\|_2$ decreases. In other words, as we allow a poorer fit to the data, smaller norm models will suffice to fit the data. We can thus trace out a curve of minimum values of $\|\mathbf{m}\|_2$ versus δ (Fig. 4.1). It is also possible to trace out this curve by considering problems of the form

$$\begin{aligned} &\min \|\mathbf{Gm} - \mathbf{d}\|_2 \\ &\|\mathbf{m}\|_2 \leq \epsilon \ . \end{aligned} \tag{4.3}$$

As ϵ decreases, the set of feasible solutions becomes smaller, and the minimum value of $\|\mathbf{Gm} - \mathbf{d}\|_2$ increases. Again, as we adjust ϵ we trace out the curve of optimal values of $\|\mathbf{m}\|_2$ and $\|\mathbf{Gm} - \mathbf{d}\|_2$ (Fig. 4.2).

A third option is to consider the **damped least squares** problem

$$\min \|\mathbf{Gm} - \mathbf{d}\|_2^2 + \alpha^2 \|\mathbf{m}\|_2^2 \tag{4.4}$$

which arises when we apply the method of Lagrange multipliers to (4.2), where α is a **regularization parameter**. It can be shown that for appropriate choices of δ, ϵ, and α, the three problems (4.2), (4.3), and (4.4) yield the same solution [84]. We will concentrate on solving the damped least squares form of the problem (4.4). Solutions

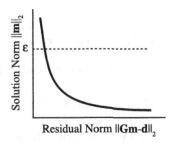

Figure 4.2 A particular model norm, ϵ, and its position on the tradeoff curve between residual misfit, $\|\mathbf{Gm} - \mathbf{d}\|_2$ and model norm, $\|\mathbf{m}\|_2$.

to (4.2) and (4.3) can be obtained using (4.4) by adjusting the regularization parameter α until the constraints are just satisfied.

When plotted on a log–log scale, the curve of optimal values of $\|\mathbf{m}\|_2$ versus $\|\mathbf{Gm} - \mathbf{d}\|_2$ often takes on a characteristic "L" shape in linear problems. This happens because $\|\mathbf{m}\|_2$ is a strictly decreasing function of α and $\|\mathbf{Gm} - \mathbf{d}\|_2$ is a strictly increasing function of α. The sharpness of the "corner" varies from problem to problem, but it is frequently well-defined. For this reason, the curve is called an **L-curve** [82]. In addition to the discrepancy principle, another popular criterion for picking the value of α is the **L-curve criterion** in which the value of α that gives the solution closest to the corner of the L-curve is selected.

In plotting the L-curve it is easy to err by using too small a range of α values, and in many cases it will be necessary to span many orders of magnitude. If nothing else is known about the problem you should confirm that at the smallest values of α, $\|\mathbf{Gm} - \mathbf{d}\|_2$ is equal to the value of the least squares solution and that at the largest values of α, $\alpha^2 \|\mathbf{m}\|_2^2$ is much larger than $\|\mathbf{Gm} - \mathbf{d}\|_2^2$.

4.2. SVD IMPLEMENTATION OF TIKHONOV REGULARIZATION

The damped least squares problem (4.4) is equivalent to the ordinary least squares problem obtained by augmenting the least squares problem for $\mathbf{Gm} = \mathbf{d}$ in the following manner

$$\min \left\| \begin{bmatrix} \mathbf{G} \\ \alpha \mathbf{I} \end{bmatrix} \mathbf{m} - \begin{bmatrix} \mathbf{d} \\ 0 \end{bmatrix} \right\|_2^2 . \tag{4.5}$$

As long as α is nonzero, the last n rows of the augmented matrix in (4.5) are obviously linearly independent. Eq. (4.5) is thus a full rank least squares problem that can be solved by the method of normal equations, i.e.

$$\begin{bmatrix} \mathbf{G}^T & \alpha \mathbf{I} \end{bmatrix} \begin{bmatrix} \mathbf{G} \\ \alpha \mathbf{I} \end{bmatrix} \mathbf{m} = \begin{bmatrix} \mathbf{G}^T & \alpha \mathbf{I} \end{bmatrix} \begin{bmatrix} \mathbf{d} \\ 0 \end{bmatrix} . \tag{4.6}$$

Eq. (4.6) simplifies to

$$(\mathbf{G}^T\mathbf{G} + \alpha^2\mathbf{I})\mathbf{m} = \mathbf{G}^T\mathbf{d} \tag{4.7}$$

which is the set of constraint equations for a **zeroth-order Tikhonov Regularization** solution of $\mathbf{Gm} = \mathbf{d}$.

Employing the SVD of $\mathbf{G}$, (4.7) can be written as

$$(\mathbf{VS}^T\mathbf{U}^T\mathbf{USV}^T + \alpha^2\mathbf{I})\mathbf{m} = \mathbf{G}^T\mathbf{d} \tag{4.8}$$

$$(\mathbf{VS}^T\mathbf{SV}^T + \alpha^2\mathbf{I})\mathbf{m} = \mathbf{VS}^T\mathbf{U}^T\mathbf{d} . \tag{4.9}$$

A simple substitution diagonalizes this system of equations and makes it straightforward to write out the solution. Let $\mathbf{x} = \mathbf{V}^T\mathbf{m}$ and $\mathbf{m} = \mathbf{Vx}$. Since $\mathbf{VV}^T = \mathbf{I}$, we can write (4.9) as

$$(\mathbf{VS}^T\mathbf{SV}^T + \alpha^2\mathbf{VV}^T)\mathbf{m} = \mathbf{VS}^T\mathbf{U}^T\mathbf{d} \tag{4.10}$$

$$\mathbf{V}(\mathbf{S}^T\mathbf{S} + \alpha^2\mathbf{I})\mathbf{V}^T\mathbf{m} = \mathbf{VS}^T\mathbf{U}^T\mathbf{d} \tag{4.11}$$

$$(\mathbf{S}^T\mathbf{S} + \alpha^2\mathbf{I})\mathbf{x} = \mathbf{S}^T\mathbf{U}^T\mathbf{d} . \tag{4.12}$$

The matrix on the left hand side of this equation is diagonal, so it is trivial to solve the system of equations

$$x_i = \frac{s_i\mathbf{U}_{.,i}^T\mathbf{d}}{s_i^2 + \alpha^2} . \tag{4.13}$$

Since $\mathbf{m} = \mathbf{Vx}$, we obtain the solution

$$\mathbf{m}_\alpha = \sum_{i=1}^{k} \frac{s_i\mathbf{U}_{.,i}^T\mathbf{d}}{s_i^2 + \alpha^2}\mathbf{V}_{.,i} \tag{4.14}$$

where $k = \min(m, n)$ so that all nonzero singular values and vectors are included. To relate this formula to (4.1), we can rewrite it slightly as

$$\mathbf{m}_\alpha = \sum_{i=1}^{k} \frac{s_i^2}{s_i^2 + \alpha^2}\frac{\mathbf{U}_{.,i}^T\mathbf{d}}{s_i}\mathbf{V}_{.,i} \tag{4.15}$$

or

$$\mathbf{m}_\alpha = \sum_{i=1}^{k} f_i \frac{\mathbf{U}_{.,i}^T\mathbf{d}}{s_i}\mathbf{V}_{.,i} . \tag{4.16}$$

Here, the **filter factors**

$$f_i = \frac{s_i^2}{s_i^2 + \alpha^2} \tag{4.17}$$

control the contribution to the sum from different terms. For $s_i \gg \alpha$, $f_i \approx 1$, and for $s_i \ll \alpha$, $f_i \approx 0$. For singular values between these two extremes, as the s_i decrease, the f_i produce a monotonically decreasing contribution of corresponding model space vectors, $\mathbf{V}_{.,i}$.

In constructing an L-curve using (4.16), it is important to make sure that the range of values included in the plot is sufficient. If the singular values of $\mathbf{G}$ are known, then it should be sufficient to use a range with the smallest value of α smaller than s_k, and a largest value of α larger than s_1.

A related method called the **damped SVD method** [84] uses the filter factors

$$\hat{f}_i = \frac{s_i}{s_i + \alpha} \; . \tag{4.18}$$

This has a similar effect to using (4.17), but transitions more slowly with the index i.

Example 4.1

We revisit the severely ill-posed Shaw problem, which was previously introduced in Examples 1.6 and 1.10, and was analyzed using the SVD in Example 3.3. The true model in this synthetic example is a spike of unit amplitude in the 10th model element, and independent $N(0, (10^{-6})^2)$ noise has been added to the data vector. We begin by computing the L-curve at 1000 points and finding its corner by estimating the point of maximum curvature. Fig. 4.3 shows the L-curve. The corner of the curve corresponds to $\alpha \approx 6.40 \times 10^{-6}$.

Next, we compute the Tikhonov regularization solution corresponding to this value of α. This solution is shown in Fig. 4.4. Note that this solution is much better than the wild solution obtained by the TSVD with $p' = 18$ (Fig. 3.24).

Alternatively, we can use the discrepancy principle to find an appropriate α to obtain a Tikhonov regularized solution. Because independent $N(0, (1 \times 10^{-6})^2)$ noise was added to these $m = 20$ data points, we search for a solution for which the norm of the residuals is $10^{-6} \cdot \sqrt{20}$, or norm $\|\mathbf{Gm} - \mathbf{d}\|_2 \approx 4.47 \times 10^{-6}$.

The discrepancy principle results in a somewhat larger value of the regularization parameter, $\alpha = 4.29 \times 10^{-5}$, than that obtained using the L-curve technique above. The corresponding solution, shown in Fig. 4.5, thus has a smaller model norm, but the two models are quite close.

It is interesting to note that the norm of the residual of the true (spike) model in this particular case, 3.86×10^{-6}, is actually slightly smaller than the tolerance that we specified via the discrepancy principle (4.47×10^{-6}). The discrepancy principle method did not recover the original spike model because the spike model has a norm of 1, whereas the solution obtained by the discrepancy principle has a norm of only 0.67.

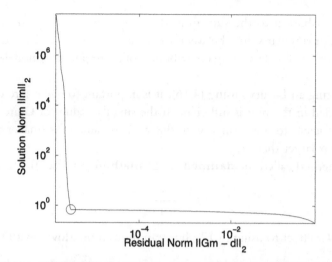

Figure 4.3 Zeroth-order Tikhonov regularization L-curve for the Shaw problem, with corner estimated using maximum functional curvature.

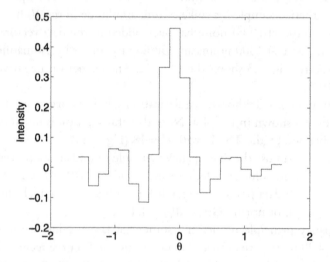

Figure 4.4 Recovery of the spike model with noise, zeroth-order Tikhonov solution ($\alpha = 6.40 \times 10^{-6}$ determined from the L-curve corner).

Plotting the singular values s_i, the values of $|(\mathbf{U}_{\cdot,i})^T \mathbf{d}|$, and the ratios $|(\mathbf{U}_{\cdot,i})^T \mathbf{d}|/s_i$ allows us to examine the discrete Picard condition for this problem (Fig. 4.6). $|(\mathbf{U}_{\cdot,i})^T \mathbf{d}|$ reaches a noise floor of about 1×10^{-6} after $i = 11$. The singular values continue to

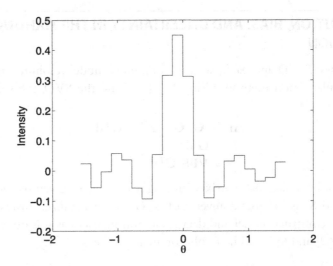

Figure 4.5 Recovery of the spike model with noise, zeroth-order Tikhonov solution ($\alpha = 4.29 \times 10^{-5}$ determined from the discrepancy principle).

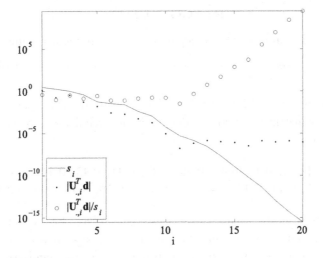

Figure 4.6 Picard plot for the Shaw problem of Example 4.1.

decay. As a consequence, the ratios increase rapidly. It is clear from this plot that we cannot expect to obtain useful information from the singular values beyond $i = 11$. The 11th singular value is $\approx 5.1 \times 10^{-6}$, which is comparable to the values of α in Figs. 4.4 and 4.5.

4.3. RESOLUTION, BIAS, AND UNCERTAINTY IN THE TIKHONOV SOLUTION

As in our earlier TSVD approach, we can compute a model resolution matrix for the Tikhonov regularization method. Using Eq. (4.7) and the SVD, the solution can be written as

$$\begin{aligned}
\mathbf{m}_\alpha &= (\mathbf{G}^T\mathbf{G} + \alpha^2\mathbf{I})^{-1}\mathbf{G}^T\mathbf{d} \\
&= \mathbf{G}^\sharp\mathbf{d} \\
&= \mathbf{VFS}^\dagger\mathbf{U}^T\mathbf{d}
\end{aligned} \tag{4.19}$$

where $\mathbf{F}$ is an n by n diagonal matrix with diagonal elements given by the filter factors f_i of (4.17), and $\mathbf{S}^\dagger$ is the pseudoinverse of $\mathbf{S}$. $\mathbf{G}^\sharp$ is a generalized inverse matrix that can be used to construct model and data resolution matrices as was done for the SVD solution in (3.62) and (3.76). The resolution matrices are

$$\mathbf{R}_\mathrm{m} = \mathbf{G}^\sharp\mathbf{G} = \mathbf{VFV}^T \tag{4.20}$$

and

$$\mathbf{R}_\mathrm{d} = \mathbf{GG}^\sharp = \mathbf{UFU}^T . \tag{4.21}$$

Note that the resolution matrices are dependent on the particular value of α (4.19).

Example 4.2

In Example 4.1, with $\alpha = 4.29 \times 10^{-5}$ as selected using the discrepancy principle, the model resolution matrix for zeroth-order Tikhonov regularization has the following diagonal elements

$$\begin{aligned}
\mathrm{diag}(\mathbf{R}_\mathrm{m}) \approx [&0.91, \ 0.49, \ 0.45, \ 0.39, \ 0.42, \ 0.41, \ 0.43, \ 0.44, \ 0.44, \dots \\
&0.45, \ 0.45, \ 0.44, \ 0.44, \ 0.43, \ 0.41, \ 0.42, \ 0.39, \ 0.45, \ 0.49, \ 0.91]^T
\end{aligned} \tag{4.22}$$

indicating that most model parameters are not well resolved. Fig. 4.7 displays the effect of this limited resolution by applying $\mathbf{R}_\mathrm{m}$ to the (true) spike model (3.62) or, equivalently, inverting noise-free spike model data using (4.6) for the regularization parameter value estimated using the discrepancy principle. Note that the result of limited resolution is that the true model "leaks" or "smears" into adjacent model parameters and is reduced in its maximum amplitude in the recovered model. In this example, the noise-free spike model recovery obtained in this resolution test is nearly identical to the recovery from the noisy spike model data using zeroth-order Tikhonov regularization (Fig. 4.5), indicating that noise has only a very slight effect on model recovery accuracy. Thus, the differences between the true and recovered models here are essentially entirely

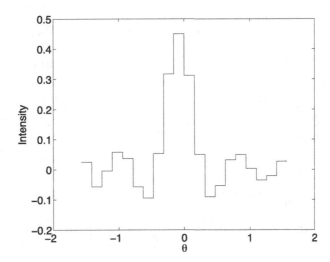

Figure 4.7 Resolution test using the spike model ($\alpha = 4.29 \times 10^{-5}$ determined from the discrepancy principle). Note that this model is nearly equivalent to that shown in Fig. 4.5.

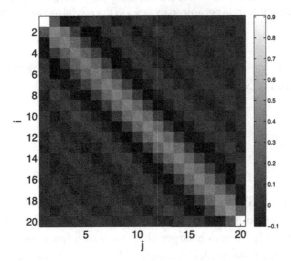

Figure 4.8 Resolution matrix for the Shaw problem, $\alpha = 4.29 \times 10^{-5}$. Note that the noise-free inversion of a spike model e_i will produce the ith column or row of $\mathbf{R}_m$ (e.g., Fig. 4.7 is a plot of the 10th column/row).

due to the regularization that was necessary to stabilize the solution, rather than from noise propagation from data to model (see Example 4.3). Fig. 4.7 displays just a single row or column from the (symmetric) $\mathbf{R}_m$, but effects of limited resolution can be examined more comprehensively by imaging the entire resolution matrix (Fig. 4.8).

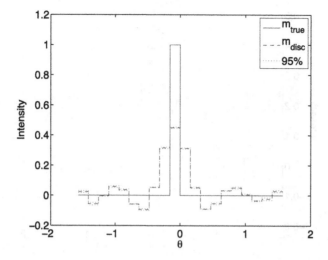

Figure 4.9 Tikhonov solution and confidence intervals for the Shaw problem, estimated using (4.24), where the true model is a spike and the data noise is independent and $N(0, (1 \times 10^{-6})^2)$. The regularization parameter $\alpha = 4.29 \times 10^{-5}$ was chosen using the discrepancy principle. The confidence interval is barely visible at this scale because inaccuracy in the model recovery is dominated by limited resolution.

As in Chapter 2, we can compute a covariance matrix for the estimated model parameters using (B.64). Since

$$\mathbf{m}_\alpha = \mathbf{G}^\sharp \mathbf{d} \tag{4.23}$$

the model covariance is

$$\text{Cov}(\mathbf{m}_\alpha) = \mathbf{G}^\sharp \text{Cov}(\mathbf{d})(\mathbf{G}^\sharp)^T . \tag{4.24}$$

Note that, as with the TSVD solution of Chapter 3, the Tikhonov regularized solution will generally be biased, and differences between the regularized solution values and the true model may actually be much larger than the confidence intervals obtained from the covariance matrix of the model parameters. Rather, the confidence intervals reflect the difference between $\mathbf{m}_\alpha$ and $\mathbf{R}_\mathbf{m}\mathbf{m}_{\text{true}}$. See Fig. 4.9.

Example 4.3

Recall our earlier example of the Shaw problem with the true spike model. Fig. 4.9 shows the true model, the solution obtained using $\alpha = 4.29 \times 10^{-5}$ chosen using the discrepancy principle, and 95% confidence intervals for the estimated parameters. Note that the confidence intervals are extremely tight, and that very few of the true model parameters are included within the confidence intervals. In this case, the regularization bias, which is *not* estimated by the covariance matrix, is far larger than the propagated data uncertainty. In other words, the inaccuracy in model recovery in this case is dominated by limited resolution rather than by noise propagation from data to model. The solution shown in Fig. 4.9 is essentially identical to the product of $\mathbf{R}_m$ and $\mathbf{m}_{\text{true}}$ shown in Fig. 4.7 or the 10th column of the resolution matrix of Fig. 4.8.

4.4. HIGHER-ORDER TIKHONOV REGULARIZATION

So far in our discussions of Tikhonov regularization we have minimized an objective function involving $\|\mathbf{m}\|_2$. In many situations, we would prefer to obtain a solution that minimizes some other measure of $\mathbf{m}$, expressed as $\mathbf{Lm}$, such as the norm of the first or second derivative of $\mathbf{m}$, reflecting a preference for a "flat" or "smooth" model. We then solve the regularized least squares problem

$$\min \|\mathbf{Gm} - \mathbf{d}\|_2^2 + \alpha^2 \|\mathbf{Lm}\|_2^2 . \tag{4.25}$$

As with (4.5), this can be rewritten as a standard linear least squares problem.

$$\min \left\| \begin{bmatrix} \mathbf{G} \\ \alpha\mathbf{L} \end{bmatrix} \mathbf{m} - \begin{bmatrix} \mathbf{d} \\ \mathbf{0} \end{bmatrix} \right\|_2^2 . \tag{4.26}$$

For example, if we have discretized our problem using simple collocation and our model is one-dimensional, then we can approximate, to a multiplicative constant, the first derivative of the model by a finite-difference operation $\mathbf{L}_1\mathbf{m}$, where

$$\mathbf{L}_1 = \begin{bmatrix} -1 & 1 & & & \\ & -1 & 1 & & \\ & & \ddots & \ddots & \\ & & & -1 & 1 \\ & & & & -1 & 1 \end{bmatrix} . \tag{4.27}$$

The matrices that are used to differentiate $\mathbf{m}$ for the purposes of regularization are commonly referred to as **roughening matrices**. In (4.27), $\mathbf{L}_1\mathbf{m}$ is a finite-difference approximation that is proportional to the first derivative of $\mathbf{m}$. By penalizing $\|\mathbf{L}_1\mathbf{m}\|_2$, we will favor solutions that are relatively flat (i.e., in the limit, constant). Note that $\|\mathbf{L}_1\mathbf{m}\|_2$ is a seminorm because it is zero for any constant model, not just for $\mathbf{m} = \mathbf{0}$.

In applying **second-order Tikhonov regularization** to a one dimensional problem, we use a roughening matrix of the form

$$\mathbf{L}_2 = \begin{bmatrix} 1 & -2 & 1 & & & \\ & 1 & -2 & 1 & & \\ & & \ddots & \ddots & \ddots & \\ & & & 1 & -2 & 1 & \\ & & & & 1 & -2 & 1 \end{bmatrix}. \tag{4.28}$$

Here, $\mathbf{L}_2\mathbf{m}$ is a finite-difference approximation that is proportional to the second derivative of $\mathbf{m}$, and the seminorm term $\|\mathbf{L}_2\mathbf{m}\|_2$ in (4.25) penalizes solutions that are rough in a squared second derivative sense. We will refer to regularization strategies of the form of (4.25) that use $\mathbf{L}_1$ and $\mathbf{L}_2$ roughening matrices as **first- and second-order Tikhonov regularization**, respectively.

If our model is higher-dimensional (e.g., two or three dimensions), then the roughening matrices described here would not be appropriate. In such cases, second-order Tikhonov regularization is often implemented using a finite-difference approximation to the Laplacian operator of appropriate dimensionality (see Exercise 4.3).

To ensure that the least squares problem (4.26) has a unique solution, we will require that the matrix

$$\mathbf{A} = \begin{bmatrix} \mathbf{G} \\ \alpha\mathbf{L} \end{bmatrix} \tag{4.29}$$

has full column rank or equivalently that $N(\mathbf{G}) \cap N(\mathbf{L}) = \{\mathbf{0}\}$.

We have already seen how to apply zeroth-order Tikhonov regularization to solve (4.25), with $\mathbf{L} = \mathbf{I}$, using the singular value decomposition (4.16). To solve and analyze higher-order Tikhonov regularization problems, we employ the **generalized singular value decomposition**, or GSVD [73,81,84]. The GSVD enables the solution to (4.25) to be expressed as a sum of filter factors times generalized singular vectors in a manner that is analogous to the series representation of the generalized inverse solution (4.16).

Unfortunately, the definition of the GSVD and associated notation are not presently standardized. In the following, we will adhere to the conventions used by the MATLAB **gsvd** command where $\mathbf{G}$ is an m by n matrix and $\mathbf{L}$ is a p by n matrix. Although MATLAB's implementation of the GSVD can handle matrices $\mathbf{G}$ and $\mathbf{L}$ that have overlapping null spaces, we assume in the following derivation that the matrix in (4.29) has full column rank so that the solution to (4.26) will be unique. To further simplify the derivation we will also assume that $\text{rank}(\mathbf{L}) = p$. This is certainly true for the $\mathbf{L}$ matrices in (4.27) and (4.28). In general we can eliminate redundant rows from $\mathbf{L}$ to make $\mathbf{L}$ have full row rank.

Under the above assumptions there exist matrices $\mathbf{U}$, $\mathbf{V}$, $\mathbf{X}$, Λ, and $\mathbf{M}$ with the following properties and relationships:

- **U** is m by m and orthogonal.
- **V** is p by p and orthogonal.
- **X** is n by n and nonsingular.
- Λ is an m by n matrix with diagonal entries that may be shifted from the main diagonal of the matrix. The diagonal entries are

$$0 \leq \Lambda_{1,k+1} \leq \Lambda_{2,k+2} \leq \cdots \leq \Lambda_{m,k+m} \leq 1 \qquad (4.30)$$

where $k = 0$ when $m > n$, and $k = n - m$ when $m \leq n$.

- **M** is a p by n diagonal matrix with

$$M_{1,1} \geq M_{2,2} \geq \cdots \geq M_{p,p} \geq 0 . \qquad (4.31)$$

-

$$\mathbf{M}^T\mathbf{M} + \mathbf{\Lambda}^T\mathbf{\Lambda} = \mathbf{I} . \qquad (4.32)$$

- The matrices **G** and **L** can be written as

$$\mathbf{G} = \mathbf{U}\mathbf{\Lambda}\mathbf{X}^T \qquad (4.33)$$

and

$$\mathbf{L} = \mathbf{V}\mathbf{M}\mathbf{X}^T . \qquad (4.34)$$

The **generalized singular values** of **G** and **L** are

$$\gamma_i = \frac{\lambda_i}{\mu_i} \qquad (4.35)$$

where

$$\lambda = \sqrt{\operatorname{diag}(\mathbf{\Lambda}^T\mathbf{\Lambda})} \qquad (4.36)$$

and

$$\mu = \sqrt{\operatorname{diag}(\mathbf{M}^T\mathbf{M})} . \qquad (4.37)$$

These definitions may seem somewhat odd, since the diagonal elements of $\mathbf{\Lambda}^T\mathbf{\Lambda}$ and $\mathbf{M}^T\mathbf{M}$ are simply squares of the diagonal elements of $\mathbf{\Lambda}$ and $\mathbf{M}$. The issue here is that the diagonals of $\mathbf{\Lambda}$ and $\mathbf{M}$ are not of the same length. The effect of these definitions is to create vectors λ and μ that are of length n, padding with zeros as needed.

Because of the ordering of the λ and μ elements, the generalized singular values appear in ascending order with

$$\gamma_1 \leq \gamma_2 \leq \cdots \leq \gamma_n . \qquad (4.38)$$

Also note that if $\mu_i = 0$, then the corresponding generalized singular value γ_i is infinite or undefined.

Letting $\mathbf{Y} = \mathbf{X}^{-T}$, we can apply (4.33) and the orthogonality of $\mathbf{U}$ to show that

$$\mathbf{Y}^T \mathbf{G}^T \mathbf{G} \mathbf{Y} = \boldsymbol{\Lambda}^T \boldsymbol{\Lambda} . \tag{4.39}$$

Notice that

$$\lambda_i = \sqrt{\mathbf{Y}_{\cdot,i}^T \mathbf{G}^T \mathbf{G} \mathbf{Y}_{\cdot,i}} = \|\mathbf{G}\mathbf{Y}_{\cdot,i}\|_2 . \tag{4.40}$$

Whenever λ_i is 0, this means that the corresponding column of $\mathbf{Y}$ is in $N(\mathbf{G})$. However, when λ_i is nonzero, the corresponding column of $\mathbf{Y}$ is not in $N(\mathbf{G})$. Since $\mathbf{Y}$ is nonsingular, the columns of $\mathbf{Y}$ are linearly independent. If we pick r so that $\lambda_r = 0$, but $\lambda_{r+1} \neq 0$, then r is the dimension of $N(\mathbf{G})$, and the vectors $\mathbf{Y}_{\cdot,1}$, $\mathbf{Y}_{\cdot,2}$, ..., $\mathbf{Y}_{\cdot,r}$ form a basis for $N(\mathbf{G})$. Note that because $\text{rank}(\mathbf{G}) + \dim(N(\mathbf{G})) = n$, we have determined that $\text{rank}(\mathbf{G}) = n - r$.

Similarly, it is easy to use (4.34) and the orthogonality of $\mathbf{V}$ to show that

$$\mathbf{Y}^T \mathbf{L}^T \mathbf{L} \mathbf{Y} = \mathbf{M}^T \mathbf{M} . \tag{4.41}$$

Since $\text{rank}(\mathbf{L}) = p$, the dimension of the null space of $\mathbf{L}$ is $n - p$. Thus μ_{p+1}, μ_{p+2}, ..., μ_n are 0, and the vectors $\mathbf{Y}_{\cdot,p+1}$, $\mathbf{Y}_{\cdot,p+2}$, ..., $\mathbf{Y}_{\cdot,n}$ form a basis for $N(\mathbf{L})$. Note that the columns of $\mathbf{Y}$ are not generally orthogonal, so we have not found orthonormal bases for the null spaces of $\mathbf{G}$ and $\mathbf{L}$.

The GSVD may be computationally expensive (or intractable for very large problems) to evaluate. However, as with the SVD, once the decomposition matrices are computed, solutions to least-squares problems may be calculated very easily. We begin by introducing the change of variables

$$\mathbf{Y}\mathbf{x} = \mathbf{m} . \tag{4.42}$$

With this substitution, the normal equations for (4.26) become

$$\left(\mathbf{G}^T \mathbf{G} + \alpha^2 \mathbf{L}^T \mathbf{L}\right) \mathbf{Y}\mathbf{x} = \mathbf{G}^T \mathbf{d} . \tag{4.43}$$

Using (4.33) and (4.34), and $\mathbf{Y} = \mathbf{X}^{-T}$, we have

$$\left(\mathbf{Y}^{-T} \boldsymbol{\Lambda}^T \mathbf{U}^T \mathbf{U} \boldsymbol{\Lambda} \mathbf{Y}^{-1} + \alpha^2 \mathbf{Y}^{-T} \mathbf{M}^T \mathbf{V}^T \mathbf{V} \mathbf{M} \mathbf{Y}^{-1}\right) \mathbf{Y}\mathbf{x} = \mathbf{Y}^{-T} \boldsymbol{\Lambda}^T \mathbf{U}^T \mathbf{d} \tag{4.44}$$

$$\left(\mathbf{Y}^{-T} \boldsymbol{\Lambda}^T \boldsymbol{\Lambda} \mathbf{Y}^{-1} + \alpha^2 \mathbf{Y}^{-T} \mathbf{M}^T \mathbf{M} \mathbf{Y}^{-1}\right) \mathbf{Y}\mathbf{x} = \mathbf{Y}^{-T} \boldsymbol{\Lambda}^T \mathbf{U}^T \mathbf{d} \tag{4.45}$$

and, finally,

$$\left(\boldsymbol{\Lambda}^T \boldsymbol{\Lambda} + \alpha^2 \mathbf{M}^T \mathbf{M}\right) \mathbf{x} = \boldsymbol{\Lambda}^T \mathbf{U}^T \mathbf{d} . \tag{4.46}$$

The matrix on the left hand side of this system of equations is diagonal, so the solution is particularly easy to see:

$$x_i = \frac{\lambda_i \mathbf{U}_{\cdot,i+k}^T \mathbf{d}}{\lambda_i^2 + \alpha^2 \mu_i^2} \tag{4.47}$$

where, as in (4.30), $k = 0$ when $m > n$, and $k = n - m$ when $m \leq n$. In terms of the generalized singular values, this can be written as

$$x_i = \frac{\gamma_i^2}{\gamma_i^2 + \alpha^2} \frac{\mathbf{U}_{\cdot,i+k}^T \mathbf{d}}{\lambda_i}. \tag{4.48}$$

Substituting this expression for x_i into $\mathbf{m} = \mathbf{Yx}$, we obtain the summation formula

$$\mathbf{m}_{\alpha, L} = \sum_{i=k+1}^{n} \frac{\gamma_i^2}{\gamma_i^2 + \alpha^2} \frac{\mathbf{U}_{\cdot,i-k}^T \mathbf{d}}{\lambda_i} Y_{\cdot,i} \tag{4.49}$$

where

$$f_i = \frac{\gamma_i^2}{\gamma_i^2 + \alpha^2} \tag{4.50}$$

are **GSVD filter factors** that are analogous to the filter factors obtained in the series expression for the zeroth-order Tikhonov regularized solution (4.17). In evaluating this sum, we sometimes encounter situations in which γ_i is infinite. In those cases, the filter factor f_i should be set to 1. Similarly, there are situations in which $\lambda_i = 0$ and $\gamma_i = 0$, producing an expression of the form $0^2/0$ in the sum. These terms should be treated as 0.

When $\mathbf{G}$ comes from an IFK, the GSVD typically has two properties that were also characteristic of the SVD. First, the m nonzero generalized singular values γ_n, γ_{n-1}, γ_{n-m+1} from (4.35) trend towards zero without any obvious break. Second, the vectors $\mathbf{U}_{\cdot,i}$, $\mathbf{V}_{\cdot,i}$, $\mathbf{X}_{\cdot,i}$, and $\mathbf{Y}_{\cdot,i}$ tend to become rougher as γ_i decreases.

Example 4.4

We return to the vertical seismic profiling example previously discussed in Examples 1.3 and 1.9. Here, consider a 1-km deep borehole experiment discretized using $m = n = 50$ observation and model points, corresponding to sensors every 20 m, and 20 m thick, constant slowness model layers. Fig. 4.10 shows the test model that we will try to recover. A synthetic data set was generated with $N(0, (2 \times 10^{-4})^2)$ (in units of seconds) noise added.

The discretized system of equations $\mathbf{Gm} = \mathbf{d}$ has a small condition number (64). This happens in part because we have chosen a very coarse discretization, which effectively

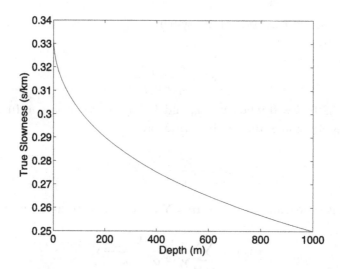

Figure 4.10 A smooth test model for the VSP problem.

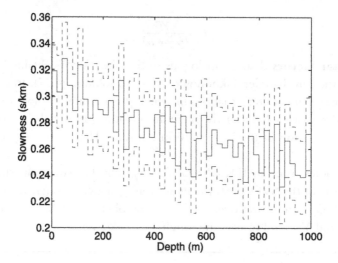

Figure 4.11 Least squares solution for the VSP problem, with 95% confidence intervals.

regularizes the problem by discretization. Another reason is that the vertical seismic profiling problem is only mildly ill-posed [57]. Fig. 4.11 shows the least squares solution, together with 95% confidence intervals.

From the statistical point of view, this solution is completely acceptable. However, suppose that from other information, we believe the slowness should vary smoothly

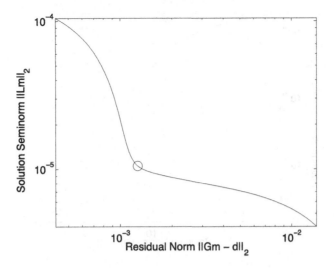

Figure 4.12 L-curve and corner for the VSP problem, first-order regularization.

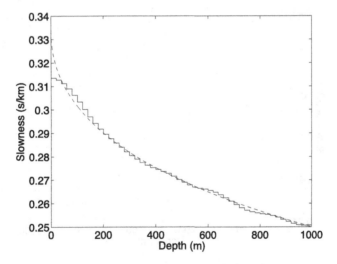

Figure 4.13 Tikhonov solution for the VSP problem, first-order regularization, $\alpha = 122$, shown in comparison with the true model (Fig. 4.10).

with depth. We will next apply higher-order Tikhonov regularization to obtain smooth solutions to this problem.

Fig. 4.12 shows the first-order Tikhonov regularization L-curve for this problem. The L-curve has a distinct corner near $\alpha \approx 122$. Fig. 4.13 shows the corresponding

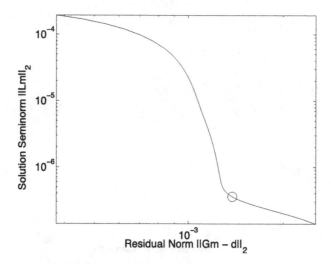

Figure 4.14 L-curve and corner for the VSP problem, second-order regularization.

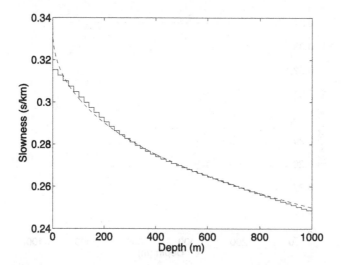

Figure 4.15 Tikhonov solution for the VSP problem, second-order regularization, $\alpha = 2341$, shown in comparison with the true model (Fig. 4.10).

solution. The first-order regularized solution is much smoother than the least squares solution, and is much closer to the true solution.

Fig. 4.14 shows the L-curve for second-order Tikhonov regularization, which has a corner near $\alpha \approx 1965$. Fig. 4.15 shows the corresponding solution. This solution is smoother still compared to the first-order regularized solution. Both the first- and

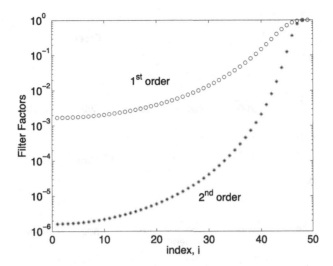

Figure 4.16 Filter factors (4.17) for optimal first- and second-order Tikhonov solutions to the VSP problem shown in Figs. 4.13 and 4.15.

second-order solutions depart most from the true solution at shallow depths where the true slowness has the greatest slope and curvature. This happens because the first- and second-order Tikhonov regularized solutions are biased towards flatness and second-derivative smoothness, respectively.

Fig. 4.16 shows filter factors corresponding to these first- and second-order solutions. Higher-order terms in (4.49) are severely downweighted in both cases, particularly in the second-order case. Because of the smoothness of the true model, the model seminorms can be reduced considerably through the selection of relatively large regularization parameters, α, without large data misfit increases. In this example the 2-norms of the difference between the first- and second-order solutions and the true model (discretized into 50 values) are approximately 1.2×10^{-5} s/km and 1.0×10^{-5} s/km, respectively.

4.5. RESOLUTION IN HIGHER-ORDER TIKHONOV REGULARIZATION

As with zeroth-order Tikhonov regularization, we can compute a resolution matrix for higher-order Tikhonov regularization. For a particular roughening matrix $\mathbf{L}$ and value of α, the Tikhonov regularization solution can be written as

$$\mathbf{m}_{\alpha,L} = \mathbf{G}^{\sharp}\mathbf{d} \qquad (4.51)$$

where

$$\mathbf{G}^{\sharp} = (\mathbf{G}^{T}\mathbf{G} + \alpha^{2}\mathbf{L}^{T}\mathbf{L})^{-1}\mathbf{G}^{T} . \qquad (4.52)$$

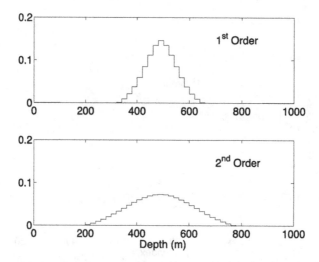

Figure 4.17 The model resolution matrix $\mathbf{R}_{\alpha,L}$ multiplied times the spike model for the first- and second-order regularized solutions of Example 4.4.

Using the GSVD, we can write this expression as

$$\mathbf{G}^{\sharp} = (\mathbf{X}\mathbf{\Lambda}^T\mathbf{U}^T\mathbf{U}\mathbf{\Lambda}\mathbf{X}^T + \alpha^2(\mathbf{X}\mathbf{M}^T\mathbf{V}^T\mathbf{V}\mathbf{M}\mathbf{X}^T))^{-1}\mathbf{X}\mathbf{\Lambda}^T\mathbf{U}^T \qquad (4.53)$$

$$= (\mathbf{X}\mathbf{\Lambda}^T\mathbf{\Lambda}\mathbf{X}^T + \alpha^2(\mathbf{X}\mathbf{M}^T\mathbf{M}\mathbf{X}^T))^{-1}\mathbf{X}\mathbf{\Lambda}^T\mathbf{U}^T \qquad (4.54)$$

$$= (\mathbf{X}(\mathbf{\Lambda}^T\mathbf{\Lambda} + \alpha^2\mathbf{M}^T\mathbf{M})\mathbf{X}^T)^{-1}\mathbf{X}\mathbf{\Lambda}^T\mathbf{U}^T \qquad (4.55)$$

$$= \mathbf{X}^{-T}(\mathbf{\Lambda}^T\mathbf{\Lambda} + \alpha^2\mathbf{M}^T\mathbf{M})^{-1}\mathbf{\Lambda}^T\mathbf{U}^T . \qquad (4.56)$$

The model resolution matrix is then

$$\mathbf{R}_{\mathrm{m}} = \mathbf{G}^{\sharp}\mathbf{G} \qquad (4.57)$$

$$= \mathbf{X}^{-T}(\mathbf{\Lambda}^T\mathbf{\Lambda} + \alpha^2\mathbf{M}^T\mathbf{M})^{-1}\mathbf{\Lambda}^T\mathbf{U}^T\mathbf{U}\mathbf{\Lambda}\mathbf{X}^T \qquad (4.58)$$

$$= \mathbf{X}^{-T}\mathbf{F}\mathbf{X}^T \qquad (4.59)$$

where

$$\mathbf{F} = (\mathbf{\Lambda}^T\mathbf{\Lambda} + \alpha^2\mathbf{M}^T\mathbf{M})^{-1}\mathbf{\Lambda}^T\mathbf{\Lambda} \qquad (4.60)$$

is a diagonal matrix of GSVD filter factors (4.50).

Example 4.5

To examine the resolution of the Tikhonov regularized inversions of Example 4.4, we perform a spike test using (4.59). Fig. 4.17 shows the effect of multiplying $\mathbf{R}_{\mathrm{m}}$ times a unit amplitude spike model (at depth 500 m) under first- and second-order Tikhonov regularization using α values of 122 and 2341, respectively. These curves can equivalently be conceptualized as rows/columns of the full resolution matrix at the index corresponding to 500 m. The spike test results indicate that these Tikhonov regularized solutions are smoothed versions of the spike model. Under first- or second-order regularization, the resolution of various model features will depend critically on how smooth or rough these features are in the true model. In Figs. 4.13 and 4.15, the higher-order solutions recover the true model better because the true model is smooth. Conversely, the spike model is not well recovered because of its rapid variation.

4.6. THE TGSVD METHOD

In the discussion of the SVD in Chapter 3, we examined the TSVD method of regularization, which rejects model space basis vectors associated with smaller singular values. Equivalently, this can be thought of as a damped SVD solution in which filter factors of one are used for basis vectors associated with larger singular values and filter factors of zero are used for basis vectors associated with smaller singular values. This approach can be extended to the GSVD solution (4.49) to produce a **truncated generalized singular value decomposition** or **TGSVD** solution. In the TGSVD solution we simply assign filter factors (4.50) of one to the q largest generalized singular values terms in the sum to obtain

$$\mathbf{m}_{q,L} = \sum_{i=n-q+1}^{n} \frac{\mathbf{U}^{T}_{\cdot,i-k}\mathbf{d}}{\lambda_{i}} Y_{\cdot,i} . \tag{4.61}$$

Example 4.6

Applying the TGSVD method to the VSP problem, we find L-curve corners near $q = 8$ in the first-order case shown in Fig. 4.18, and $q = 7$ in the second-order case shown in Fig. 4.19. Examining the filter factors obtained for the corresponding Tikhonov solutions shown in Fig. 4.16, we find that they decay precipitously with decreasing index near these locations. Figs. 4.20 and 4.21 show the corresponding TGSVD solutions. The model recovery is comparable to that obtained with the Tikhonov method. The 2-norms of the difference between the first- and second-order solutions and the true model are approximately 1.0×10^{-2} s/km and 7.1×10^{-3} s/km, respectively, which are similar to the Tikhonov solutions in Example 4.4.

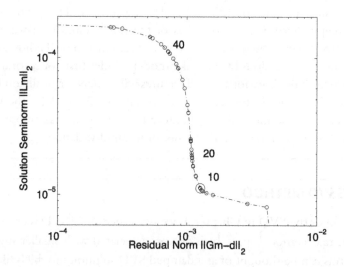

Figure 4.18 TGSVD L-curve for the VSP problem as a function of q for first-order regularization with the $q = 8$ solution indicated.

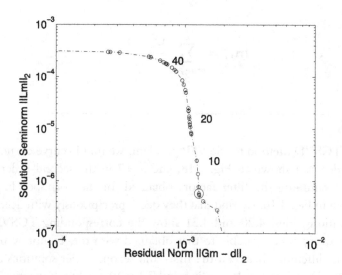

Figure 4.19 TGSVD L-curve for the VSP problem as a function of q for second-order regularization with the $q = 7$ solution indicated.

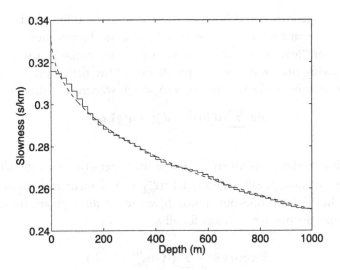

Figure 4.20 TGSVD solution of the VSP problem, $q = 8$, first-order regularization, shown in comparison with the true model.

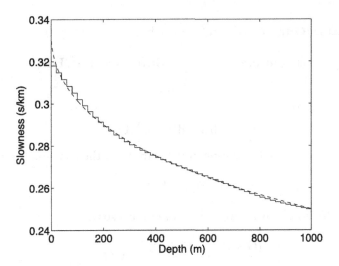

Figure 4.21 TGSVD solution of the VSP problem, $q = 7$, second-order regularization.

4.7. GENERALIZED CROSS-VALIDATION

Generalized cross-validation (GCV) is an alternative method for selecting a regularization parameter, α, that has a number of desirable statistical properties.

In ordinary or "leave-one-out" cross-validation, we consider the models that are obtained by leaving one of the m data points out of the fitting process. Consider the modified Tikhonov regularization problem in which we ignore a data point d_k,

$$\min \sum_{i \neq k} ((\mathbf{Gm})_i - d_i)^2 + \alpha^2 \|\mathbf{Lm}\|_2^2 . \tag{4.62}$$

Call the solution to this problem $\mathbf{m}_{\alpha,L}^{[k]}$, where the superscript indicates that d_k was left out of the computation. Ideally, the model $\mathbf{m}_{\alpha,L}^{[k]}$ would accurately predict the missing data value d_k. In the leave-one-out approach, we select the regularization parameter α so as to minimize the predictive errors for all k

$$\min g(\alpha) = \frac{1}{m} \sum_{k=1}^{m} ((\mathbf{Gm}_{\alpha,L}^{[k]})_k - d_k)^2 . \tag{4.63}$$

Unfortunately, computing $g(\alpha)$ involves solving m problems of the form (4.62). Generalized cross-validation is a way to speed up this computation.

First, let

$$\tilde{d}_i = \begin{cases} (\mathbf{Gm}_{\alpha,L}^{[k]})_k & i = k \\ d_i & i \neq k . \end{cases} \tag{4.64}$$

Note that because $(\mathbf{Gm}_{\alpha,L}^{[k]})_k = \tilde{d}_k$, $\mathbf{m}_{\alpha,L}^{[k]}$ also solves

$$\min ((\mathbf{Gm})_k - \tilde{d}_k)^2 + \sum_{i \neq k} ((\mathbf{Gm})_i - \tilde{d}_i)^2 + \alpha^2 \|\mathbf{Lm}\|_2^2 \tag{4.65}$$

which is equivalent to

$$\min \|\mathbf{Gm} - \tilde{\mathbf{d}}\|_2^2 + \alpha^2 \|\mathbf{Lm}\|_2^2 . \tag{4.66}$$

This result is known as the **leave-one-out lemma**. By the leave-one-out lemma,

$$\mathbf{m}_{\alpha,L}^{[k]} = \mathbf{G}^\sharp \tilde{\mathbf{d}} . \tag{4.67}$$

We will use (4.67) to eliminate $\mathbf{m}_{\alpha,L}^{[k]}$ from (4.63), because

$$\frac{(\mathbf{GG}^\sharp \tilde{\mathbf{d}})_k - (\mathbf{GG}^\sharp \mathbf{d})_k}{\tilde{d}_k - d_k} = (\mathbf{GG}^\sharp)_{k,k} \tag{4.68}$$

where $(\mathbf{GG}^\sharp)_{k,k}$ are the diagonal elements of the data resolution matrix (4.21). Subtracting both sides of the equation from one gives

$$\frac{\tilde{d}_k - d_k - (\mathbf{GG}^\sharp \tilde{\mathbf{d}})_k + (\mathbf{GG}^\sharp \mathbf{d})_k}{\tilde{d}_k - d_k} = 1 - (\mathbf{GG}^\sharp)_{k,k}. \tag{4.69}$$

Since $(\mathbf{GG}^\sharp \mathbf{d})_k = (\mathbf{Gm}_{\alpha,L})_k$, $(\mathbf{GG}^\sharp \tilde{\mathbf{d}})_k = \tilde{d}_k$, and $(\mathbf{Gm}_{\alpha,L}^{[k]})_k = \tilde{d}_k$, (4.69) simplifies to

$$\frac{(\mathbf{Gm}_{\alpha,L})_k - d_k}{(\mathbf{Gm}_{\alpha,L}^{[k]})_k - d_k} = 1 - (\mathbf{GG}^\sharp)_{k,k}. \tag{4.70}$$

Rearranging this formula and substituting into (4.63), we obtain

$$g(\alpha) = \frac{1}{m} \sum_{k=1}^{m} \left(\frac{(\mathbf{Gm}_{\alpha,L})_k - d_k}{1 - (\mathbf{GG}^\sharp)_{k,k}} \right)^2. \tag{4.71}$$

We can simplify the formula further by replacing the $(\mathbf{GG}^\sharp)_{k,k}$ with the average value

$$(\mathbf{GG}^\sharp)_{k,k} \approx \frac{1}{m} \mathrm{Tr}(\mathbf{GG}^\sharp) \tag{4.72}$$

which gives an expression for (4.63) that can be evaluated as a function of α

$$g(\alpha) \approx \frac{1}{m} \sum_{k=1}^{m} \frac{((\mathbf{Gm}_{\alpha,L})_k - d_k)^2}{(\frac{1}{m}(m - \mathrm{Tr}(\mathbf{GG}^\sharp)))^2} \tag{4.73}$$

$$= \frac{m \|\mathbf{Gm}_{\alpha,L} - \mathbf{d}\|_2^2}{\mathrm{Tr}(\mathbf{I} - \mathbf{GG}^\sharp)^2}. \tag{4.74}$$

It can be shown that under reasonable assumptions regarding the noise and smoothness of $\mathbf{m}_{\mathrm{true}}$, the value of α that minimizes (4.74) approaches the value that minimizes $E[\mathbf{Gm}_{\alpha,L} - \mathbf{d}_{\mathrm{true}}]$ as the number of data points m goes to infinity, and that under the same assumptions, $E[\|\mathbf{m}_{\mathrm{true}} - \mathbf{m}_{\alpha,L}\|_2]$ goes to 0 as m goes to infinity [48,217]. In practice, the size of the data set is fixed in advance, so the limit is not directly applicable. However, these results provide a theoretical justification for using the GCV method to select the Tikhonov regularization parameter.

Example 4.7

Figs. 4.22 and 4.23 show $g(\alpha)$ for the VSP test problem, using first- and second-order Tikhonov regularization, respectively. Respective GCV function (4.74) minima occur near $\alpha = 76.3$ and $\alpha = 981$, which are somewhat smaller than the α values estimated previously using the L-curve (Example 4.4). The corresponding models (Figs. 4.24 and 4.25) thus have somewhat larger seminorms.

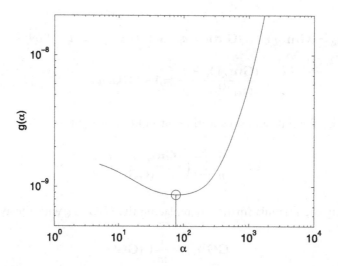

Figure 4.22 GCV curve for the VSP problem and its minimum, first-order regularization.

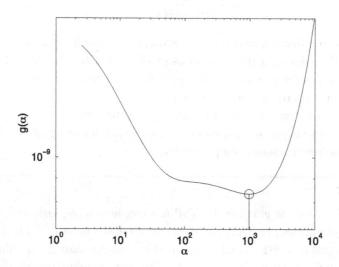

Figure 4.23 GCV curve for the VSP problem and its minimum, second-order regularization.

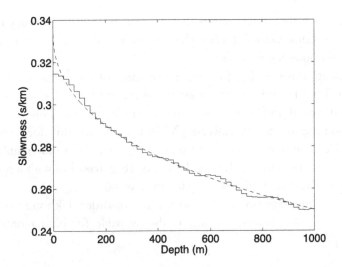

Figure 4.24 GCV solution for the VSP problem, first-order, $\alpha = 76.3$, shown in comparison with the true model.

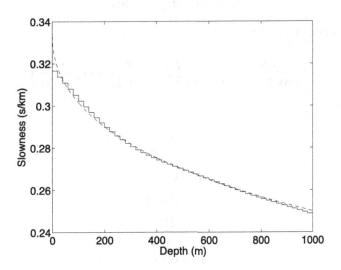

Figure 4.25 GCV solution for the VSP problem, second-order, $\alpha = 981$, shown in comparison with the true model.

4.8. ERROR BOUNDS

We next present two theoretical results that help to address the accuracy of Tikhonov regularization solutions. We will present these results in a simplified form, covering only zeroth-order Tikhonov regularization.

The first question is whether for a particular value of the regularization parameter, α, we can establish a bound on the sensitivity of the regularized solution to the noise in the observed data $\mathbf{d}$ and/or errors in the system matrix $\mathbf{G}$. This would provide a sort of condition number for the inverse problem. Note that this does not tell us how far the regularized solution is from the true model, since Tikhonov regularization has introduced a bias in the solution. Under Tikhonov regularization with a nonzero α, we would not obtain the true model even if the noise was 0.

The following theorem gives a bound for zeroth-order Tikhonov regularization. A somewhat more complicated formula is also available for higher-order Tikhonov regularization [84].

Theorem 4.1. Suppose that $0 < \alpha \le 1$ and the problems

$$\min \|\mathbf{Gm} - \mathbf{d}\|_2^2 + \alpha^2 \|\mathbf{m}\|_2^2 \tag{4.75}$$

and

$$\min \|\bar{\mathbf{G}}\mathbf{m} - \bar{\mathbf{d}}\|_2^2 + \alpha^2 \|\mathbf{m}\|_2^2 \tag{4.76}$$

are solved to obtain $\mathbf{m}_\alpha$ and $\bar{\mathbf{m}}_\alpha$. Then

$$\frac{\|\mathbf{m}_\alpha - \bar{\mathbf{m}}_\alpha\|_2}{\|\mathbf{m}_\alpha\|_2} \le \frac{\bar{\kappa}_\alpha}{1 - \epsilon \bar{\kappa}_\alpha} \left(2\epsilon + \frac{\|\mathbf{e}\|_2}{\|\mathbf{d}_\alpha\|_2} + \epsilon \bar{\kappa}_\alpha \frac{\|\mathbf{r}_\alpha\|_2}{\|\mathbf{d}_\alpha\|_2} \right) \tag{4.77}$$

where

$$\bar{\kappa}_\alpha = \frac{\|\mathbf{G}\|_2}{\alpha} \tag{4.78}$$

$$\mathbf{E} = \mathbf{G} - \bar{\mathbf{G}} \tag{4.79}$$

$$\mathbf{e} = \mathbf{d} - \bar{\mathbf{d}} \tag{4.80}$$

$$\epsilon = \frac{\|\mathbf{E}\|_2}{\|\mathbf{G}\|_2} \tag{4.81}$$

$$\mathbf{d}_\alpha = \mathbf{Gm}_\alpha \tag{4.82}$$

and

$$\mathbf{r}_\alpha = \mathbf{d} - \mathbf{d}_\alpha . \tag{4.83}$$

In the particular case when $\mathbf{G} = \bar{\mathbf{G}}$, and the only difference between the two problems is $\mathbf{e} = \mathbf{d} - \bar{\mathbf{d}}$, the inequality becomes even simpler

$$\frac{\|\mathbf{m}_\alpha - \bar{\mathbf{m}}_\alpha\|_2}{\|\mathbf{m}_\alpha\|_2} \leq \bar{\kappa}_\alpha \frac{\|\mathbf{e}\|_2}{\|\mathbf{d}_\alpha\|_2} . \tag{4.84}$$

The condition number $\bar{\kappa}_\alpha$ is inversely proportional to α. Thus increasing α will decrease the sensitivity of the solution to perturbations in the data. Of course, increasing α also increases the error in the solution due to regularization bias and decreases resolution.

The second question is whether we can establish any sort of bound on the norm of the difference between the regularized solution and the true model. This bound would incorporate both sensitivity to noise and the bias introduced by Tikhonov regularization. Such a bound must of course depend on the magnitude of the noise in the data. It must also depend on the particular regularization parameter chosen. Tikhonov developed a beautiful theorem that addresses this question in the context of inverse problems involving IFK's [206]. More recently, Neumaier has developed a version of Tikhonov's theorem that can be applied directly to discretized problems [147].

Recall that in a discrete ill-posed linear inverse problem, the matrix $\mathbf{G}$ commonly has a smoothing effect, in that when we multiply $\mathbf{Gm}$, the result is smoother than $\mathbf{m}$. Similarly, if we multiply $\mathbf{G}^T$ times $\mathbf{Gm}$, the result will be even smoother than $\mathbf{Gm}$. This smoothing in such problems is a consequence of the fact that the singular vectors corresponding to the larger singular values of $\mathbf{G}$ are smooth. Note, however, that this is not a universal property of all matrices; for example, if $\mathbf{G}$ is a matrix that approximates the differentiation operator, then $\mathbf{Gm}$ will be rougher than $\mathbf{m}$.

For discrete ill-posed problems, models in the range of $\mathbf{G}^T$ will form a relatively smooth subspace of all possible models in R^n. Models in this subspace can be written using the range of $\mathbf{G}^T$ as a basis as $\mathbf{m} = \mathbf{G}^T\mathbf{w}$, using the coefficients $\mathbf{w}$. Furthermore, models in the range of $\mathbf{G}^T\mathbf{G}$ form a subspace of $R(\mathbf{G}^T)$, since any model in $R(\mathbf{G}^T\mathbf{G})$ can be written as $\mathbf{m} = \mathbf{G}^T(\mathbf{Gw})$, which is a linear combination of columns of $\mathbf{G}^T$. Because of the smoothing effect of $\mathbf{G}$ and $\mathbf{G}^T$, we would expect these models to be even smoother than the models in $R(\mathbf{G}^T)$. We could construct smaller subspaces of R^n that contain even smoother models, but it turns out that with zeroth-order Tikhonov regularization these are the only subspaces of interest.

There is another way to see that models in $R(\mathbf{G}^T)$ will be relatively smooth. Recall that the vectors $\mathbf{V}_{\cdot,1}, \mathbf{V}_{\cdot,2}, \ldots, \mathbf{V}_{\cdot,p}$ from the SVD of $\mathbf{G}$ form an orthonormal basis for $R(\mathbf{G}^T)$. For discrete ill-posed problems, we know from Chapter 3 that these basis vectors will be relatively smooth, so linear combinations of these vectors in $R(\mathbf{G}^T)$ should be smooth.

The following theorem gives a bound on the total error including bias due to regularization and error due to noise in the data for zeroth-order Tikhonov regularization [147].

Theorem 4.2. Suppose that we use zeroth-order Tikhonov regularization with regularization parameter α to solve $\mathbf{Gm} = \mathbf{d}$, and that $\mathbf{m}_{\text{true}}$ can be expressed as one of the following distinct cases, for some $\mathbf{w}$, and as parameterized by p,

$$\mathbf{m}_{\text{true}} = \begin{cases} \mathbf{G}^T \mathbf{w} & p = 1 \\ \mathbf{G}^T \mathbf{Gw} & p = 2 \end{cases} \tag{4.85}$$

and that

$$\|\mathbf{Gm}_{\text{true}} - \mathbf{d}\|_2 \leq \Delta \|\mathbf{w}\|_2 \tag{4.86}$$

for some $\Delta > 0$. Then

$$\|\mathbf{m}_{\text{true}} - \mathbf{G}^{\sharp}\mathbf{d}\|_2 \leq \left(\frac{\Delta}{2\alpha} + \gamma \alpha^p \right) \|\mathbf{w}\|_2 \tag{4.87}$$

where

$$\gamma = \begin{cases} 1/2 & p = 1 \\ 1 & p = 2 \end{cases} . \tag{4.88}$$

Furthermore, if we begin with the bound

$$\|\mathbf{Gm}_{\text{true}} - \mathbf{d}\|_2 \leq \delta \tag{4.89}$$

we can let

$$\Delta = \frac{\delta}{\|\mathbf{w}\|_2} . \tag{4.90}$$

Under this condition the optimal value of α is

$$\hat{\alpha} = \left(\frac{\Delta}{2\gamma p} \right)^{\frac{1}{p+1}} = O(\Delta^{\frac{1}{p+1}}) . \tag{4.91}$$

With this choice of α,

$$\Delta = 2\gamma p \hat{\alpha}^{p+1} \tag{4.92}$$

and the error bound simplifies to

$$\|\mathbf{m}_{\text{true}} - \mathbf{G}_{\hat{\alpha}}^{\sharp}\mathbf{d}\|_2 \leq \gamma(p+1)\hat{\alpha}^p \|\mathbf{w}\|_2 = O(\Delta^{\frac{p}{p+1}}) . \tag{4.93}$$

This theorem tells us that the error in the Tikhonov regularization solution depends on both the noise level δ and on the regularization parameter α. For larger values of α, the error due to regularization in recovering the true model will be dominant. For smaller values of α, the error due to noise in the data will be dominant. We seek the optimal value of α that best balances these effects. Using the optimal α, (4.93) indicates that we can obtain an error bound of $O(\Delta^{2/3})$ if $p = 2$, and an error bound of $O(\Delta^{1/2})$ if $p = 1$.

Of course, the above result can only be applied when our true model lives in the restricted subspace of $R(\mathbf{G}^T)$. In practice, even if the model does lie in $R(\mathbf{G}^T)$, the vector $\mathbf{w}$ could have a very large norm, making the bound uninformative.

Thus, applying this theorem in a quantitative fashion is typically impractical. However, the theorem does provide some useful insight into the ability of Tikhonov regularization to recover a true model. The first point is that the accuracy of the regularized solution depends very much on the smoothness of the true model. If $\mathbf{m}_{\text{true}}$ is not smooth, then Tikhonov regularization simply will not give an accurate solution. Furthermore, if the model $\mathbf{m}_{\text{true}}$ is smooth, then we can hope for an error in the Tikhonov regularized solution which is $O(\delta^{1/2})$ or $O(\delta^{2/3})$. Another way of saying this is that we can hope *at best* for an answer with about two thirds as many correct significant digits as the data.

Example 4.8

Recall the Shaw problem previously considered in Examples 4.1 and 4.3. Because $\mathbf{G}^T$ is a nonsingular matrix, the spike model should lie in $R(\mathbf{G}^T)$. However, $\mathbf{G}^T$ is numerically singular due to the ill-posedness of the problem, and the spike model thus lies outside of the effective range of $\mathbf{G}^T$. Any unregularized attempt to find $\mathbf{w}$ produces a meaningless answer due to numerical instability. Because the spike model does not lie in $R(\mathbf{G}^T)$, Theorem 4.2 does not apply.

Fig. 4.26 shows a smooth model that does lie in the range of $\mathbf{G}^T$. For this model we constructed a synthetic data set with noise as before at $\delta = 4.47 \times 10^{-6}$. Eq. (4.93) suggests using $\alpha = 8.0 \times 10^{-4}$. The resulting error bound is 8.0×10^{-4}, whereas the actual norm of the model error is 6.6×10^{-4}. Here the data were accurate to roughly six digits, and the solution was accurate to roughly four digits. Fig. 4.27 shows the reconstruction of the model with $N(0, (1.0 \times 10^{-6})^2)$ noise added to the data vector, showing that the solution is well recovered. This example once again demonstrates the importance of smoothness in the true model in determining how accurately it can be reconstructed with Tikhonov regularization.

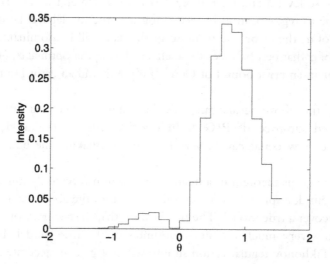

Figure 4.26 A smooth model in $R(\mathbf{G}^T)$.

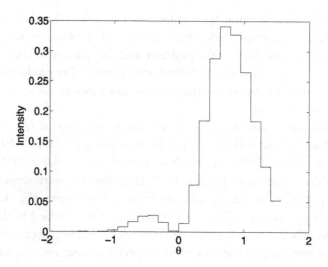

Figure 4.27 Reconstruction of the smooth model with $N(0, (1.0 \times 10^{-6})^2)$ noise.

4.9. USING BOUNDS AS CONSTRAINTS

In many physical situations, bounds exist, or can be usefully estimated, on the maximum and/or minimum values of model parameters. For example, the model parameters may represent a physical quantity such as density that is inherently nonnegative, establishing a strict lower bound for model parameters of $\mathbf{0}$. The problem of solving for a least squares solution that includes this constraint can be expressed as

$$\min \|\mathbf{Gm} - \mathbf{d}\|_2 \\ \mathbf{m} \geq \mathbf{0} \tag{4.94}$$

where $\mathbf{m} \geq \mathbf{0}$ means that every element of the vector $\mathbf{m}$ must be nonnegative. This **nonnegative least squares** problem can be solved by an algorithm called NNLS that was originally developed by Lawson and Hanson [120]. MATLAB includes a command, **lsqnonneg**, that solves the NNLS problem.

More generally, we might also declare a strict upper bound, so that model parameters may not exceed some value, for example, a density of 3500 kg/m^3 for crustal rocks in a particular region. Given the lower and upper bound vectors $\mathbf{l}$ and $\mathbf{u}$, we can pose the **bounded variables least squares** (BVLS) problem

$$\min \|\mathbf{Gm} - \mathbf{d}\|_2 \\ \mathbf{m} \geq \mathbf{l} \\ \mathbf{m} \leq \mathbf{u} . \tag{4.95}$$

Given a BVLS algorithm for solving (4.95), we can also perform Tikhonov regularization with bounds by augmenting the system of equations (e.g., (4.5)) and then solving the augmented system under bounding constraints. Stark and Parker [192] developed an algorithm for solving the BVLS problem, which we employ here as the library function **bvls**. A similar algorithm is given in the 1995 edition of Lawson and Hanson's book [120].

A related optimization problem involves minimizing or maximizing a linear function of the model for a set of n coefficients c_i, subject to bounds constraints and a constraint on the misfit. This problem can be formulated as

$$\min \mathbf{c}^T \mathbf{m} \\ \|\mathbf{Gm} - \mathbf{d}\|_2 \leq \delta \\ \mathbf{m} \geq \mathbf{l} \\ \mathbf{m} \leq \mathbf{u} . \tag{4.96}$$

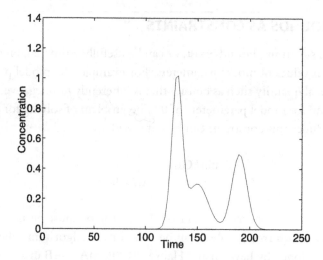

Figure 4.28 True source history.

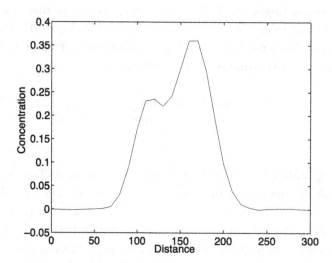

Figure 4.29 Concentration data as a function of position, x, taken at $T = 300$.

Example 4.9

Recall the source history reconstruction problem of Example 1.7, where data are taken in concentration units at spatial positions, x, at a particular time (assumed dimensionless here), T. Fig. 4.28 shows the true (smooth) source model used to generate the data, and Fig. 4.29 shows these data as a function of distance, x, at time $T = 300$, with $N(0, 0.001^2)$ noise added.

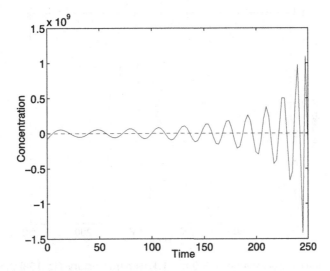

Figure 4.30 Least squares source history solution with true history (Fig. 4.28) shown as a dashed curve. Because the least squares solution has extremely large amplitudes, the true model appears as a flat line at this scale.

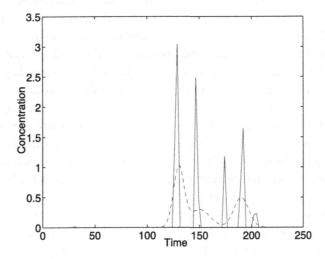

Figure 4.31 NNLS source history solution with true history (Fig. 4.28) shown as a dashed curve.

Fig. 4.30 shows the least squares solution, which has the extremely large amplitudes and oscillatory behavior characteristic of an unregularized solution to an ill-posed problem. This solution is, furthermore, physically unrealistic in having negative concentrations. Fig. 4.31 shows the nonnegative least squares solution, which, although

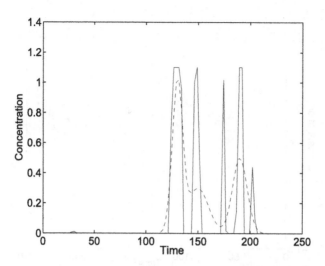

Figure 4.32 BVLS source history solution ($l = 0$, $u = 1.1$) with true history (Fig. 4.28) shown as a dashed curve.

certainly more realistic in having all concentration values nonnegative, consists of high-amplitude spikes that do not accurately reconstruct the (smooth) true source history. Suppose that the solubility limit of the contaminant is known to be 1.1 units, thus providing a natural upper bound on model parameters. Fig. 4.32 shows the corresponding BVLS solution, which exhibits spiky features similar to those of Fig. 4.31. Further regularization is indicated if we believe that the true model is smooth.

Fig. 4.33 shows the L-curve for a second-order Tikhonov regularization solution applied with bounds of $0 \leq \mathbf{m} \leq 1.1$ on the model vector elements. Fig. 4.34 shows the regularized solution for $\alpha = 0.01$. This solution correctly reveals the two major input concentration peaks. As is typical for cases of non-ideal model resolution, the solution peaks are somewhat lower and broader than those of the true model. This solution does not, however, resolve the smaller subsidiary peak near $t = 150$.

We can additionally use (4.96) to establish bounds on linear combinations of the model parameters. For example, we might want to establish bounds on the average concentration from $t = 125$ to $t = 150$. These concentrations appear in positions 51 through 60 of the model vector $\mathbf{m}$. We let c_i be zero in positions 1 through 50 and 61 through 100, and let c_i be 0.1 in positions 51 through 60 to form a 10 time-sample averaging function. The value of $\mathbf{c}^T\mathbf{m}$ is the average of model parameters 51 through 60. The value of the solution to (4.96) will then be a lower bound on the average concentration from $t = 125$ to $t = 150$. Similarly, by maximizing $\mathbf{c}^T\mathbf{m}$ or equivalently by minimizing $-\mathbf{c}^T\mathbf{m}$, we can obtain an upper bound on the average concentration from $t = 125$ to $t = 150$. Solving the minimization problems for $\mathbf{c}^T\mathbf{m}$ and $-\mathbf{c}^T\mathbf{m}$, we

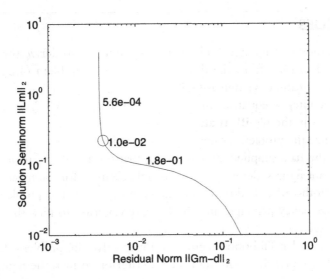

Figure 4.33 L-curve for the second-order Tikhonov solution with BVLS ($l = 0, u = 1.1$) implementation, corner at $\alpha = 0.01$.

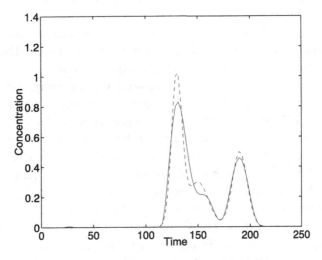

Figure 4.34 Second-order Tikhonov regularization source history solution determined from the L-curve of Fig. 4.33, with true history (Fig. 4.28) shown as a dashed curve.

obtain, respectively, a lower bound of 0.36 and an upper bound of 0.73 for the average concentration during this time period. The true concentration average over this interval (Fig. 4.28) is 0.57.

4.10. EXERCISES

1. Use the method of Lagrange Multipliers (Appendix C) to derive the damped least squares problem (4.4) from the discrepancy principle problem (4.2), and demonstrate that (4.4) can be written as (4.5).

2. Consider the integral equation and data set from Problem 3.5. You can find a copy of this data set in the file **ifk.mat**.
   a. Discretize the problem using simple collocation.
   b. Using the data supplied, and assuming that the numbers are accurate to four significant figures, determine a reasonable bound δ for the misfit.
   c. Use zeroth-order Tikhonov regularization to solve the problem. Use GCV, the discrepancy principle and the L-curve criterion to pick the regularization parameter.
   d. Use first-order Tikhonov regularization to solve the problem. Use GCV, the discrepancy principle and the L-curve criterion to pick the regularization parameter.
   e. Use second-order Tikhonov regularization to solve the problem. Use GCV, the discrepancy principle and the L-curve criterion to pick the regularization parameter.
   f. Analyze the resolution of your solutions. Are the features you see in your inverse solutions unambiguously real? Interpret your results. Describe the size and location of any significant features in the solution.

3. Consider the following problem in **cross-well tomography**. Two vertical wells are located 1600 meters apart. A seismic source is inserted in one well at depths of 50, 150, ..., 1550 m. A string of receivers is inserted in the other well at depths of 50 m, 150 m, ..., 1550 m. See Fig. 4.35. For each source-receiver pair, a travel time is recorded, with a measurement standard deviation of 0.5 ms. There are 256 ray paths and 256 corresponding data points. We wish to determine the velocity structure in the two-dimensional plane between the two wells.
   Discretizing the problem into a 16 by 16 grid of 100 meter by 100 meter blocks gives 256 model parameters. The **G** matrix and noisy data **d** for this problem (assuming straight ray paths) are in the file **crosswell.mat**. The order of parameter indexing from the slowness grid to the model vector is row-by-row (e.g., Example 1.12).
   a. Use the TSVD to solve this inverse problem using an L-curve. Plot the result.
   b. Use zeroth-order Tikhonov regularization to solve this problem and plot your solution. Explain why it is hard to use the discrepancy principle to select the regularization parameter. Use the L-curve criterion to select your regularization parameter. Plot the L-curve as well as your solution.
   c. Use second-order Tikhonov regularization to solve this problem and plot your solution. Because this is a two-dimensional problem, you will need to imple-

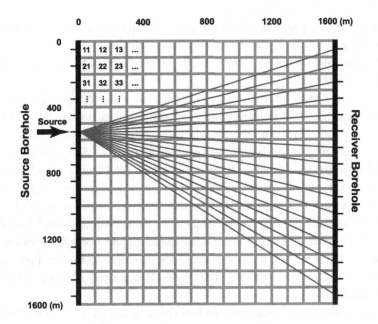

Figure 4.35 Cross-well tomography problem, showing block discretization, block numbering convention, and one set of straight source-receiver ray paths.

ment a finite–difference approximation to the Laplacian (second derivative in the horizontal direction plus the second derivative in the vertical direction) in the roughening matrix. The **L** matrix can be generated using the following MATLAB code:

```
L=zeros(14*14,256);
k=1;
for i=2:15,
  for j=2:15,
    M=zeros(16,16);
    M(i,j)=-4;
    M(i,j+1)=1;
    M(i,j-1)=1;
    M(i+1,j)=1;
    M(i-1,j)=1;
    L(k,:)=reshape(M,256,1)';
    k=k+1;
  end
end
```

What, if any, problems did you have in using the L-curve criterion on this problem? Plot the L-curve as well as your solution.

d. Discuss your results. If vertical bands appeared in some of your solutions, can you explain why?

4. Apply second-order Tikhonov regularization to solve the problem of finding a smooth curve that approximately fits a set of noisy data points. Write a MATLAB program to find a function specified at the 191 points $x = 1, 1.1, 1.2, \ldots, 20$ that approximately fits the 20 data points specified at $x = 1, 2, 3, \ldots, 20$ given in the file **interpdata.mat**. Construct an appropriate 20 by 191 **G** matrix, and use the library function **get_l_rough** to obtain the necessary second-order roughening matrix, **L**. Produce solutions for regularization parameter values of $\alpha = 0.2, 0.4, 0.6, \ldots, 10$ and show the tradeoff curve between 2-norm data misfit and model seminorm on a linear-linear plot. If the data noise is independent and normally distributed with a standard deviation of 0.3, use the discrepancy principle to find and plot an optimal interpolated curve along with the data points. What is the χ^2 value of this solution? Is it reasonable?

5. In some situations it is appropriate to bias the regularized solution towards a particular model $\mathbf{m}_0$. In this case, we would solve

$$\min \|\mathbf{Gm} - \mathbf{d}\|_2^2 + \alpha^2 \|\mathbf{L}(\mathbf{m} - \mathbf{m}_0)\|_2^2 . \tag{4.97}$$

Write this as an ordinary linear least squares problem. What are the normal equations? Can you find a solution for this problem using the GSVD?

6. Using the method of Lagrange multipliers, develop a formula that can be used to solve

$$\min \mathbf{c}^T \mathbf{m}$$
$$\|\mathbf{Gm} - \mathbf{d}\|_2 \le \delta . \tag{4.98}$$

7. In many cases, it is valuable to implement multiple regularization constraints, such as smoothness combined with bounds. Consider the gravity anomaly data in the vector **dn** (units of m/s^2) found in the file **gdata.mat**. The data are taken at the surface above a 1000 m section of a partially debris-blocked buried pipe located with its center at a depth of 25 m. Use the methodology of Example 1.4 to invert these data for subsurface density perturbations, assuming that the pipe has a cross-sectional area of 1 m^2. Anomalous density along the pipe relative to background density is parameterized as point masses located at 100 equally spaced points along the section, as specified by the vector **rhox**. The noisy surface gravity anomaly observations contained in the vector **dn** are made at 500 points along the same 1000 m span as the buried pipe section of interest at points specified in the vector **x**. Invert the data for density perturbations along the buried pipe by assuming that all mass in each segment can be concentrated at the points specified by **rhox**.

a. Invert for density perturbations along the pipe transect in kg/m^3 using least squares.

b. Invert for density perturbations along the pipe transect in kg/m^3 using second-order Tikhonov regularization and an L–curve analysis.

c. Solve the problem using second-order Tikhonov regularization combined with BVLS and an L–curve analysis. Employ the knowledge that density variations for $x < 91$ and $x > 899$ m are negligible, and that the density variations overall are bounded by -150 kg/m$^3 \le \delta\rho(x) \le 150$ kg/m^3.

d. Compare and discuss your three solutions. Where are the prominent maxima and minima located?

4.11. NOTES AND FURTHER READING

Hansen's book [84] is a very complete reference on the linear algebra of Tikhonov regularization. Arnold Neumaier's tutorial [147] is also a very useful reference. Two other surveys of Tikhonov regularization are [56,57]. Vogel [216] includes an extensive discussion of methods for selecting the regularization parameter. Hansen's MATLAB Regularization Tools [83] is a collection of software for performing regularization within MATLAB.

The GSVD was first defined by Van Loan [127]. References on the GSVD and algorithms for computing the GSVD include [3,72,81,84].

Selecting the regularization parameter is an important problem in both theory and practice. Much of the literature on functional analytic approaches assumes that the noise level is known. When the noise level is known, the discrepancy principle provides a scheme for selecting the regularization parameter for ill-posed problems that is convergent in the sense that, in the limit as the noise level goes to zero, the regularized solution goes to $\mathbf{m}_{\text{true}}$ [57].

In practice, the noise level is often unknown, so there has been a great deal of interest in schemes for selecting the regularization parameter without its prior knowledge. The two most popular approaches are the L–curve method and GCV. The L–curve method was introduced by Hansen [82,84]. GCV was introduced by Craven and Wahba [48, 217]. The formula for GCV given here is very expensive to compute for large problems. A GCV algorithm for large scale problems is given by Golub and von Matt [73]. Vogel has shown that the L–curve method can fail to converge [215]. It can be shown that no scheme that depends only on noisy data without knowledge of the noise level can be convergent in the limit as the noise level goes to zero [57]. A Bayesian method of regularization optimization for Tikhonov-style problems can be found in [213].

Within statistics, poorly conditioned linear regression problems are said to suffer from "multicollinearity." A method called "Ridge Regression," which is identical to Tikhonov regularization, is often used to deal with such problems [52]. Statisticians also use a method called "Principal Components Regression" (PCR) which is identical to the TSVD method [142].

Methods for bounded variables least squares problems and minimizing a linear function subject to a bound on the misfit are given in [192]. Some applications of these techniques can be found in [92,158,161,190,191].

CHAPTER FIVE

Discretizing Inverse Problems Using Basis Functions

Synopsis

Techniques for discretizing continuous inverse problems characterized by Fredholm integral equations of the first kind using continuous basis functions are discussed, both for general basis functions and for representers. The Gram matrix is defined and used in analyzing these problems. A method of creating an orthonormal basis to solve problems where the Gram matrix is ill-conditioned is shown. The method of Backus and Gilbert is also introduced.

5.1. DISCRETIZATION BY EXPANSION OF THE MODEL

To this point, we have discretized continuous inverse problems using the collocation scheme discussed in Chapter 1. In this chapter, we discuss an alternative approach in which we expand the unknown model $m(x)$ in terms of a set of basis functions.

We begin with a Fredholm integral equation of the first kind,

$$d(x) = \int_a^b g(x, \, \xi) m(\xi) \, d\xi \, . \tag{5.1}$$

Recall from Section A.10 that we can generalize linear algebra, including the concepts of linear independence, dot products, orthogonality, and the 2-norm, to spaces defined by functions. We will assume that our model $m(x)$ lies in the space $L_2(a, \, b)$ of functions that are square integrable on $[a, b]$.

In $L_2(a, \, b)$, the dot product of two functions $f(x)$ and $g(x)$ is defined as

$$\langle f(x), g(x) \rangle = \int_a^b f(x)g(x) \, dx \tag{5.2}$$

the 2-norm of $f(x)$ is

$$\| f(x) \|_2 = \sqrt{\int_a^b f(x)^2 \, dx} \tag{5.3}$$

and the distance between two functions under this norm is

$$\| f(x) - g(x) \|_2 = \sqrt{\int_a^b (f(x) - g(x))^2 \, dx} \, . \tag{5.4}$$

Parameter Estimation and Inverse Problems
https://doi.org/10.1016/B978-0-12-804651-7.00010-9

Ideally, we could try to solve the Tikhonov regularization problem

$$\min \|m(x)\|_2$$

$$\left\| \int_a^b g(x, \xi) m(\xi) d\xi - d(x) \right\|_2 \leq \Delta .$$
(5.5)

However, in practical cases, we do not fully know $d(x)$ in the sense that we normally have a finite number of data points $d_1 = d(x_1)$, $d_2 = d(x_2)$, $\ldots$, $d_m = d(x_m)$ arranged in a vector $\mathbf{d}$. An appropriate approach is thus to minimize the model norm $\|m(x)\|_2$ subject to a constraint on the misfit at the data points

$$\min \|m(x)\|_2$$
$$\|\mathbf{g}(m(x)) - \mathbf{d}\|_2 \leq \delta$$
(5.6)

where

$$\mathbf{g}(m(x)) = \begin{bmatrix} \int_a^b g(x_1, \xi) m(\xi) \, d\xi \\ \int_a^b g(x_2, \xi) m(\xi) \, d\xi \\ \vdots \\ \int_a^b g(x_m, \xi) m(\xi) \, d\xi \end{bmatrix} .$$
(5.7)

Consider a finite dimensional subspace H of $L_2(a, b)$, with the basis

$$h_1(x), h_2(x), \ldots, h_n(x) .$$
(5.8)

Any model $m(x)$ in the subspace H can be written uniquely as

$$m(x) = \sum_{j=1}^n \beta_j h_j(x) .$$
(5.9)

We will seek the model $m(x)$ in H that has minimum norm and adequately fits the data.

Substituting (5.9) into (5.1), and considering only those points x_i where we have data, we obtain

$$d(x_i) = \int_a^b g(x_i, \xi) \left(\sum_{j=1}^n \beta_j h_j(\xi) \right) d\xi \quad i = 1, 2, \ldots, m .$$
(5.10)

To simplify the notation, let

$$g_i(\xi) = g(x_i, \xi) \quad i = 1, 2, \ldots, m$$
(5.11)

and

$$d_i = d(x_i) \quad i = 1, 2, \ldots, m . \tag{5.12}$$

We interchange the integral and sum to get

$$d_i = \sum_{j=1}^{n} \beta_j \int_a^b g_i(x) h_j(x) \, dx \quad i = 1, 2, \ldots, m . \tag{5.13}$$

If we let $\mathbf{G}$ be the m by n matrix with entries

$$G_{i,j} = \int_a^b g_i(x) h_j(x) \, dx \tag{5.14}$$

then the discretized inverse problem can be written as

$$\mathbf{G}\boldsymbol{\beta} = \mathbf{d} \tag{5.15}$$

which can be solved to find the functional expansion coefficients $\boldsymbol{\beta}$ in (5.9).

Because of noise in the data, we do not normally desire, nor will we be generally able to solve (5.15) exactly. Furthermore, we need to be able to regularize our solution to the inverse problem. We could try to solve

$$\begin{aligned} \min \|\boldsymbol{\beta}\|_2 \\ \|\mathbf{G}\boldsymbol{\beta} - \mathbf{d}\|_2 \le \delta . \end{aligned} \tag{5.16}$$

However, $\|\boldsymbol{\beta}\|_2$ is generally not proportional to $\|m(x)\|_2$. Instead, note that

$$\begin{aligned} \|m(x)\|_2^2 &= \int_a^b m(x)^2 \, dx \\ &= \int_a^b \left(\sum_{i=1}^{n} \beta_i h_i(x) \right) \left(\sum_{j=1}^{n} \beta_j h_j(x) \right) dx \\ &= \sum_{i=1}^{n} \sum_{j=1}^{n} \beta_i \beta_j \int_a^b h_i(x) h_j(x) \, dx . \end{aligned} \tag{5.17}$$

If we let $\mathbf{H}$ be the n by n matrix with

$$H_{i,j} = \int_a^b h_i(x) h_j(x) \, dx \tag{5.18}$$

then

$$\|m(x)\|_2^2 = \boldsymbol{\beta}^T \mathbf{H} \boldsymbol{\beta} . \tag{5.19}$$

The matrix $\mathbf{H}$ is called the **Gram matrix** of the functions $h_i(x)$. It is relatively easy to show that $\mathbf{H}$ is a symmetric and positive definite matrix. Since $\mathbf{H}$ is symmetric and positive definite, it has a Cholesky factorization

$$\mathbf{H} = \mathbf{R}^T \mathbf{R} \tag{5.20}$$

so that

$$\|m(x)\|_2^2 = \boldsymbol{\beta}^T \mathbf{R}^T \mathbf{R} \boldsymbol{\beta}$$
$$= \|\mathbf{R}\boldsymbol{\beta}\|_2^2 . \tag{5.21}$$

We can write the Tikhonov regularization problem as

$$\min \|\mathbf{R}\boldsymbol{\beta}\|_2$$
$$\|\mathbf{G}\boldsymbol{\beta} - \mathbf{d}\|_2 \leq \delta . \tag{5.22}$$

We can use the Lagrange multiplier technique (Appendix C) to turn this into an unconstrained minimization problem

$$\min \|\mathbf{G}\boldsymbol{\beta} - \mathbf{d}\|_2^2 + \alpha^2 \|\mathbf{R}\boldsymbol{\beta}\|_2^2 \tag{5.23}$$

where the regularization parameter α is selected so that we obtain a solution that adequately fits the data. This problem can easily be solved using the Cholesky factorization of the normal equations, the generalized singular value decomposition, or iterative methods discussed in Chapter 6.

Note that if we choose an orthonormal basis of functions, then

$$H_{i,j} = \int_a^b h_i(x) h_j(x) \, dx = \begin{cases} 1 & i = j \\ 0 & i \neq j \end{cases} \tag{5.24}$$

or $\mathbf{H} = \mathbf{R} = \mathbf{I}$, and the minimization problem simplifies to (5.16).

To minimize a model derivative norm (e.g., $\|m''(x)\|_2$ instead of $\|m(x)\|_2$), note that because

$$m''(x) = \sum_{j=1}^n \beta_j h_j''(x) \tag{5.25}$$

we can simply use the second derivative Gram matrix

$$H_{i,j} = \int_a^b h_i''(x) h_j''(x) \, dx \tag{5.26}$$

instead of (5.18).

The simple collocation scheme described in Chapter 1 can also be thought of as a special case of this approach, in which the $h_i(x)$ basis functions are "boxcar" functions centered at the measurement points x_i. A variety of other bases have been used in practice. For example, sine and cosine functions can be used to produce Fourier series solutions (Example 5.1), and, for problems involving spherical symmetries, the spherical harmonic functions are often used. Recently, many researchers have also investigated the use of wavelet bases in the solution of inverse problems.

The choice of the finite dimensional subspace H and its basis is critical to the success of this method in solving practical problems. As we have seen, the method finds the model $m(x)$ in the subspace H that minimizes $\|m(x)\|_2$ subject to fitting the data. There is no guarantee that the model we obtain will actually have the minimum norm over the entire space $L_2(a, b)$. However, if the subspace includes a sufficiently large class of functions, the resulting solution may be an adequate approximation to the solution of (5.6).

Example 5.1

We revisit the Shaw problem for $m = 20$ data points, choosing to represent the model as a linear combination of $n = 31$ Fourier series basis functions given by

$$
\begin{aligned}
h_1(\theta) &= 1 \\
h_j(\theta) &= \sin(2\pi(j-1)\theta/T) & j &= 2, \ldots, (n+1)/2 \\
h_j(\theta) &= \cos(2\pi(j-(n+1)/2)\theta/T) & j &= (n+1)/2+1, \ldots, n
\end{aligned}
\tag{5.27}
$$

defined on the interval $[\pi/2, \pi/2]$, where the fundamental period $T = \pi$.

We construct the appropriate 20 by 31 $\mathbf{G}$ matrix, where the column vector $\mathbf{G}_{.j}$ is the Shaw forward predicted data for the basis function h_j. We discretize the $h_j(\theta)$ into 1000 equal angular segments for the purposes of calculating the necessary integrals and for constructing the model from the $\boldsymbol{\beta}$ coefficients. Normalizing these basis functions to have equal norms on the space $L_2(-\pi/2, \pi/2)$, we can jointly fit the data and minimize the model norm, solving (5.16) using (4.5)

$$
\min \left\| \begin{bmatrix} \mathbf{G} \\ \alpha\mathbf{I} \end{bmatrix} \boldsymbol{\beta} - \begin{bmatrix} \mathbf{d} \\ \mathbf{0} \end{bmatrix} \right\|_2^2
\tag{5.28}
$$

where we select an appropriate value of α to satisfy the discrepancy principle via a suitable search algorithm. As in previous Shaw problem examples, we add independent normally distributed $N(0, (10^{-6})^2)$ noise to the data vector, and the desired residual norm δ is thus given by 10^{-6} times the square root of the number of data points, m, or about 4.47×10^{-6}. The corresponding solution is shown in Fig. 5.1.

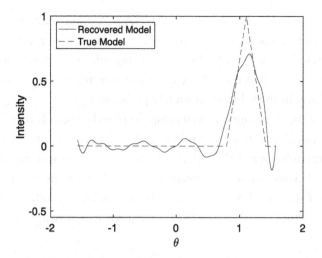

Figure 5.1 Recovery of a triangle function model with a width of 0.62 radians centered on $\theta = 1.098$, satisfying the discrepancy principle using a set of 31 Fourier series model basis functions ($\alpha = 5.97 \times 10^{-6}$, $\chi = 4.48 \times 10^{-6} \approx \delta$).

5.2. USING REPRESENTERS AS BASIS FUNCTIONS

A particularly interesting choice of basis is to use the representers (1.33) themselves as the basis functions. This approach results in a solution that does effectively minimize $\|m(x)\|_2$ over the entire space $L_2(a, b)$.

Consider the space G spanned by the representers

$$g_j(\xi) = g(x_j, \xi) \quad j = 1, 2, \ldots, m . \tag{5.29}$$

Let $G^\perp$ be the perpendicular complement of G. $G^\perp$ consists of the functions $f(x)$ in $L_2(a, b)$ such that

$$\int_a^b f(x)g_j(x) \, dx = 0 \quad j = 1, 2, \ldots, m . \tag{5.30}$$

Thus every function $f(x)$ in $G^\perp$ is in the null space of the forward operator. Every function $m(x)$ in $L_2(a, b)$ can be written uniquely as

$$m(x) = m_1(x) + m_2(x) \tag{5.31}$$

where $m_1(x)$ is in G and $m_2(x)$ is in $G^\perp$. By the Pythagorean theorem,

$$\|m(x)\|_2^2 = \|m_1(x)\|_2^2 + \|m_2(x)\|_2^2 . \tag{5.32}$$

Since any nonzero function $m_2(x)$ would have no effect on the fit to the data but would increase $\|m(x)\|_2$, the optimal norm-minimizing solution to (5.6) will always have

$m_2(x) = 0$. This means that minimizing $\|m(x)\|_2$ over our subspace G will be equivalent to minimizing $\|m(x)\|_2$ over $L_2(a, b)$.

The Gram matrix of the representers, $\mathbf{\Gamma}$, has entries

$$\Gamma_{i,j} = \int_a^b g_i(x)g_j(x)\ dx \tag{5.33}$$

In terms of this matrix,

$$\|m(x)\|_2^2 = \boldsymbol{\beta}^T \mathbf{\Gamma} \boldsymbol{\beta} \tag{5.34}$$

so that the discretized Tikhonov regularization problem (5.22) becomes

$$\begin{aligned}\min \boldsymbol{\beta}^T \mathbf{\Gamma} \boldsymbol{\beta} \\ \|\mathbf{\Gamma}\boldsymbol{\beta} - \mathbf{d}\|_2 \le \delta\ .\end{aligned} \tag{5.35}$$

We can reformulate this as a damped least squares problem

$$\min \|\mathbf{\Gamma}\boldsymbol{\beta} - \mathbf{d}\|_2^2 + \alpha^2 \boldsymbol{\beta}^T \mathbf{\Gamma} \boldsymbol{\beta}\ . \tag{5.36}$$

This damped least squares problem has the normal equations

$$\left(\mathbf{\Gamma}^T \mathbf{\Gamma} + \alpha^2 \mathbf{\Gamma}\right) \boldsymbol{\beta} = \mathbf{\Gamma}^T \mathbf{d}\ . \tag{5.37}$$

5.3. REFORMULATION IN TERMS OF AN ORTHONORMAL BASIS

This linear system of equations (5.37) has a condition number that is always greater than or equal to the condition number of $\mathbf{\Gamma}$. However, if $\mathbf{\Gamma}$ is ill-conditioned, then this method may not produce a numerically stable solution.

Parker [157] developed an approach to solving the inverse problem in this situation via reformulation using an orthonormal basis. Since $\mathbf{\Gamma}$ is symmetric and positive definite, we can write it (A.80) as

$$\mathbf{\Gamma} = \mathbf{Q}\mathbf{\Lambda}\mathbf{Q}^T \tag{5.38}$$

where $\mathbf{\Lambda}$ is a diagonal matrix of nonnegative eigenvalues $\lambda_1 \ge \lambda_2 \ge \cdots \ge \lambda_n$ and $\mathbf{Q}$ is an orthogonal matrix.

Next, define a set of orthonormal basis functions

$$h_j(x) = \sum_{i=1}^m \frac{1}{\sqrt{\lambda_j}} Q_{i,j} g_i(x) \quad j = 1, 2, \dots, m\ . \tag{5.39}$$

We demonstrate that these functions are orthonormal by evaluating the inner product of $h_j(x)$ and $h_k(x)$

$$
\begin{aligned}
\int_a^b h_j(x)h_k(x)\,dx &= \int_a^b \left(\sum_{i=1}^m \frac{1}{\sqrt{\lambda_j}} Q_{i,j}g_i(x)\right)\left(\sum_{l=1}^m \frac{1}{\sqrt{\lambda_k}} Q_{l,k}g_l(x)\right)\\
&= \frac{1}{\sqrt{\lambda_j\lambda_k}} \sum_{i=1}^m \sum_{l=1}^m Q_{i,j}Q_{l,k} \int_a^b g_i(x)g_l(x)\,dx\\
&= \frac{1}{\sqrt{\lambda_j\lambda_k}} \sum_{i=1}^m Q_{i,j} \sum_{l=1}^m Q_{l,k}\Gamma_{l,i}\\
&= \frac{1}{\sqrt{\lambda_j\lambda_k}} \sum_{i=1}^m Q_{i,j} \left(Q^T\Gamma\right)_{k,i}\\
&= \frac{1}{\sqrt{\lambda_j\lambda_k}} (Q^T\Gamma Q)_{j,k}\\
&= \frac{1}{\sqrt{\lambda_j\lambda_k}} \Lambda_{j,k}\\
&= \delta_{j,k}
\end{aligned}
\tag{5.40}
$$

where $\delta_{j,k} = 1$ if $j = k$ and 0 otherwise.

In practice, when Γ is nearly singular, some of the eigenvalues might be effectively zero. In that case, let $p < m$ be the effective rank of Γ, and let

$$
h_j(x) = \sum_{i=1}^m \frac{1}{\sqrt{\lambda_j}} Q_{i,j}g_i(x) \quad j = 1, 2, \ldots, p .
\tag{5.41}
$$

Let Q_p consist of the first p columns of Q and let Λ_p be the diagonal matrix of eigenvalues $\lambda_1, \lambda_2, \ldots, \lambda_p$.

The set of orthonormal basis functions $h_1(x), h_2(x), \ldots, h_p(x)$ spans the same space G as the original linearly dependent set of representers $g_1(x), g_2(x), \ldots, g_m(x)$. We can now express the solution to the inverse problem in terms of the new orthonormal basis as

$$
f(x) = \sum_{i=1}^m \beta_i g_i(x)
\tag{5.42}
$$

or, as

$$
f(x) = \sum_{i=1}^p \gamma_i h_i(x)
\tag{5.43}
$$

where

$$\boldsymbol{\gamma} = \boldsymbol{\Lambda}_p^{1/2} \mathbf{Q}_p^T \boldsymbol{\beta} \qquad (5.44)$$

and

$$\boldsymbol{\beta} = \mathbf{Q}_p \boldsymbol{\Lambda}_p^{-1/2} \boldsymbol{\gamma} \ . \qquad (5.45)$$

With respect to this new basis, the damped least squares problem (5.36) becomes

$$\min \| \boldsymbol{\Gamma} \mathbf{Q}_p \boldsymbol{\Lambda}_p^{-1/2} \boldsymbol{\gamma} - \mathbf{d} \|_2^2 + \alpha^2 \boldsymbol{\gamma}^T \boldsymbol{\gamma} \ . \qquad (5.46)$$

Can we extend this approach to minimizing a derivative-based model roughness measure (e.g., $\|m''(x)\|$ as we did in (5.26))? It is still true that adding a nonzero function $m_2(x)$ from $G^\perp$ to $m(x)$ will have no effect on the fit to the data. Unfortunately, derivatives of models in G and $G^\perp$ will no longer generally satisfy the orthogonality relationship (5.32), e.g.,

$$\|m''(x)\|_2^2 \neq \|m_1''(x)\|_2^2 + \|m_2''(x)\|_2^2 \qquad (5.47)$$

because orthogonality of two functions $f(x)$ and $g(x)$ does not imply orthogonality of $f''(x)$ and $g''(x)$. Since adding a nonzero function $m_2(x)$ from $G^\perp$ might actually decrease $\|m''(x)\|_2$, minimizing $\|m''(x)\|_2$ over our subspace G will not be equivalent to minimizing over $L_2(a, b)$.

5.4. THE METHOD OF BACKUS AND GILBERT

The method of Backus and Gilbert [2,159,169] is applicable to continuous linear inverse problems of the form

$$d(x) = \int_a^b g(x, \xi) m(\xi) \, d\xi \qquad (5.48)$$

where we have observations at points $x_1, x_2, \ldots, x_n$. Let

$$d_j = d(x_j) \quad j = 1, \ 2, \ \ldots, \ m \qquad (5.49)$$

and

$$g_j(\xi) = g(x_j, \xi) \quad j = 1, \ 2, \ \ldots, \ m \ . \qquad (5.50)$$

Using (5.48), we can write the d_j in terms of the true model, m_{true}, as

$$\begin{aligned} d_j &= \int_a^b g(x_j, \ \xi) m_{\text{true}}(\xi) \, d\xi \\ &= \int_a^b g_j(\xi) m_{\text{true}}(\xi) \, d\xi \ . \end{aligned} \qquad (5.51)$$

The Backus and Gilbert method estimates $m_{\text{true}}(x)$ at some point $\hat{x}$, given the m data values d_j. The method produces model estimates that are linear combinations of the data elements

$$\hat{m} = \sum_{j=1}^{m} c_j d_j \approx m_{\text{true}}(\hat{x}) \tag{5.52}$$

where the coefficients c_j are to be determined.

Combining (5.51) and (5.52) gives

$$\begin{aligned}
\hat{m} &= \sum_{j=1}^{m} c_j \int_a^b g_j(x) m_{\text{true}}(x) \; dx \\
&= \int_a^b \left(\sum_{j=1}^{m} c_j g_j(x) \right) m_{\text{true}}(x) \; dx \\
&= \int_a^b A(x) m_{\text{true}}(x) \; dx
\end{aligned} \tag{5.53}$$

where

$$A(x) = \sum_{j=1}^{m} c_j g_j(x) \; . \tag{5.54}$$

The function $A(x)$ is called an **averaging kernel**. Ideally, we would like the averaging kernel to closely approximate a delta function

$$A(x) = \delta(x - \hat{x}) \tag{5.55}$$

because, assuming exact data, (5.53) would then produce exact agreement between the true and estimated model ($\hat{m} = m_{\text{true}}(\hat{x})$). Since this is not possible, we will instead select the coefficients so that the area under the averaging kernel is one, and so that the width of the averaging kernel around $\hat{x}$ is as small as possible.

In terms of the coefficients $\mathbf{c}$, the unit area constraint can be written as

$$\begin{aligned}
\int_a^b A(x) \; dx &= \\
\int_a^b \sum_{j=1}^{m} c_j g_j(x) \; dx &= \\
\sum_{j=1}^{m} c_j \left(\int_a^b g_j(x) \; dx \right) &= 1 \; .
\end{aligned} \tag{5.56}$$

Letting

$$q_j = \int_a^b g_j(x) \, dx \tag{5.57}$$

the unit area constraint (5.56) can be written as

$$\mathbf{q}^T\mathbf{c} = 1 . \tag{5.58}$$

Averaging kernel widths can be usefully characterized in a variety of ways [159]. The most commonly used measure is the second moment of $A(x)$ about $\hat{x}$

$$w = \int_a^b A(x)^2 (x - \hat{x})^2 \, dx . \tag{5.59}$$

In terms of the coefficients $\mathbf{c}$, this can be written using (5.56) as

$$w = \mathbf{c}^T\mathbf{Hc} \tag{5.60}$$

where

$$H_{j,k} = \int_a^b g_j(x) g_k(x) (x - \hat{x})^2 \, dx . \tag{5.61}$$

Now, the problem of finding the optimal coefficients can be written as

$$\begin{aligned} \min \mathbf{c}^T\mathbf{Hc} \\ \mathbf{c}^T\mathbf{q} = 1 \end{aligned} \tag{5.62}$$

which can be solved using the Lagrange multiplier technique.

In practice, the data may be noisy, and the solution may thus be unstable. For measurements with independent errors, the standard deviation of the estimate is given by

$$\text{Var}(\hat{m}) = \sum_{j=1}^m c_j^2 \sigma_j^2 \tag{5.63}$$

where σ_j is the standard deviation of the jth observation.

The solution can be stabilized by adding a constraint on the variance to (5.62)

$$\begin{aligned} \min \mathbf{c}^T\mathbf{Hc} \\ \mathbf{q}^T\mathbf{c} = 1 \\ \sum_{j=1}^n c_j^2 \sigma_j^2 \le \delta . \end{aligned} \tag{5.64}$$

Again, this problem can be solved by the method of Lagrange multipliers. Smaller values of δ decrease the variance of the estimate, but restrict the choice of coefficients so that the width of the averaging kernel increases. There is thus a tradeoff between stability of the solution and the width of the averaging kernel.

The method of Backus and Gilbert produces a model estimate at a particular point $\hat{x}$. It is possible to use the method to compute estimates on a grid of points $x_1, x_2, \ldots, x_n$. However, since the averaging kernels may not be well localized around their associated grid points, and may vary and overlap in complicated ways, this is not equivalent to the simple collocation method of model representation introduced in Chapter 1. Furthermore, such an approach requires the computationally intensive solution of (5.64) for each point. For these reasons, the method does not receive wide use.

Example 5.2

For a spherically symmetric Earth model, the mass M_e and moment of inertia I_e are determined by the radial density $\rho(r)$, where

$$M_e = \int_0^{R_e} \left(4\pi r^2\right) \rho(r) \, dr \tag{5.65}$$

and

$$I_e = \int_0^{R_e} \left(\frac{8}{3}\pi r^4\right) \rho(r) \, dr . \tag{5.66}$$

Using $R_e = 6.3708 \times 10^6$ m as the radius of a spherical Earth, and supposing that from astronomical measurements we can infer that $M_e = 5.973 \pm 0.0005 \times 10^{24}$ kg and $I_e = 8.02 \pm 0.005 \times 10^{37}$ kg·m², we will estimate the density of the Earth in the lower mantle (e.g., at $r = 5000$ km), and core (e.g., at $r = 1000$ km).

Eqs. (5.65) and (5.66) include terms that span an enormous numerical range. Scaling so that $\hat{r} = r/R_e$, $\hat{\rho} = \rho/1000$, $\hat{I}_e = I_e/10^{37}$, and $\hat{M}_e = M_e/10^{24}$ gives

$$\hat{M}_e = 0.2586 \int_0^1 \left(4\pi\hat{r}^2\right) \hat{\rho}(\hat{r}) \, d\hat{r} \tag{5.67}$$

and

$$\hat{I}_e = 1.0492 \int_0^1 \left(\frac{8}{3}\pi\hat{r}^4\right) \hat{\rho}(\hat{r}) \, d\hat{r} . \tag{5.68}$$

Applying (5.62) for $r = 5000$ km gives the coefficient values $\mathbf{c}^T = [1.1809, \ -0.1588]$ and a corresponding model density of 5.8 g/cm³. This is a fairly accurate estimate of density for this radius, where standard Earth models estimated using seismological methods [122] infer densities of approximately 5 g/cm³. The associated standard deviation (5.63) is 0.001 g/cm³, so there is very little sensitivity to data uncertainty.

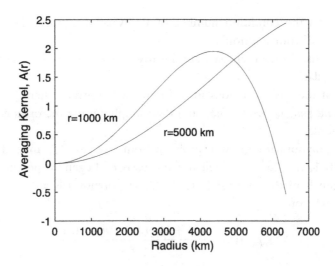

Figure 5.2 Averaging kernels for target radii of 1000 and 5000 km.

At $r = 1000$ km, we obtain the coefficients $\mathbf{c}^T = [2.5537, \ -1.0047]$, and a corresponding density estimate of 7.2 g/cm^3. This is not an accurate estimate for the density of the inner core, where standard Earth models have densities of around 13 g/cm^3. The corresponding standard deviation is just 0.005 g/cm^3, so this inaccuracy is not related to instability in the inverse problem.

Fig. 5.2 shows the averaging kernels corresponding to these model element estimates, and explains both the successful mantle and failed core density estimates. In the mantle case, the averaging kernel has much of its area near the targeted radius of 5000 km. In the core case, however, the averaging kernel has most of its area at much greater r, and little area near the target radius of 1000 km. The fundamental reason for this situation is that both the mass and moment of inertia are relatively insensitive to the density of the innermost Earth.

5.5. EXERCISES

1. Consider the Earth density estimation problem of Example 5.1. Apply the generalized inverse to estimate the density of the Earth as a function of radius, using the given values of mass and moment of inertia. Obtain a density model composed of 20 spherical shells of equal thickness, and compare your results to a standard model.

2. Use the Gram matrix technique to discretize the integral equation from Example 5.1.

   a. Solve the resulting linear system of equations, and plot the resulting model.

    b. What was the condition number of $\mathbf{\Gamma}$? What does this tell you about the accuracy of your solution?

3. Show that if the representers $g_i(t)$ are linearly independent, then the Gram matrix $\mathbf{\Gamma}$ is nonsingular.

4. Show that if the basis functions in (5.9) are orthonormal, then $\|m(x)\|_2 = \|\boldsymbol{\beta}\|_2$. Hint: Expand $\|m(x)\|_2^2$ using (5.3), and then simplify using the orthogonality of the basis functions.

5. Recall the polynomial regression problem from Exercise 2.5. Instead of using the polynomials $1, x, \ldots, x^{19}$, we will use the basis of Legendre polynomials, which are orthogonal on the interval $[-1, 1]$. These polynomials are generated by the recurrence relation

$$p_{n+1}(x) = \frac{(2n+1)xp_n(x) - np_{n-1}(x)}{n+1} \tag{5.69}$$

starting with

$$p_0(x) = 1 \tag{5.70}$$

and

$$p_1(x) = x . \tag{5.71}$$

The next two Legendre polynomials are thus $p_2(x) = (3x^2 - 1)/2$ and $p_3(x) = (5x^3 - 3x)/2$. This recurrence relation can be used both to compute coefficients of the Legendre polynomials and to compute values of the polynomials for particular values of x.

Use the first 20 Legendre polynomials to fit a polynomial of degree 19 to the data from Exercise 2.5. Express your solution as a linear combination of the Legendre polynomials and also as a regular polynomial. How well-conditioned was this system of equations? Plot your solution and compare it with your solution to Exercise 2.5.

6. Construct specific examples of functions $m_1(x)$ and $m_2(x)$ in $L_2(-1, 1)$, such that $m(x) = m_1(x) + m_2(x)$, $m_1(x) \perp m_2(x)$, but $\|m''(x)\|_2^2 \neq \|m_1''(x)\|_2^2 + \|m_2''(x)\|_2^2$.

5.6. NOTES AND FURTHER READING

Techniques for discretizing integral equations are discussed in [57,159,211,220]. A variety of basis functions have been used to discretize integral equations including sines and cosines, spherical harmonics, B-splines, and wavelets. In selecting the basis functions, it is important to select a basis that can reasonably represent likely models. The basis functions must be linearly independent, so that a function can be written in terms of the basis functions in exactly one way, and (5.9) is thus unique. As we have seen, the use of an orthonormal basis has the further advantage that $\|\boldsymbol{\beta}\|_2 = \|m(x)\|_2$.

The selection of an appropriate basis for a particular problem is a fine art that requires detailed knowledge of the problem as well as of the behavior of the basis functions. Beware that a poorly selected basis may not adequately approximate the solution, resulting in an estimated model $m(x)$ that is very wrong. The choice of basis can also have a very large effect on the condition number of the discretized problem, potentially making it very ill-conditioned.

An important theoretical question is whether the solutions to discretized versions of a continuous inverse problem with noise-free data will converge to a solution of the continuous inverse problem. Engl, Hanke, and Neubauer provide an explicit example showing that nonconvergence of a discretization scheme is possible [57]. They also provide conditions under which convergence is guaranteed. For Fredholm integral equations of the first kind, the Gram matrix approach using representers can be shown to be convergent [57].

A comparison of the method of Backus and Gilbert with Tikhonov regularization when applied to problems in helioseismology can be found in [41]. Many authors have suggested variations on the method of Backus and Gilbert in which other measures of the quality of the averaging kernel are optimized [119,129,164,165], with one motivation being to reduce resolution artifacts in seismic tomography [221,222].

CHAPTER SIX

Iterative Methods

Synopsis

We present several methods for solving linear inverse problems that may be far too large for the methods previously discussed to be practical. These methods are iterative, in that a sequence of trial solutions is generated that converges to a final solution. Kaczmarz's algorithm and stochastic gradient descent methods require access to only one row of $\mathbf{G}$ at a time. Other methods, including gradient descent and the method of conjugate gradients, are based on matrix–vector multiplications. When the method of conjugate gradients is applied to the normal equations, the resulting conjugate gradient least squares (CGLS) method regularizes the solution to the inverse problem. Iterative and stochastic methods for estimating model resolution for very large model spaces are described. Illustrative examples involving tomography and image deblurring are given.

6.1. INTRODUCTION

SVD-based pseudoinverse and Tikhonov regularization solutions become impractical when we consider larger problems in which $\mathbf{G}$ has, for example, tens of thousands of rows and columns. Storing the elements of such a large $\mathbf{G}$ matrix can require a great deal of computer memory. If many of the elements in the $\mathbf{G}$ matrix are zero, as for example in ray-theory tomography, then $\mathbf{G}$ is a **sparse matrix**, and we can obviate this problem by only storing the nonzero elements of $\mathbf{G}$ and their locations. The **density** of $\mathbf{G}$ is the percentage of nonzero elements in the matrix. MATLAB can store sparse matrices and most of its built-in functions operate efficiently on sparse matrices. **Dense matrices** contain enough nonzero elements that sparse storage schemes are not efficient.

Methods for the solution of linear systems of equations based on matrix factorizations such as the Cholesky factorization, QR factorization, or SVD do not tend to work well with sparse matrices. The problem is that the matrices that occur in the factorization of $\mathbf{G}$ are generally more dense than $\mathbf{G}$ itself. In particular, the $\mathbf{U}$ and $\mathbf{V}$ matrices in the SVD and the $\mathbf{Q}$ matrix in the QR factorization are orthogonal matrices, which typically makes these matrices dense.

The iterative methods discussed in this chapter do not require the storage of large dense matrices. Instead, they work by generating a sequence of models $\mathbf{m}^{(0)}$, $\mathbf{m}^{(1)}$, ..., that converge to an optimal solution. Here and throughout this chapter, we will use superscripts to denote algorithm iterations. Steps in these algorithms typically involve multiplying $\mathbf{G}$ and $\mathbf{G}^T$ times vectors, which can be done without additional storage. Because iterative methods can also take advantage of the sparsity commonly found in the $\mathbf{G}$ matrix, they are often used for very large problems.

Parameter Estimation and Inverse Problems
https://doi.org/10.1016/B978-0-12-804651-7.00011-0

For example, consider a two-dimensional tomography problem where the model is of size 256 by 256 (65,536 model elements) and there are 100,000 ray paths. Each ray path will pass through less than 1% of the model cells, so the vast majority of the elements (more than 99%) in $\mathbf{G}$ will be zero. If we stored $\mathbf{G}$ as a full double precision matrix, it would require 100,000 rows times 65,536 columns times 8 bytes per entry or about 50 gigabytes of storage. Furthermore, if we wished to apply the SVD, the $\mathbf{U}$ matrix would be of size 100,000 by 100,000 and require 80 gigabytes of storage, while the $\mathbf{V}$ matrix would be of size 65,536 by 65,536 and require about 35 gigabytes. Because of the extreme sparsity of $\mathbf{G}$, such a matrix can be stored much more compactly. We have roughly 1000 entries per row and 65,536 rows, and allowing 16 bytes per entry (for the entry and its row and column indices) this would still require less than 1 gigabyte of storage using a sparse scheme.

The point at which it becomes necessary to use sparse matrix storage depends on the computer that we are using. Although the memory capacity of computers has been increasing steadily for many years, it is safe to say that there will always be problems for which sparse matrix storage will be highly desireable or required. Furthermore, programs that use sparse matrix storage often run much faster than programs that do not take advantage of sparsity. Thus it is wise to consider using sparse storage when appropriate even in cases where sufficient memory is available to store the full matrix.

6.2. ROW ACTION METHODS FOR TOMOGRAPHY PROBLEMS

The approaches discussed in this section are known as **row action methods**, because they work with one row of $\mathbf{G}$ at a time. For extremely large problems, it is sometimes necessary to store the $\mathbf{G}$ matrix externally and bring rows of $\mathbf{G}$ into main memory one at a time for processing. In comparison, the methods discussed later in this chapter make use of the entire $\mathbf{G}$ matrix in each iteration.

Kaczmarz's algorithm is an easy to implement method for solving a linear system of equations $\mathbf{Gm} = \mathbf{d}$ [103,104]. To understand the algorithm, note that each of the m equations $\mathbf{G}_{i,}\,\mathbf{m} = d_i$ defines an n-dimensional hyperplane in R^m. Kaczmarz's algorithm starts with an initial solution $\mathbf{m}^{(0)}$, and then moves to a solution $\mathbf{m}^{(1)}$ by projecting the initial solution onto the hyperplane defined by the first row in $\mathbf{G}$. Next $\mathbf{m}^{(1)}$ is projected onto the hyperplane defined by the second row in $\mathbf{G}$, and so forth. The process is repeated until the solution has been projected onto all m hyperplanes defined by the system of equations. At that point, a new cycle of projections begins. These cycles are repeated until the solution has converged sufficiently. Fig. 6.1 shows an example in which Kaczmarz's algorithm is used to solve the system of equations

$$y = 1$$
$$-x + y = -1 \; .$$

$$(6.1)$$

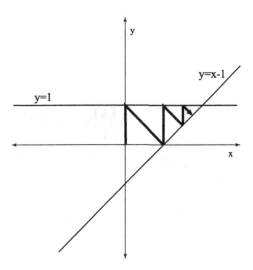

Figure 6.1 Kaczmarz's algorithm on a system of two equations.

To implement the algorithm, we need a formula to compute the projection of a vector onto the hyperplane defined by equation i. Let $\mathbf{G}_{i,\cdot}$ be the ith row of $\mathbf{G}$. Consider the hyperplane defined by $\mathbf{G}_{i+1,\cdot}\mathbf{m} = d_{i+1}$. Because the vector $\mathbf{G}_{i+1,\cdot}^T$ is perpendicular to this hyperplane, the update to $\mathbf{m}^{(i)}$ from the constraint due to row $i+1$ of $\mathbf{G}$ will be proportional to $\mathbf{G}_{i+1,\cdot}^T$.

$$\mathbf{m}^{(i+1)} = \mathbf{m}^{(i)} + \beta \mathbf{G}_{i+1,\cdot}^T \tag{6.2}$$

Using the fact that $\mathbf{G}_{i+1,\cdot}\mathbf{m}^{(i+1)} = d_{i+1}$ to solve for β, we obtain

$$\mathbf{G}_{i+1,\cdot}\left(\mathbf{m}^{(i)} + \beta \mathbf{G}_{i+1,\cdot}^T\right) = d_{i+1} \tag{6.3}$$

$$\mathbf{G}_{i+1,\cdot}\mathbf{m}^{(i)} - d_{i+1} = -\beta \mathbf{G}_{i+1,\cdot}\mathbf{G}_{i+1,\cdot}^T \tag{6.4}$$

$$\beta = -\frac{\mathbf{G}_{i+1,\cdot}\mathbf{m}^{(i)} - d_{i+1}}{\mathbf{G}_{i+1,\cdot}\mathbf{G}_{i+1,\cdot}^T}. \tag{6.5}$$

Thus, the update formula is

$$\mathbf{m}^{(i+1)} = \mathbf{m}^{(i)} - \frac{\mathbf{G}_{i+1,\cdot}\mathbf{m}^{(i)} - d_{i+1}}{\mathbf{G}_{i+1,\cdot}\mathbf{G}_{i+1,\cdot}^T}\mathbf{G}_{i+1,\cdot}^T. \tag{6.6}$$

Kaczmarz showed that if the system of equations $\mathbf{Gm} = \mathbf{d}$ has a unique solution, then the algorithm will converge to this solution [104].

However, if the system of equations is inconsistent, then the algorithm must ultimately fail to converge, because there will always be at least one equation that is not satisfied. For example, if the system of equations $\mathbf{Gm} = \mathbf{d}$ is defined by two parallel

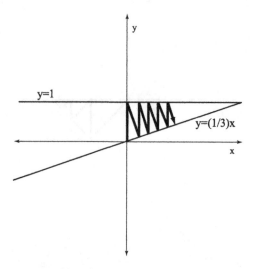

Figure 6.2 Slow convergence occurs when hyperplanes are nearly parallel.

hyperplanes, then the iteration will simply bounce back and forth between the two hyperplanes. A technique called underrelaxation can be used to improve the convergence of the algorithm. Rather than taking the full step specified in (6.6), we multiply the step by a factor $\omega < 1$, Over time this factor ω is reduced to 0. It can be shown that a version of Kaczmarz's algorithm with controlled underrelaxation and $\mathbf{m}^{(0)} = 0$ converges to a minimum norm least squares solution $\mathbf{m} = \mathbf{G}^\dagger \mathbf{d}$ [39,80].

Algorithm 6.1 Kaczmarz's Algorithm

Given a system of equations $\mathbf{Gm} = \mathbf{d}$.

1. Let $\mathbf{m}^{(0)} = \mathbf{0}$.
2. Let $\omega = 1$.
3. For $i = 0, 1, \ldots, m - 1$, let

$$\mathbf{m}^{(i+1)} = \mathbf{m}^{(i)} - \omega \frac{\mathbf{G}_{i+1,\cdot}\mathbf{m}^{(i)} - d_{i+1}}{\mathbf{G}_{i+1,\cdot}\mathbf{G}_{i+1,\cdot}^T}\mathbf{G}_{i+1,\cdot}^T . \qquad (6.7)$$

4. If the solution has not yet sufficiently converged, reduce ω (e.g. by $\omega = 0.95\omega$) and return to step 3 to perform another sweep through the constraints.

A second important question is how quickly Kaczmarz's algorithm will converge to a solution. If the hyperplanes described by the system of equations are nearly orthogonal, then the algorithm will converge very quickly. However, if two or more hyperplanes are nearly parallel to each other, convergence can be extremely slow. Fig. 6.2 shows a typical situation in which the algorithm zigzags back and forth with-

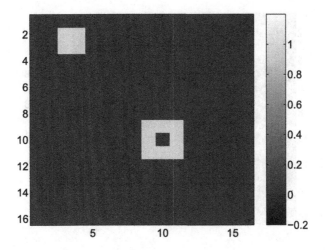

Figure 6.3 True model.

out making much progress towards a solution. As the two lines become more nearly parallel, the problem becomes worse. This problem can be alleviated by picking an ordering of the equations such that adjacent equations describe hyperplanes that are nearly orthogonal to each other. In the context of tomography, this can be done by ordering the equations so that successive equations do not share common model cells. Another approach that is easy to implement is to process the rows of **G** in random order.

Example 6.1

Consider a tomographic reconstruction problem with the same geometry used in Exercise 3.4, in which the slowness structure is parameterized in homogeneous blocks of size l by l. A background velocity of 3000 m/s has been subtracted from the travel times, and we are inverting for perturbations from this background model. The true slowness perturbation model is shown in Fig. 6.3. Synthetic data were generated, with normally distributed random noise added. The random noise had standard deviation 0.01. Fig. 6.4 shows the TSVD solution. The two anomalies are apparent, but it is not possible to distinguish the small hole within the larger of the two.

Fig. 6.5 shows the solution obtained after 200 iterations of Kaczmarz's algorithm. This solution is extremely similar to the TSVD solution and both solutions are about the same distance from the true model.

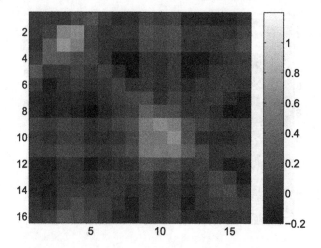

Figure 6.4 Truncated SVD solution.

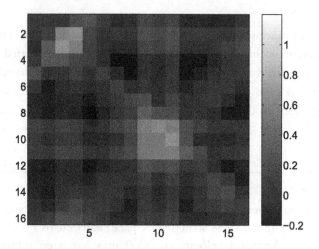

Figure 6.5 Kaczmarz's algorithm solution.

6.3. THE GRADIENT DESCENT METHOD

A very general iterative approach to minimizing a function $f(\mathbf{m})$ is to compute the gradient of f and then step from the current iterate in the direction of the negative of the gradient.

$$\mathbf{m}^{(k+1)} = \mathbf{m}^{(k)} - \alpha_k \nabla f(\mathbf{m}^{(k)}) \ . \tag{6.8}$$

The step size parameter α_k is critical. If α_k is too small, then the algorithm can make very slow progress towards a minimum point. If α_k is too large, then the algorithm can step to a point with a larger function value and ultimately may fail to converge.

There is a large body of research on ways to select the step size parameter for various classes of functions f, but in this section we will consider only the linear least squares problem

$$\min f(\mathbf{m}) = \|\mathbf{Gm} - \mathbf{d}\|_2^2 \ . \tag{6.9}$$

We can write $f(\mathbf{m})$ as

$$f(\mathbf{m}) = (\mathbf{Gm} - \mathbf{d})^T (\mathbf{Gm} - \mathbf{d}) = \mathbf{m}^T \mathbf{G}^T \mathbf{Gm} - 2(\mathbf{G}^T \mathbf{d})^T \mathbf{m} + \mathbf{d}^T \mathbf{d} \ . \tag{6.10}$$

The gradient is

$$\nabla f(\mathbf{m}) = 2\mathbf{G}^T \mathbf{Gm} - 2\mathbf{G}^T \mathbf{d} \ . \tag{6.11}$$

The factor of 2 makes some of the following formulas unnecessarily complicated, so we will fold the factor of 2 into the constant α_k and write the gradient descent iteration as

$$\mathbf{m}^{(k+1)} = \mathbf{m}^{(k)} - \alpha_k (\mathbf{G}^T \mathbf{Gm}^{(k)} - \mathbf{G}^T \mathbf{d}) \ . \tag{6.12}$$

The factor $(\mathbf{G}^T \mathbf{Gm}^{(k)} - \mathbf{G}^T \mathbf{d})$ occurs often, so we will let

$$\mathbf{p}^{(k)} = \mathbf{G}^T \mathbf{Gm}^{(k)} - \mathbf{G}^T \mathbf{d} \ . \tag{6.13}$$

We will also need the residual vector

$$\mathbf{r}^{(k)} = \mathbf{d} - \mathbf{Gm}^{(k)} \ . \tag{6.14}$$

In implementing the method it is important to avoid the expensive computation of the matrix–matrix product $\mathbf{G}^T \mathbf{G}$. This can be done with two matrix–vector products by first computing $\mathbf{r}^{(k)}$, and then letting $\mathbf{p}^{(k)} = -\mathbf{G}^T \mathbf{r}^{(k)}$.

We can now compute $f(\mathbf{m}^{(k+1)})$

$$f(\mathbf{m}^{(k+1)}) = (\mathbf{Gm}^{(k+1)} - \mathbf{d})^T (\mathbf{Gm}^{(k+1)} - \mathbf{d}) \tag{6.15}$$

$$= (\mathbf{G}(\mathbf{m}^{(k)} - \alpha_k \mathbf{p}^{(k)}) - \mathbf{d})^T (\mathbf{G}(\mathbf{m}^{(k)} - \alpha_k \mathbf{p}^{(k)}) - \mathbf{d}) \tag{6.16}$$

$$= (-\mathbf{r}^{(k)} - \alpha_k \mathbf{Gp}^{(k)})^T (-\mathbf{r}^{(k)} - \alpha_k \mathbf{Gp}^{(k)}) \tag{6.17}$$

$$= \mathbf{r}^{(k)^T} \mathbf{r}^{(k)} + 2\alpha_k \mathbf{p}^{(k)^T} \mathbf{G}^T \mathbf{r}^{(k)} + \alpha_k^2 \mathbf{p}^{(k)^T} \mathbf{G}^T \mathbf{Gp}^{(k)} \ . \tag{6.18}$$

Since $f(\mathbf{m}^{(k)}) = \mathbf{r}^{(k)^T} \mathbf{r}^{(k)}$, we have

$$f(\mathbf{m}^{(k+1)}) - f(\mathbf{m}^{(k)}) = 2\alpha_k \mathbf{p}^{(k)^T} \mathbf{G}^T \mathbf{r}^{(k)} + \alpha_k^2 \mathbf{p}^{(k)^T} \mathbf{G}^T \mathbf{Gp}^{(k)} \ . \tag{6.19}$$

Since $\mathbf{p}^{(k)} = -\mathbf{G}^T\mathbf{r}^{(k)}$,

$$f(\mathbf{m}^{(k+1)}) - f(\mathbf{m}^{(k)}) = -2\alpha_k\mathbf{p}^{(k)}{}^T\mathbf{p}^{(k)} + \alpha_k^2\mathbf{p}^{(k)}{}^T\mathbf{G}^T\mathbf{G}\mathbf{p}^{(k)} \ . \tag{6.20}$$

We can factor $\mathbf{p}^{(k)}$ out of this expression to get

$$f(\mathbf{m}^{(k+1)}) - f(\mathbf{m}^{(k)}) = \mathbf{p}^{(k)}{}^T(-2\alpha_k\mathbf{I} + \alpha_k^2\mathbf{G}^T\mathbf{G})\mathbf{p}^{(k)} \ . \tag{6.21}$$

The decrease in the objective function value from iteration k to iteration $k+1$ is thus

$$f(\mathbf{m}^{(k)}) - f(\mathbf{m}^{(k+1)}) = \mathbf{p}^{(k)}{}^T(2\alpha_k\mathbf{I} - \alpha_k^2\mathbf{G}^T\mathbf{G})\mathbf{p}^{(k)} \ . \tag{6.22}$$

Let $\lambda_{\max}(\mathbf{G}^T\mathbf{G})$ be the largest eigenvalue of $\mathbf{G}^T\mathbf{G}$. Consider the matrix $\mathbf{H} = 2\alpha_k\mathbf{I} - \alpha_k^2\mathbf{G}^T\mathbf{G}$. If

$$2\alpha_k \geq \alpha_k^2\lambda_{\max}(\mathbf{G}^T\mathbf{G}) \tag{6.23}$$

then $\mathbf{H}$ is positive definite and $f(\mathbf{m}^{(k+1)}) < f(\mathbf{m}^{(k)})$. The associated condition on α_k is thus

$$\alpha_k \leq \frac{2}{\lambda_{\max}(\mathbf{G}^T\mathbf{G})} \ . \tag{6.24}$$

However, if α_k is larger than the critical value $2/(\lambda_{\max}(\mathbf{G}^T\mathbf{G}))$ then $\mathbf{H}$ will not be positive definite, and depending on the value of $\mathbf{p}^{(k)}$, the objective function value might increase from iteration k to $k+1$. One simple approach to selecting α_k is to pick a constant value $\alpha_k = \omega$ such that

$$\alpha_k = \omega < \frac{2}{\lambda_{\max}(\mathbf{G}^T\mathbf{G})} \ . \tag{6.25}$$

This ensures that the objective function f is reduced at each iteration. With a bit more work it can be shown that $\mathbf{p}^{(k)}$ converges to $\mathbf{0}$, and that $\mathbf{m}^{(k)}$ converges to a least squares solution [218]. This version of the gradient descent algorithm for minimizing $\|\mathbf{Gm} - \mathbf{d}\|_2^2$ is known as the Landweber iteration. The Landweber iteration has some interesting regularization properties that are explored in Exercise 6.2.

Another approach to selecting α_k, called **exact line search**, is to find the value of α_k that achieves the greatest possible decrease in the objective function [218].

$$\max_\alpha g(\alpha) = f(\mathbf{m}^{(k)}) - f(\mathbf{m}^{(k+1)}) = 2\alpha\mathbf{p}^{(k)}{}^T\mathbf{p}^{(k)} - \alpha^2\mathbf{p}^{(k)}\mathbf{G}^T\mathbf{G}\mathbf{p}^{(k)} \tag{6.26}$$

Differentiating with respect to α and setting $g'(\alpha) = 0$ gives

$$\alpha_k = \frac{\mathbf{p}^{(k)}{}^T\mathbf{p}^{(k)}}{\mathbf{p}^{(k)}{}^T\mathbf{G}^T\mathbf{G}\mathbf{p}^{(k)}} \ . \tag{6.27}$$

In implementation it is important to avoid the expensive multiplication of $\mathbf{G}^T\mathbf{G}$. Thus the formula for α_k should be evaluated as

$$\alpha_k = \frac{\mathbf{p}^{(k)T}\mathbf{p}^{(k)}}{(\mathbf{G}\mathbf{p}^{(k)})^T(\mathbf{G}\mathbf{p}^{(k)})} \tag{6.28}$$

where the product $\mathbf{G}\mathbf{p}^{(k)}$ only needs to be computed once.

It can be shown that the gradient descent method with exact line search converges linearly to the optimal solution to the least squares problem. That is, the distance to the optimal solution is reduced by a constant factor less than one in each iteration. However, the rate of convergence depends on the 2-norm condition number of $\mathbf{G}^T\mathbf{G}$ [150]. In particular,

$$\lim_{k\to\infty} \frac{\|\mathbf{x}^{(k+1)} - \mathbf{x}^*\|_2}{\|\mathbf{x}^{(k)} - \mathbf{x}^*\|_2} \leq \frac{\kappa_2(\mathbf{G}^T\mathbf{G}) - 1}{\kappa_2(\mathbf{G}^T\mathbf{G}) + 1} \tag{6.29}$$

where $\kappa_2(\mathbf{G}^T\mathbf{G})$ is the 2-norm condition number of $\mathbf{G}^T\mathbf{G}$. When $\mathbf{G}^T\mathbf{G}$ is well conditioned, this ratio can be small and the convergence is rapid. However, if $\mathbf{G}^T\mathbf{G}$ has a large condition number, then this ratio can be very close to one and the convergence can be extremely slow.

The method of **stochastic gradient descent (SGD)** is an algorithm that uses gradient descent within a row action method [18,114]. We write the least squares objective function as

$$f(\mathbf{m}) = \|\mathbf{G}\mathbf{m} - \mathbf{d}\|_2^2 = \sum_{i=1}^m f_i(\mathbf{m}) = \sum_{i=1}^m (\mathbf{G}\mathbf{m} - \mathbf{d})_i^2 . \tag{6.30}$$

The gradients of the individual terms are

$$\nabla f_i(\mathbf{m}) = 2\mathbf{G}_{i,.}^T\mathbf{G}_{i,.}\mathbf{m} - 2d_i\mathbf{G}_{i,.}^T . \tag{6.31}$$

We then apply the method of gradient descent to each of the functions $f_i(\mathbf{m})$, $i = 1, 2, \ldots, m$. As with Kaczmarz's algorithm, it can be helpful to process the rows of $\mathbf{G}$ in random order. The resulting algorithm is below.

Algorithm 6.2 Stochastic Gradient Descent (SGD)

Given an initial solution $\mathbf{m}^{(0)}$ and a step size parameter α. For $k = 0, 1, \ldots$,

1. Pick a random row i of $\mathbf{G}$.
2. Let

$$\mathbf{m}^{(k+1)} = \mathbf{m}^{(k)} - \alpha(\mathbf{G}_{i,.}^T\mathbf{G}_{i,.}\mathbf{m} - d_i\mathbf{G}_{i,.}^T) . \tag{6.32}$$

3. Let $k = k + 1$.
4. Repeat until convergence.

Because we don't have access to the entirety of $\mathbf{G}$, it is not possible to perform an exact line search in this algorithm. Rather, a fixed parameter α is used. Because $\mathbf{m}^{(k)}$ is updated in this algorithm after each row is processed, rather than with one update combining the gradients of all rows, it is not exactly equivalent to the method of gradient descent. However, the algorithm is very easy to parallelize, and works well for extremely large scale problems.

6.4. THE CONJUGATE GRADIENT METHOD

We next consider the **conjugate gradient** (CG) method for solving a symmetric and positive definite system of equations $\mathbf{Ax} = \mathbf{b}$. We will later apply the CG method to solving the normal equations, $\mathbf{G}^T\mathbf{Gm} = \mathbf{G}^T\mathbf{d}$. In that case, $\mathbf{A} = \mathbf{G}^T\mathbf{G}$.

Consider the quadratic optimization problem

$$\min \phi(\mathbf{x}) = \frac{1}{2}\mathbf{x}^T\mathbf{Ax} - \mathbf{b}^T\mathbf{x} \tag{6.33}$$

where $\mathbf{A}$ is an n by n symmetric and positive definite matrix. We require $\mathbf{A}$ be positive definite so that the function $\phi(\mathbf{x})$ will be convex and have a unique minimum. We can calculate $\nabla\phi(\mathbf{x}) = \mathbf{Ax} - \mathbf{b}$ and set it equal to zero to find the minimum, which will be equivalent to solving the system of equations $\mathbf{Ax} = \mathbf{b}$.

The conjugate gradient method approaches the problem of minimizing $\phi(\mathbf{x})$ by constructing a basis for R^n in which the minimization problem is extremely simple. The basis vectors $\mathbf{p}^{(0)}, \mathbf{p}^{(1)}, \ldots, \mathbf{p}^{(n-1)}$ are selected so that

$$\mathbf{p}^{(i)^T}\mathbf{Ap}^{(j)} = 0 \text{ when } i \neq j . \tag{6.34}$$

A collection of vectors with this property is said to be **mutually conjugate** with respect to $\mathbf{A}$. We express $\mathbf{x}$ in terms of these basis vectors as

$$\mathbf{x} = \sum_{i=0}^{n-1}\alpha_i\mathbf{p}^{(i)} \tag{6.35}$$

so that

$$\phi(\boldsymbol{\alpha}) = \frac{1}{2}\left(\sum_{i=0}^{n-1}\alpha_i\mathbf{p}^{(i)}\right)^T\mathbf{A}\left(\sum_{j=0}^{n-1}\alpha_j\mathbf{p}^{(j)}\right) - \mathbf{b}^T\left(\sum_{i=0}^{n-1}\alpha_i\mathbf{p}^{(i)}\right) . \tag{6.36}$$

The product $\mathbf{x}^T\mathbf{Ax}$ can be written as a double sum, so the equation becomes

$$\phi(\boldsymbol{\alpha}) = \frac{1}{2}\sum_{i=0}^{n-1}\sum_{j=0}^{n-1}\alpha_i\alpha_j\mathbf{p}^{(i)^T}\mathbf{Ap}^{(j)} - \mathbf{b}^T\left(\sum_{i=0}^{n-1}\alpha_i\mathbf{p}^{(i)}\right) . \tag{6.37}$$

Since the vectors are mutually conjugate with respect to $\mathbf{A}$, this simplifies to

$$\phi(\boldsymbol{\alpha}) = \frac{1}{2} \sum_{i=0}^{n-1} \alpha_i^2 \mathbf{p}^{(i)^T} \mathbf{A} \mathbf{p}^{(i)} - \mathbf{b}^T \left(\sum_{i=0}^{n-1} \alpha_i \mathbf{p}^{(i)} \right) \tag{6.38}$$

or

$$\phi(\boldsymbol{\alpha}) = \frac{1}{2} \sum_{i=0}^{n-1} \left(\alpha_i^2 \mathbf{p}^{(i)^T} \mathbf{A} \mathbf{p}^{(i)} - 2\alpha_i \mathbf{b}^T \mathbf{p}^{(i)} \right) . \tag{6.39}$$

Eq. (6.39) shows that $\phi(\boldsymbol{\alpha})$ consists of n terms, each of which is independent of the other terms. Thus we can minimize $\phi(\boldsymbol{\alpha})$ by selecting each α_i to minimize the ith term,

$$\alpha_i^2 \mathbf{p}^{(i)^T} \mathbf{A} \mathbf{p}^{(i)} - 2\alpha_i \mathbf{b}^T \mathbf{p}^{(i)} . \tag{6.40}$$

Differentiating with respect to α_i and setting the derivative equal to zero, we find that the optimal value for α_i is

$$\alpha_i = \frac{\mathbf{b}^T \mathbf{p}^{(i)}}{\mathbf{p}^{(i)^T} \mathbf{A} \mathbf{p}^{(i)}} . \tag{6.41}$$

This shows that if we have a basis of vectors that are mutually conjugate with respect to $\mathbf{A}$, then minimizing $\phi(\mathbf{x})$ is very easy. However, we have not yet shown how to construct the mutually conjugate basis vectors.

Our algorithm will actually construct a sequence of solution vectors $\mathbf{x}^{(k)}$, residual vectors $\mathbf{r}^{(k)} = \mathbf{b} - \mathbf{A}\mathbf{x}^{(k)}$, and basis vectors $\mathbf{p}^{(k)}$. The algorithm begins with $\mathbf{x}^{(0)} = \mathbf{0}$, $\mathbf{r}^{(0)} = \mathbf{b}$, $\mathbf{p}^{(0)} = \mathbf{r}^{(0)}$, and $\alpha_0 = (\mathbf{r}^{(0)^T} \mathbf{r}^{(0)}) / (\mathbf{p}^{(0)^T} \mathbf{A} \mathbf{p}^{(0)})$.

Suppose that at the start of iteration k of the algorithm we have constructed $\mathbf{x}^{(0)}$, $\mathbf{x}^{(1)}, \ldots, \mathbf{x}^{(k)}, \mathbf{r}^{(0)}, \mathbf{r}^{(1)}, \ldots, \mathbf{r}^{(k)}, \mathbf{p}^{(0)}, \mathbf{p}^{(1)}, \ldots, \mathbf{p}^{(k)}$ and $\alpha_0, \alpha_1, \ldots, \alpha_k$. For the induction step, we assume that the first $k+1$ basis vectors $\mathbf{p}^{(i)}$ are mutually conjugate with respect to $\mathbf{A}$, the first $k+1$ residual vectors $\mathbf{r}^{(i)}$ are mutually orthogonal, and that $\mathbf{r}^{(i)^T} \mathbf{p}^{(j)} = 0$ when $i \neq j$.

We let

$$\mathbf{x}^{(k+1)} = \mathbf{x}^{(k)} + \alpha_k \mathbf{p}^{(k)} . \tag{6.42}$$

This effectively adds one more term of (6.35) into the solution. Next, we let

$$\mathbf{r}^{(k+1)} = \mathbf{r}^{(k)} - \alpha_k \mathbf{A} \mathbf{p}^{(k)} . \tag{6.43}$$

This correctly updates the residual, because

$$\mathbf{r}^{(k+1)} = \mathbf{b} - \mathbf{A}\mathbf{x}^{(k+1)} \tag{6.44}$$

$$= \mathbf{b} - \mathbf{A} \left(\mathbf{x}^{(k)} + \alpha_k \mathbf{p}^{(k)} \right) \tag{6.45}$$

$$= \left(\mathbf{b} - \mathbf{Ax}^{(k)} \right) - \alpha_k \mathbf{Ap}^{(k)} \tag{6.46}$$

$$= \mathbf{r}^{(k)} - \alpha_k \mathbf{Ap}^{(k)} . \tag{6.47}$$

We let

$$\beta_{k+1} = \frac{\|\mathbf{r}^{(k+1)}\|_2^2}{\|\mathbf{r}^{(k)}\|_2^2} = \frac{\mathbf{r}^{(k+1)^T} \mathbf{r}^{(k+1)}}{\mathbf{r}^{(k)^T} \mathbf{r}^{(k)}} \tag{6.48}$$

and

$$\mathbf{p}^{(k+1)} = \mathbf{r}^{(k+1)} + \beta_{k+1} \mathbf{p}^{(k)} . \tag{6.49}$$

This formula for $\mathbf{p}^{(k+1)}$ seems mysterious at first, but as we shall soon see, it ensures that $\mathbf{p}$ vectors will be conjugate.

In the following calculations, it will be useful to know that $\mathbf{b}^T \mathbf{p}^{(k)} = \mathbf{r}^{(k)^T} \mathbf{r}^{(k)}$. This is shown by

$$\mathbf{b}^T \mathbf{p}^{(k)} = (\mathbf{r}^{(k)} + \mathbf{Ax}^{(k)})^T \mathbf{p}^{(k)} \tag{6.50}$$

$$= \mathbf{r}^{(k)^T} \mathbf{p}^{(k)} + \mathbf{x}^{(k)^T} \mathbf{Ap}^{(k)} . \tag{6.51}$$

Since $\mathbf{A}$ is symmetric, $\mathbf{x}^{(k)^T} \mathbf{Ap}^{(k)} = \mathbf{p}^{(k)^T} \mathbf{Ax}^{(k)}$, so

$$\mathbf{b}^T \mathbf{p}^{(k)} = \mathbf{r}^{(k)^T} \mathbf{p}^{(k)} + \mathbf{p}^{(k)^T} \mathbf{Ax}^{(k)} \tag{6.52}$$

$$= \mathbf{r}^{(k)^T} (\mathbf{r}^{(k)} + \beta_k \mathbf{p}^{(k-1)}) + \mathbf{p}^{(k)^T} \mathbf{Ax}^{(k)} \tag{6.53}$$

$$= \mathbf{r}^{(k)^T} \mathbf{r}^{(k)} + \beta_k \mathbf{r}^{(k)^T} \mathbf{p}^{(k-1)} + \mathbf{p}^{(k)^T} \mathbf{A} \sum_{l=0}^{k-1} \alpha_l \mathbf{p}^{(l)} \tag{6.54}$$

$$= \mathbf{r}^{(k)^T} \mathbf{r}^{(k)} + 0 + 0 \tag{6.55}$$

$$= \mathbf{r}^{(k)^T} \mathbf{r}^{(k)} . \tag{6.56}$$

We will now show that $\mathbf{r}^{(k+1)}$ is orthogonal to $\mathbf{r}^{(i)}$ for $i \le k$. For every $i < k$,

$$\mathbf{r}^{(k+1)^T} \mathbf{r}^{(i)} = (\mathbf{r}^{(k)} - \alpha_k \mathbf{Ap}^{(k)})^T \mathbf{r}^{(i)} \tag{6.57}$$

$$= \mathbf{r}^{(k)^T} \mathbf{r}^{(i)} - \alpha_k \mathbf{p}^{(k)^T} \mathbf{Ar}^{(i)} \tag{6.58}$$

$$= \mathbf{r}^{(k)^T} \mathbf{r}^{(i)} - \alpha_k \mathbf{r}^{(i)^T} \mathbf{Ap}^{(k)} . \tag{6.59}$$

Since $\mathbf{r}^{(k)}$ is orthogonal to all earlier $\mathbf{r}^{(i)}$ vectors,

$$\mathbf{r}^{(k+1)^T} \mathbf{r}^{(i)} = 0 - \alpha_k \mathbf{p}^{(k)^T} \mathbf{Ar}^{(k)} . \tag{6.60}$$

Also, since $\mathbf{p}^{(i)} = \mathbf{r}^{(i)} + \beta_i \mathbf{p}^{(i-1)}$,

$$\mathbf{r}^{(k+1)^T} \mathbf{r}^{(i)} = 0 - \alpha_k \left(\mathbf{p}^{(i)} - \beta_i \mathbf{p}^{(i-1)} \right)^T \mathbf{A} \mathbf{p}^{(k)} . \tag{6.61}$$

Both $\mathbf{p}^{(i)}$ and $\mathbf{p}^{(i-1)}$ are conjugate with $\mathbf{p}^{(k)}$. Thus for $i < k$,

$$\mathbf{r}^{(k+1)^T} \mathbf{r}^{(i)} = 0 . \tag{6.62}$$

We also have to show that $\mathbf{r}^{(k+1)^T} \mathbf{r}^{(k)} = 0$.

$$\mathbf{r}^{(k+1)^T} \mathbf{r}^{(k)} = (\mathbf{r}^{(k)} - \alpha_k \mathbf{A} \mathbf{p}^{(k)})^T \mathbf{r}^{(k)} \tag{6.63}$$

$$= \mathbf{r}^{(k)^T} \mathbf{r}^{(k)} - \alpha_k \mathbf{p}^{(k)^T} \mathbf{A} \mathbf{r}^{(k)} \tag{6.64}$$

$$= \mathbf{r}^{(k)^T} \mathbf{r}^{(k)} - \alpha_k \mathbf{p}^{(k)^T} \mathbf{A} (\mathbf{p}^{(k)} - \beta_k \mathbf{p}^{(k-1)}) \tag{6.65}$$

$$= \mathbf{r}^{(k)^T} \mathbf{r}^{(k)} - \alpha_k (\mathbf{p}^{(k)} - \beta_k \mathbf{p}^{(k-1)})^T \mathbf{A} \mathbf{p}^{(k)} \tag{6.66}$$

$$= \mathbf{r}^{(k)^T} \mathbf{r}^{(k)} - \alpha_k \mathbf{p}^{(k)^T} \mathbf{A} \mathbf{p}^{(k)} + \alpha_k \beta_k \mathbf{p}^{(k-1)^T} \mathbf{A} \mathbf{p}^{(k)} \tag{6.67}$$

$$= \mathbf{r}^{(k)^T} \mathbf{r}^{(k)} - \mathbf{r}^{(k)^T} \mathbf{r}^{(k)} + \alpha_k \beta_k 0 \tag{6.68}$$

$$= 0 . \tag{6.69}$$

Next, we will show that $\mathbf{r}^{(k+1)}$ is orthogonal to $\mathbf{p}^{(i)}$ for $i \le k$.

$$\mathbf{r}^{(k+1)^T} \mathbf{p}^{(i)} = \mathbf{r}^{(k+1)^T} \left(\mathbf{r}^{(i)} + \beta_i \mathbf{p}^{(i-1)} \right) \tag{6.70}$$

$$= \mathbf{r}^{(k+1)^T} \mathbf{r}^{(i)} + \beta_i \mathbf{r}^{(k+1)^T} \mathbf{p}^{(i-1)} \tag{6.71}$$

$$= 0 + \beta_i \mathbf{r}^{(k+1)^T} \mathbf{p}^{(i-1)} \tag{6.72}$$

$$= \beta_i \left(\mathbf{r}^{(k)} - \alpha_k \mathbf{A} \mathbf{p}^{(k)} \right)^T \mathbf{p}^{(i-1)} \tag{6.73}$$

$$= \beta_i (\mathbf{r}^{(k)^T} \mathbf{p}^{(i-1)} - \alpha_k \mathbf{p}^{(i-1)^T} \mathbf{A} \mathbf{p}^{(k)}) \tag{6.74}$$

$$= \beta_i (0 - 0) = 0 . \tag{6.75}$$

Finally, we need to show that $\mathbf{p}^{(k+1)^T} \mathbf{A} \mathbf{p}^{(i)} = 0$ for $i \le k$. For $i < k$,

$$\mathbf{p}^{(k+1)^T} \mathbf{A} \mathbf{p}^{(i)} = (\mathbf{r}^{(k+1)} + \beta_{k+1} \mathbf{p}^{(k)})^T \mathbf{A} \mathbf{p}^{(i)} \tag{6.76}$$

$$= \mathbf{r}^{(k+1)^T} \mathbf{A} \mathbf{p}^{(i)} + \beta_{k+1} \mathbf{p}^{(k)^T} \mathbf{A} \mathbf{p}^{(i)} \tag{6.77}$$

$$= \mathbf{r}^{(k+1)^T} \mathbf{A} \mathbf{p}^{(i)} + 0 \tag{6.78}$$

$$= \mathbf{r}^{(k+1)^T} \left(\frac{1}{\alpha_i} (\mathbf{r}^{(i)} - \mathbf{r}^{(i+1)}) \right) \tag{6.79}$$

$$= \frac{1}{\alpha_i} \left(\mathbf{r}^{(k+1)^T} \mathbf{r}^{(i)} - \mathbf{r}^{(k+1)^T} \mathbf{r}^{(i+1)} \right) = 0 . \tag{6.80}$$

For $i = k$,

$$\mathbf{p}^{(k+1)^T} \mathbf{A} \mathbf{p}^{(k)} = \left(\mathbf{r}^{(k+1)} + \beta_{k+1} \mathbf{p}^{(k)} \right)^T \left(\frac{1}{\alpha_k} (\mathbf{r}^{(k)} - \mathbf{r}^{(k+1)}) \right) \tag{6.81}$$

$$= \frac{1}{\alpha_k} \left(\beta_{k+1} \left(\mathbf{r}^{(k)} + \beta_k \mathbf{p}^{(k-1)} \right)^T \mathbf{r}^{(k)} - \mathbf{r}^{(k+1)^T} \mathbf{r}^{(k+1)} \right) \tag{6.82}$$

$$= \frac{1}{\alpha_k} \left(\beta_{k+1} \mathbf{r}^{(k)^T} \mathbf{r}^{(k)} + \beta_{k+1} \beta_k \mathbf{p}^{(k-1)^T} \mathbf{r}^{(k)} - \mathbf{r}^{(k+1)^T} \mathbf{r}^{(k+1)} \right) \tag{6.83}$$

$$= \frac{1}{\alpha_k} \left(\mathbf{r}^{(k+1)^T} \mathbf{r}^{(k+1)} + \beta_{k+1} \beta_k \cdot 0 - \mathbf{r}^{(k+1)^T} \mathbf{r}^{(k+1)} \right) \tag{6.84}$$

$$= 0 . \tag{6.85}$$

We have now shown that the algorithm generates a sequence of mutually conjugate basis vectors. In theory, the algorithm will find an exact solution to the system of equations in n iterations. In practice, due to roundoff errors in the computation, the exact solution may not be obtained in n iterations. In practical implementations of the algorithm, we iterate until the residual is smaller than some tolerance that we specify. The algorithm can be summarized as follows.

Algorithm 6.3 Conjugate Gradient Method

Given a positive definite and symmetric system of equations $\mathbf{A}\mathbf{x} = \mathbf{b}$, and an initial solution $\mathbf{x}^{(0)}$, let $\beta_0 = 0$, $\mathbf{p}^{(-1)} = \mathbf{0}$, $\mathbf{r}^{(0)} = \mathbf{b} - \mathbf{A}\mathbf{x}^{(0)}$, and $k = 0$.

1. If $k > 0$, let $\beta_k = \frac{\mathbf{r}^{(k)^T} \mathbf{r}^{(k)}}{\mathbf{r}^{(k-1)^T} \mathbf{r}^{(k-1)}}$.
2. Let $\mathbf{p}^{(k)} = \mathbf{r}^{(k)} + \beta_k \mathbf{p}^{(k-1)}$.
3. Let $\alpha_k = \frac{\|\mathbf{r}^{(k)}\|_2^2}{\mathbf{p}^{(k)^T} \mathbf{A} \mathbf{p}^{(k)}}$.
4. Let $\mathbf{x}^{(k+1)} = \mathbf{x}^{(k)} + \alpha_k \mathbf{p}^{(k)}$.
5. Let $\mathbf{r}^{(k+1)} = \mathbf{r}^{(k)} - \alpha_k \mathbf{A} \mathbf{p}^{(k)}$.
6. Let $\beta_k = \frac{\|\mathbf{r}^{(k+1)}\|_2^2}{\|\mathbf{r}^{(k)}\|_2^2}$.
7. Let $k = k + 1$.
8. Repeat the previous steps until convergence.

A major advantage of the CG method is that it requires storage only for the vectors $\mathbf{x}^{(k)}$, $\mathbf{p}^{(k)}$, $\mathbf{r}^{(k)}$ and the symmetric matrix $\mathbf{A}$. If $\mathbf{A}$ is large and sparse, then sparse matrix techniques can be used to store $\mathbf{A}$ more efficiently. Unlike factorization methods such as QR, SVD, or Cholesky factorization, there will be no fill-in of the zero elements in $\mathbf{A}$ at any stage in the solution process. Thus it is possible to solve extremely large systems using CG in cases where direct factorization would require far too much storage. In fact, the only way in which the algorithm uses $\mathbf{A}$ is in one multiplication of $\mathbf{A}\mathbf{p}^{(k)}$ for each iteration.

In some applications of the CG method, it is possible to perform these matrix vector multiplications without explicitly constructing $\mathbf{A}$. For example, if $\mathbf{A} = \mathbf{G}^T\mathbf{G}$, then we can write the matrix–vector product $\mathbf{Ap}^{(k)}$ as

$$\mathbf{Ap}^{(k)} = \mathbf{G}^T(\mathbf{Gp}^{(k)}) \tag{6.86}$$

using two matrix–vector products and avoiding the matrix–matrix product $\mathbf{G}^T\mathbf{G}$.

As with the method of gradient descent, the rate at which the conjugate gradient method converges to an optimal solution depends on the condition number of the matrix $\mathbf{A}$. [218]. For the conjugate gradient method,

$$\lim_{k\to\infty} \frac{\|\mathbf{x}^{(k+1)} - \mathbf{x}^*\|_2}{\|\mathbf{x}^{(k)} - \mathbf{x}^*\|_2} \le \frac{\sqrt{\kappa_2(\mathbf{A})} - 1}{\sqrt{\kappa_2(\mathbf{A})} + 1} . \tag{6.87}$$

Compare this convergence rate with the convergence rate for the gradient descent method given by (6.29). Since this limit depends on the square root of the condition number of $\mathbf{A}$, the method is somewhat less sensitive to ill-conditioning than the method of gradient descent.

6.5. THE CGLS METHOD

The CG method by itself can only be applied to positive definite systems of equations, and is thus not directly applicable to general least squares problems. In the **conjugate gradient least squares** (CGLS) method, we solve a least squares problem

$$\min \|\mathbf{Gm} - \mathbf{d}\|_2 \tag{6.88}$$

by applying CG to the normal equations

$$\mathbf{G}^T\mathbf{Gm} = \mathbf{G}^T\mathbf{d} . \tag{6.89}$$

In implementing this algorithm it is important to avoid roundoff errors. One important source of error is the evaluation of the residual, $\mathbf{G}^T\mathbf{d} - \mathbf{G}^T\mathbf{Gm}$. It turns out that this calculation is more accurate when we factor out $\mathbf{G}^T$ and compute $\mathbf{G}^T(\mathbf{d} - \mathbf{Gm})$. We will use the notation $\mathbf{s}^{(k)} = \mathbf{d} - \mathbf{Gm}^{(k)}$, and $\mathbf{r}^{(k)} = \mathbf{G}^T\mathbf{s}^{(k)}$. Note that we can compute $\mathbf{s}^{(k+1)}$ recursively from $\mathbf{s}^{(k)}$ as follows

$$\mathbf{s}^{(k+1)} = \mathbf{d} - \mathbf{Gm}^{(k+1)} \tag{6.90}$$
$$= \mathbf{d} - \mathbf{G}(\mathbf{m}^{(k)} + \alpha_k\mathbf{p}^{(k)}) \tag{6.91}$$
$$= (\mathbf{d} - \mathbf{Gm}^{(k)}) - \alpha_k\mathbf{Gp}^{(k)} \tag{6.92}$$
$$= \mathbf{s}^{(k)} - \alpha_k\mathbf{Gp}^{(k)} . \tag{6.93}$$

With this trick, we can now state the CGLS algorithm.

Algorithm 6.4 CGLS

Given a least squares problem $\min \|\mathbf{Gm} - \mathbf{d}\|_2$, let $k = 0$, $\mathbf{m}^{(0)} = \mathbf{0}$, $\mathbf{p}^{(-1)} = \mathbf{0}$, $\beta_0 = 0$, $\mathbf{s}^{(0)} = \mathbf{d} - \mathbf{Gm}^{(0)}$, and $\mathbf{r}^{(0)} = \mathbf{G}^T \mathbf{s}^{(0)}$.

1. If $k > 0$, let $\beta_k = \frac{\mathbf{r}^{(k)T} \mathbf{r}^{(k)}}{\mathbf{r}^{(k-1)T} \mathbf{r}^{(k-1)}}$.

2. Let $\mathbf{p}^{(k)} = \mathbf{r}^{(k)} + \beta_k \mathbf{p}^{(k-1)}$.

3. Let $\alpha_k = \frac{\|\mathbf{r}^{(k)}\|_2^2}{(\mathbf{p}^{(k)T} \mathbf{G}^T)(\mathbf{Gp}^{(k)})}$.

4. Let $\mathbf{m}^{(k+1)} = \mathbf{m}^{(k)} + \alpha_k \mathbf{p}^{(k)}$.

5. Let $\mathbf{s}^{(k+1)} = \mathbf{s}^{(k)} - \alpha_k \mathbf{Gp}^{(k)}$.

6. Let $\mathbf{r}^{(k+1)} = \mathbf{G}^T \mathbf{s}^{(k+1)}$.

7. Let $k = k + 1$.

8. Repeat the previous steps until convergence.

Notice that this algorithm only requires one multiplication of $\mathbf{Gp}^{(k)}$ and one multiplication of $\mathbf{G}^T \mathbf{s}^{(k+1)}$ per iteration. We never explicitly compute $\mathbf{G}^T \mathbf{G}$, which might require considerable time, and which might have far more nonzero elements than $\mathbf{G}$ itself. In practice, this algorithm typically gives good solutions after a very small number of iterations. We include a library function **cgls** that implements the algorithm.

The CGLS algorithm has an important property that is useful for solving ill-posed problems. It can be shown that, for exact arithmetic, $\|\mathbf{m}^{(k)}\|_2$ increases monotonically and $\|\mathbf{d} - \mathbf{Gm}^{(k)}\|_2$ decreases monotonically with increasing iterations [79,84]. We can use the discrepancy principle together with this property to obtain a regularized solution by simply terminating the CGLS algorithm once $\|\mathbf{d} - \mathbf{Gm}^{(k)}\|_2 < \delta$.

An alternative way to use CGLS is to explicitly solve the Tikhonov regularized problem (4.4) by applying the algorithm to

$$\min \left\| \begin{bmatrix} \mathbf{G} \\ \alpha \mathbf{L} \end{bmatrix} \mathbf{m} - \begin{bmatrix} \mathbf{d} \\ \mathbf{0} \end{bmatrix} \right\|_2^2 . \tag{6.94}$$

For nonzero values of the regularization parameter α, this least squares problem should be reasonably well-conditioned. By solving this problem for several values of α, we can compute an L-curve. The disadvantage of this approach is that the number of CGLS iterations for each value of α may be large, and we need to solve the problem for several values of α. Thus the computational effort is far greater.

Figure 6.6 Blurred image.

Example 6.2

A commonly used mathematical model of image blurring involves the two-dimensional convolution of the true image $I_{\text{true}}(x, y)$ with a **point spread function**, $\Psi(u, v)$ [17].

$$I_{\text{blurred}}(x, \ y) = \int_{-\infty}^{\infty} \int_{-\infty}^{\infty} I_{\text{true}}(x - u, \ y - v)\Psi(u, \ v) \ du \ dv. \qquad (6.95)$$

Here the point spread function shows how a point in the true image is altered in the blurred image. A point spread function that is commonly used to represent the blurring that occurs because an image is out of focus is the **Gaussian point spread function**

$$\Psi(u, \ v) = e^{\frac{u^2+v^2}{2\sigma^2}}. \qquad (6.96)$$

Here the parameter σ controls the relative width of the point spread function. In practice, the blurred image and point spread function are discretized into pixels. In theory, Ψ is nonzero for all values of u and v. However, it becomes small quickly as u and v increase. If we set small values of Ψ to 0, then the $\mathbf{G}$ matrix in the discretized problem will be sparse.

Fig. 6.6 shows an image that has been blurred and also has a small amount of added noise. This image is of size 200 pixels by 200 pixels, so the $\mathbf{G}$ matrix for the blurring operator is of size 40,000 by 40,000. Fortunately, the blurring matrix $\mathbf{G}$ is quite sparse, with less than 0.1% nonzero elements. The sparse matrix requires about 12 megabytes of

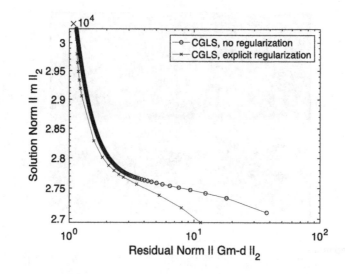

Figure 6.7 L-curves for CGLS deblurring, with and without explicit regularization.

storage. A dense matrix of this size would require about 13 gigabytes of storage. Using the SVD approach to Tikhonov regularization would require far more storage than many current computers have. However, CGLS works quite well on this problem.

Fig. 6.7 shows the L-curve for the solution of this problem by CGLS without explicit regularization with points for successive iterations plotting from right to left. Also shown is the L-curve for CGLS with explicit regularization and varying values of α. For the first 30 or so iterations of CGLS without explicit regularization, $\|\mathbf{Gm} - \mathbf{d}\|_2$ decreases quickly. After that point, the improvement in misfit slows down, while $\|\mathbf{m}\|_2$ increases rapidly. Fig. 6.8 shows the CGLS solution without explicit regularization after 30 iterations. The blurring has been greatly improved. Note that 30 iterations is far less than the size of the matrix ($n = 40,000$). Unfortunately, further CGLS iterations do not significantly improve the image. In fact, noise builds up rapidly, both because of the accumulation of roundoff errors and because the algorithm is converging slowly towards an unregularized least squares solution. Fig. 6.9 shows the CGLS solution after 100 iterations. In this image the noise has been greatly amplified, with little or no improvement in the clarity of the image.

The CGLS solutions with explicit Tikhonov regularization in Fig. 6.7 were obtained for 22 values of α. For each value of α, 200 CGLS iterations were performed. The resulting L-curve is slightly better than the L-curve from the CGLS solution without explicit regularization in that corresponding values of $\|\mathbf{m}\|_2$ or $\|\mathbf{Gm} - \mathbf{d}\|_2$ are smaller

Figure 6.8 CGLS solution after 30 iterations, without explicit regularization.

Figure 6.9 CGLS solution after 100 iterations, without explicit regularization.

along the curve. However, its generation required approximately 40 times as much computational effort. The corner solution for $\alpha = 7.0 \times 10^{-4}$ is shown in Fig. 6.10, and is similar to the one shown in Fig. 6.8.

We have focused on the CGLS method because it is relatively easy to derive and because of its useful implicit regularization properties. However, many other iterative methods have been developed for large-scale least squares problems [5,64,174], and an analysis of the implicit regularizing effects of these methods can be found in the book

Figure 6.10 CGLS solution, with explicit regularization, $\alpha = 7.0 \times 10^{-4}$.

by Hanke [79]. In particular, the LSQR method of Paige and Saunders has also been very widely used in geophysical applications [154,155]. MATLAB has an efficient and robust implementation of LSQR in its **lsqr** routine.

6.6. RESOLUTION ANALYSIS FOR ITERATIVE METHODS

Analyzing the resolution of a regularized solution computed by the iterative methods described in this chapter is a challenging problem. The formulas developed in Chapter 4 for the model resolution matrix under zeroth-order Tikhonov regularization were based on the SVD of the $\mathbf{G}$ matrix, but as we have seen it is often infeasible to compute the SVD of a large matrix. For higher-order regularization, we have formulas for the resolution matrix based on the GSVD of the $\mathbf{G}$ matrix, such as (4.59), but it is also impractical to compute the GSVD of very large matrices.

A brute force approach that offers some advantages over the computation of the SVD or GSVD is to use (4.52) with Cholesky factorization of $(\mathbf{G}^T\mathbf{G} + \alpha^2 \mathbf{L}^T\mathbf{L})$ [24]. Cholesky factorization is considerably faster in practice than computing the SVD. However, since the Cholesky factor of this matrix is typically dense, this is still a very large-scale computation that would generally need to be performed in parallel on a computer with sufficient memory to store a dense n by n matrix.

If we want to compute an individual column of the model resolution matrix, this can be accomplished by solving an additional least squares problem. Applying the expression for the least squares Tikhonov inverse operator $\mathbf{G}^\sharp$ (4.52), the ith column of $\mathbf{R}_\mathrm{m}$ can be expressed as

$$\begin{aligned}
\mathbf{R}_{.,i} &= \mathbf{R}_m \mathbf{e}_i \\
&= \mathbf{G}^\sharp \mathbf{G} \mathbf{e}_i \\
&= \mathbf{G}^\sharp \mathbf{G}_{.,i} \\
&= (\mathbf{G}^T \mathbf{G} + \alpha^2 \mathbf{L}^T \mathbf{L})^{-1} \mathbf{G}^T \mathbf{G}_{.,i} \,.
\end{aligned} \tag{6.97}$$

This can be put in the form of (4.26), a regularized least squares problem

$$\min \left\| \begin{bmatrix} \mathbf{G} \\ \alpha \mathbf{L} \end{bmatrix} \mathbf{R}_{.,i} - \begin{bmatrix} \mathbf{G}_{.,i} \\ \mathbf{0} \end{bmatrix} \right\|_2^2 \tag{6.98}$$

that can be solved for $\mathbf{R}_{.,i}$ using CGLS, LSQR, or some other iterative method. We could conceivably solve (6.98) to retrieve every column of the model resolution matrix in this manner, but this would require the solution of n least squares problems.

A number of authors have considered iterative techniques for computing low-rank approximations to the resolution matrix in combination with iterative least squares solution methods [15,16,224]. In such approaches, iterative methods are applied to compute the largest k singular values of a matrix and the associated singular vectors. For a sparse matrix, these methods are vastly more efficient than standard algorithms for computing the SVD.

Suppose that we obtain a low-rank (k singular values and vectors) approximation to the SVD

$$\mathbf{G} \approx \mathbf{U}_k \mathbf{L}_k \mathbf{V}_k^T \tag{6.99}$$

where $\mathbf{L}_k$ is a k by k nonsingular matrix, and $\mathbf{U}_k$ and $\mathbf{V}_k$ are matrices with k orthogonal columns. The pseudoinverse of $\mathbf{G}$ is then approximately

$$\mathbf{G}^\dagger \approx \mathbf{V}_k \mathbf{L}_k^{-1} \mathbf{U}_k^T \,. \tag{6.100}$$

Combining these approximations, we have

$$\mathbf{R}_m \approx \mathbf{V}_k \mathbf{V}_k^T \,. \tag{6.101}$$

Note that we would not want to explicitly multiply out $\mathbf{V}_k \mathbf{V}_k^T$ creating a huge and dense n by n resolution matrix. Rather, we could store the relatively small matrix $\mathbf{V}_k$ and use it to compute desired entries of $\mathbf{R}_m$, as needed.

A more sophisticated version of this approach is to compute a low-rank approximation to the singular value decomposition of the augmented $\mathbf{G}$ matrix [223]. Consider an augmented system of equations

$$\mathbf{A} = \begin{bmatrix} \mathbf{G} \\ \alpha \mathbf{L} \end{bmatrix} \tag{6.102}$$

using truncated SVD approximations for $\mathbf{A}$ and its pseudoinverse, $\mathbf{A}^\dagger$,

$$\mathbf{A} \approx \mathbf{U}_k \mathbf{S}_k \mathbf{V}_k^T \qquad (6.103)$$

and

$$\mathbf{A}^\dagger \approx \mathbf{V}_k \mathbf{S}_k^{-1} \mathbf{U}_k^T \qquad (6.104)$$

where $\mathbf{U}_k$ and $\mathbf{V}_k$ are composed of the first k columns of $\mathbf{U}$ and $\mathbf{V}$, and $\mathbf{S}_k$ is a square k by k matrix diagonal matrix of the largest k singular values.

It is convenient to write $\mathbf{U}_k$ as

$$\mathbf{U}_k = \begin{bmatrix} \mathbf{U}_G \\ \mathbf{U}_L \end{bmatrix} \qquad (6.105)$$

where $\mathbf{U}_G$ and $\mathbf{U}_L$ contains the rows of $\mathbf{U}_k$ corresponding to those of $\mathbf{G}$ and $\alpha \mathbf{L}$, respectively, in $\mathbf{A}$. We can thus partition (6.103) using

$$\mathbf{G} \approx \mathbf{U}_G \mathbf{S}_k \mathbf{V}_k^T . \qquad (6.106)$$

Given (6.106) and (6.104), we can calculate an approximation of the model resolution matrix

$$\mathbf{R}_m = \mathbf{A}^\dagger \begin{bmatrix} \mathbf{G} \\ \mathbf{0} \end{bmatrix} \approx \mathbf{V}_k \mathbf{S}_k^{-1} \begin{bmatrix} \mathbf{U}_G^T \mathbf{U}_L^T \end{bmatrix} \begin{bmatrix} \mathbf{U}_G \mathbf{S}_k \mathbf{V}_k^T \\ \mathbf{0} \end{bmatrix} \qquad (6.107)$$

which simplifies to

$$\mathbf{R}_m \approx \mathbf{V}_k \mathbf{S}_k^{-1} \mathbf{U}_G^T \mathbf{U}_G \mathbf{S}_k \mathbf{V}_k^T . \qquad (6.108)$$

Using the identity

$$\mathbf{U}_G^T \mathbf{U}_G = \mathbf{I} - \mathbf{U}_L^T \mathbf{U}_L \qquad (6.109)$$

we can alternatively write (6.108) as

$$\mathbf{R}_m \approx \mathbf{V}_k \left(\mathbf{I} - \mathbf{S}_k^{-1} \mathbf{U}_L^T \mathbf{U}_L \mathbf{S}_k \right) \mathbf{V}_k^T . \qquad (6.110)$$

Depending on the relative sizes of $\mathbf{U}_G$ and $\mathbf{U}_L$, one or the other of (6.108) or (6.110) might be more easily evaluated. Again, we would probably not want to, or be able to, explicitly multiply out the matrices to obtain the entire dense n by n model resolution matrix for a large problem, but could instead use (6.108) or (6.110) to compute specific desired elements of the resolution matrix.

In practice, it may not be feasible for very large problems to compute, or even store, a sufficient number of singular values and vectors to adequately approximate the model resolution matrix using low-rank SVD-based approximations [49]. We will thus consider an alternative stochastic approach based on [12] that estimates solely the resolution matrix diagonal, but which is not limited by low-rank matrix approximations.

Consider a sequence of s random vectors from $\mathbf{R}^n$, $\mathbf{v}_1, \ldots, \mathbf{v}_s$, with independent elements drawn from a standard normal distribution. The sth corresponding estimate for the diagonal of an n by n square matrix $\mathbf{A}$ is given by

$$\mathbf{q}_s = \left[\sum_{k=1}^{s} \mathbf{v}_k \odot \mathbf{A}\mathbf{v}_k \right] \oslash \left[\sum_{k=1}^{s} \mathbf{v}_k \odot \mathbf{v}_k \right] \tag{6.111}$$

where $\odot$ denotes element-wise multiplication and $\oslash$ denotes element-wise division of vectors. Note that the only way in which $\mathbf{A}$ is used within this algorithm is in matrix–vector multiplications, so as long as we have a function that can implement this operation, then we can implement the algorithm. As we have already seen, the multiplication of $\mathbf{R}_m$ with a vector can be effected by solving a least squares problem (6.98). Specifically, to solve for the kth product $\mathbf{y} = \mathbf{R}_m\mathbf{v}_k$, we solve [131]

$$\min \left\| \begin{bmatrix} \mathbf{G} \\ \alpha\mathbf{L} \end{bmatrix} \mathbf{y} - \begin{bmatrix} \mathbf{G}\mathbf{v}_k \\ \mathbf{0} \end{bmatrix} \right\|_2^2. \tag{6.112}$$

It is necessary to solve s least squares systems of equations (6.112) to calculate (6.111). In practice, relatively small values of s in (6.111) (e.g. 100 to 1000) are adequate even when there are tens or hundreds of thousands of model parameters [131].

The stochastic algorithm (6.111) can also be used to implement generalized cross-validation. In (4.74), the most difficult computation is evaluation of the trace of $\mathbf{I} - \mathbf{G}\mathbf{G}^\sharp$. Computing this matrix explicitly and then evaluating the trace would require excessive or infeasible time and storage. Using (6.111) and evaluating the necessary $\mathbf{G}^\sharp$ matrix–vector multiplications by solving an associated least squares problem using iterative method such as LSQR or CGLS makes it tractable to compute an approximation to the GCV function (4.74), where we solve s least squares problems for each value of α [131].

Example 6.3

We will examine the solution to a large-scale tomography problem involving 30,000 rays passing through a cubic volume discretized into a 30 by 30 by 30 model of 27,000 cells (Fig. 6.11).

Ray path start and end points are uniformly randomly distributed across the sides of the volume with 10,000 rays passing between each pair of parallel sides. The corresponding $\mathbf{G}$ matrix is 30,000 by 27,000, but only 0.2% of its entries are nonzero.

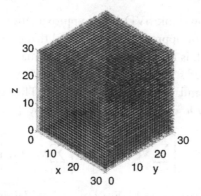

Figure 6.11 A 30 by 30 by 30 block tomography model interrogated by 30,000 ray paths.

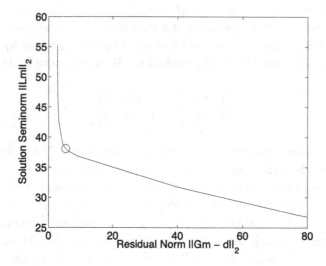

Figure 6.12 L-curve for the tomography example of Fig. 6.11 and its corner near $\alpha = 0.5$.

Synthetic travel time data (with travel time differences from the background slowness model of between approximately -20 and 30 time units) were generated from a zero-mean smooth model with $N(0, (0.5)^2)$ noise added. We implemented second-order Tikhonov regularization with a small amount of additional zeroth-order regularization to regularize the solution.

Fig. 6.12 shows the L-curve for this problem. Fig. 6.13 shows a plot of the GCV function (4.74) computed using stochastic estimates of the diagonal of $\mathbf{I} - \mathbf{GG}^{\sharp}$. The L-curve corner and GCV minimum both occur at about $\alpha = 0.5$. For the solution obtained using $\alpha = 0.5$, we computed a stochastic estimate of the diagonal of the reso-

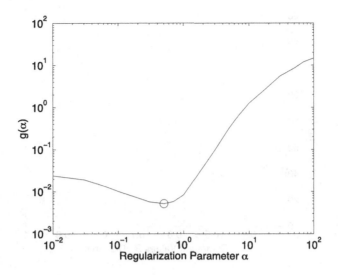

Figure 6.13 GCV plot for the tomography example of Fig. 6.11 and its minimum near $\alpha = 0.5$.

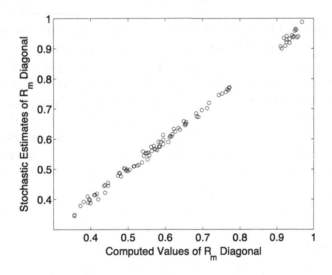

Figure 6.14 Stochastic estimates versus computed values of the diagonal of $\mathbf{R}_m$.

lution matrix using (6.111). For comparison, we also computed exact resolution matrix diagonal element values for 100 randomly selected indices using (6.97). Fig. 6.14 shows a scatter plot of the stochastic estimates compared with corresponding exact determinations, and indicates that the stochastic estimates are accurate to a few percent. It is common to use counts of the number of rays transecting each cell as a proxy for as-

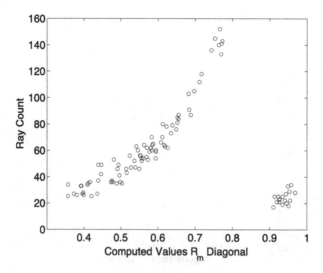

Figure 6.15 Comparison of ray densities and resolution.

sessing the diagonal of the resolution matrix in tomographic problems. In general, the association between ray density and resolution in a tomography problem will depend on the specific ray path geometry. Fig. 6.15 shows a scatter plot of the ray densities versus the accurately computed values of the diagonal elements of $\mathbf{R}_m$ for the 100 random indices plotted in Fig. 6.14. The relationship shows an increasing trend between ray density and the resolution diagonal element values, but there are also anomalously well-resolved outliers in the plot with fewer than 40 ray paths and resolution diagonal element values greater than 0.9. Further examination shows that the anomalously well-resolved outliers correspond to cells that reside on the model exterior and are thus constrained by two-dimensional ray path geometries.

6.7. EXERCISES

1. Apply Kaczmarz's algorithm to the cross-well tomography problem of Exercise 4.3.
2. The gradient descent method with a constant step size is known as the **Landweber iteration** [86]. The algorithm begins with $\mathbf{m}^{(0)} = \mathbf{0}$, and then follows the iteration

$$\mathbf{m}^{(k+1)} = \mathbf{m}^{(k)} - \omega \mathbf{G}^T(\mathbf{G}\mathbf{m}^{(k)} - \mathbf{d}) \ . \tag{6.113}$$

To ensure convergence, the parameter ω must be selected so that $0 < \omega < 2/\lambda_{\max}(\mathbf{G}^T\mathbf{G}) = 2/s_1^2$, where s_1 is the largest singular value of $\mathbf{G}$.

In practice, the CGLS method generally works better than the Landweber iteration. However, it is easier to analyze the performance of the Landweber iteration. It can be shown that the kth iterate of the Landweber iteration is exactly the same as the SVD solution with filter factors of

$$f_i^{(k)} = 1 - (1 - \omega s_i^2)^k .\tag{6.114}$$

   a. Implement the Landweber iteration and apply it to the Shaw problem from Example 3.3.
   b. Verify that $\mathbf{m}^{(10)}$ from the Landweber iteration matches the SVD solution with filter factors given by (6.114).
   c. Derive (6.114).

3. The Landweber iteration described in the previous exercise converges under the condition that $0 < \omega < 2/s_1^2$ where s_1 is the largest singular value of $\mathbf{G}$ (or equivalently, $s_1 = \|\mathbf{G}\|_2$). As a practical matter we typically cannot compute the full SVD of $\mathbf{G}$. However, it is possible to rapidly estimate this quantity using an iterative method that we will derive in this exercise. Recall from Appendix A that $\|\mathbf{G}\|_2 = s_1$ is the square root of the largest eigenvalue of the matrix $\mathbf{G}^T\mathbf{G}$.

   a. Using (A.77), diagonalize the matrix $\mathbf{A} = \mathbf{G}^T\mathbf{G}$, and use the diagonalization to show that, for the kth power of $\mathbf{A}$,

$$\mathbf{A}^k = \mathbf{P}\mathbf{\Lambda}^k\mathbf{P}^{-1} .\tag{6.115}$$

   Assume that the eigenvalues of $\mathbf{A}$ are sorted so that $\lambda_1 \geq \lambda_2 \geq \cdots \geq \lambda_n \geq 0$.
   b. Take an arbitrary vector $\mathbf{x}$ in $\mathbf{R}^n$, and write it in terms of the eigenvectors of $\mathbf{A}$ as

$$\mathbf{x} = \alpha_1\mathbf{v}_1 + \cdots + \alpha_n\mathbf{v}_n .\tag{6.116}$$

   Then show that for $k \geq 1$,

$$\mathbf{A}^k\mathbf{x} = \alpha_1\lambda_1^k\mathbf{v}_1 + \cdots + \alpha_n\lambda_n^k\mathbf{v}_n .\tag{6.117}$$

   c. Starting with a random vector $\mathbf{x}^{(0)}$, and assuming that $\alpha_1 \neq 0$ (a reasonable assumption for a random vector), show that

$$\lim_{k \to \infty} \frac{\|\mathbf{A}^k\mathbf{x}^{(0)}\|_2}{\|\mathbf{A}^{k-1}\mathbf{x}^{(0)}\|_2} = \lambda_1 .\tag{6.118}$$

This leads to a simple iterative algorithm for estimating $s_1 = \sqrt{\lambda_1}$. Start with a random vector $\mathbf{x}^{(0)}$. In each iteration, let

$$\mathbf{x}^{(k+1)} = \frac{\mathbf{G}^T(\mathbf{G}\mathbf{x}^{(k)})}{\|\mathbf{x}^{(k)}\|_2}\tag{6.119}$$

and let

$$\rho_{k+1} = \sqrt{\|\mathbf{x}^{(k+1)}\|_2} \; . \tag{6.120}$$

The sequence ρ_k converges to s_1. This function is implemented in MATLAB as the **normest** function. Write your own implementation of this function using the formulas above, and compare the results you obtain with the MATLAB function.

4. Use the Landweber iteration discussed in Exercise 6.2 to deblur the image from Example 6.3. In order to use the Landweber iteration, you will need to estimate $s_1 = \|\mathbf{G}\|_2$, which can be done using the **normest** command discussed in Exercise 6.3.

5. The file **blur.mat** contains a blur-inducing sparse $\mathbf{G}$ matrix and a data vector $\mathbf{d}$, where the corresponding 100 by 100 pixel gray scale image can be obtained using **reshape(d,100,100)**.

    a. How large is the $\mathbf{G}$ matrix? How many nonzero elements does it have? How much storage would be required for the $\mathbf{G}$ matrix if all of its elements were nonzero? How much storage would the SVD of $\mathbf{G}$ require?

    b. Plot the raw image.

    c. Using the **cgls** algorithm with 100 iterations, deblur the image by solving $\mathbf{Gm} = \mathbf{d}$ and interpret the evolution of your solutions with increasing iterations.

6. Show that if $\mathbf{p}^{(0)}$, $\mathbf{p}^{(1)}$, ..., $\mathbf{p}^{(n-1)}$ are nonzero and mutually conjugate with respect to an n by n symmetric and positive definite matrix $\mathbf{A}$, then the vectors are also linearly independent. Hint: Use the definition of linear independence.

7. Recall the iteratively reweighted least squares method introduced in Chapter 2. The system of equations (2.98) solved in each iteration of the IRLS algorithm might well be dense, even if the matrix $\mathbf{G}$ is sparse. However, this system of equations can also be thought of as the normal equations for a particular least squares problem that will be sparse when $\mathbf{G}$ is sparse. Write down this least squares problem. Write a MATLAB function that uses the **lsqr** routine to solve these least squares problems in an IRLS scheme to minimize $\|\mathbf{Gm} - \mathbf{d}\|_1$. Hint: use the fact that $\mathbf{R}$ is a diagonal matrix with positive elements and you can thus take its square root.

8. Implement the method of gradient descent with exact line search for minimizing $\|\mathbf{Gm} - \mathbf{d}\|_2^2$. Your implementation should be careful to avoid computing $\mathbf{G}^T\mathbf{G}$. Modify Example 6.3 to use your implementation of gradient descent with explicit regularization. How does your solution compare with the solution obtained using CGLS, both in terms of solution quality and running time?

9. Implement the method of stochastic gradient descent and apply it to the problem in Example 6.1.

6.8. NOTES AND FURTHER READING

Iterative methods for the solution of linear systems of equations are an important topic in numerical analysis. Some basic references include [5,109,174,181,218].

Several variations on Kaczmarz's algorithm appeared early on in the development of computed tomography scanning for medical applications. Two of these algorithms, ART and SIRT, achieve storage savings by approximating the entries of the $\mathbf{G}$ matrix rather than storing them explicitly. ART and SIRT are discussed in [106,144,208], and parallel algorithms based on these methods are discussed in [40]. There is a striking similarity between Kaczmarz's algorithm with the hyperplanes taken in random order and the stochastic gradient descent method. Recent work has focused on the analysis of convergence rates for these methods [145,197].

In practice, the conjugate gradient and LSQR methods generally provide better performance than the row action methods. There are some interesting connections between SIRT and the conjugate gradient method discussed in [151,187,188].

Hestenes and Stiefel are generally credited with the invention of the conjugate gradient method [91]. However, credit is also due to Lanczos [116]. The history of the conjugate gradient method is discussed in [71,90]. Shewchuk's technical report [181] provides an introduction to the conjugate gradient method with illustrations that help to make the geometry of the method very clear. Filter factors for the CGLS method similar to those in Exercise 6.2 can be determined. These are derived in [84].

The resolution of LSQR solutions is discussed in [15,16]. Schemes have been developed for using CGLS with explicit regularization and dynamic adjustment of the regularization parameter α [107,108,135]. This can potentially remove the computational burden of solving the problem for many values of α. An alternative approach can be used to compute regularized solutions for several values of α at once [65].

The performance of the CG algorithm degrades dramatically on poorly conditioned systems of equations. In such situations **preconditioning** can be used to improve the performance of CG. Essentially, preconditioning involves a change of variables $\bar{\mathbf{x}} = \mathbf{Cx}$. The matrix $\mathbf{C}$ is selected so that the resulting system of equations will be better conditioned than the original system of equations [50,72,209,218]. For ill-conditioned discrete linear inverse problems, standard preconditioners have a regularization effect, but it can be difficult to analyze how a particular preconditioner will bias the solution of the least squares problem. Preconditioners can be specially designed to achieve particular regularization effects [29].

The conjugate gradient method can also be generalized to nonlinear minimization problems [76,181]. We introduce the nonlinear conjugate gradient method of Fletcher and Reeves [63] in Chapter 10.

Inverse problems in image processing are a very active area of research. Some books on inverse problems in imaging include [17,144].

CHAPTER SEVEN

Sparsity Regularization and Total Variation Techniques

Synopsis

Sparsity regularization selects solutions with the minimum number of nonzero model parameters. In compressive sensing, sparsity regularization is applied in association with a change of basis, where the basis is chosen so that the desired model will be sparse, in the sense of having only a few nonzero coefficients in the model expansion. The Iterative Soft Threshholding (ISTA) and Fast Iterative Soft Threshholding (FISTA) algorithms are introduced. Total variation regularization uses a regularization term based on the 1-norm of the model gradient. Resulting models allow for discontinuous jumps so that solutions are biased towards piecewise-constant functions. An Iteratively Reweighted Least Squares (IRLS) algorithm and the Alternating Direction Method of Multipliers (ADMM) algorithm for solving the total variation problem are introduced.

7.1. SPARSITY REGULARIZATION

Sometimes there are reasons to expect that many of the unknown model parameters will be zero. We may choose to minimize the number of nonzero entries in $\mathbf{m}$ to obtain a sparse model. This approach of regularizing for sparsity has been extensively used in signal and image processing in the field of compressive sensing, and is increasingly finding use in geophysical inverse problems as well [27,68,225].

The notation $\|\mathbf{m}\|_0$ is commonly used to denote the number of nonzero entries in $\mathbf{m}$. Note that this 0-norm definition is non-standard because it is inconsistent with the definition of the p-norm in (A.85) and does not satisfy the requirements for a vector norm in Section A.7. We can formulate a corresponding regularized inverse problem as

$$\min \|\mathbf{m}\|_0$$
$$\|\mathbf{Gm} - \mathbf{d}\|_2 \leq \delta \ . \tag{7.1}$$

Unfortunately, these kinds of optimization problems can be extremely difficult to solve.

A surprisingly effective alternative to solving (7.1) is to find the least squares solution that minimizes $\|\mathbf{m}\|_1$. To see why this is a reasonable approach, consider the set of models with $\|\mathbf{m}\|_2 = 1$. Among the models with $\|\mathbf{m}\|_2 = 1$, it turns out that the models with precisely one nonzero entry of $+1$ or -1 have the smallest 1-norms (Fig. 7.1). Thus, regularizing a least squares problem to minimize its model 1-norm will tend to produce sparse solutions. This tendency for 1-norm regularized models to be sparse becomes even more prominent in higher dimensions. This heuristic approach of minimizing $\|\mathbf{m}\|_1$ instead of $\|\mathbf{m}\|_0$ works very well in practice, and research has demonstrated

181

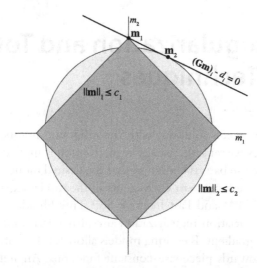

Figure 7.1 Two-dimensional demonstration of the use of model 1-norm minimization (7.2) to obtain sparsity regularization. The square shaded area shows the region $\|\mathbf{m}\|_1 \leq c_1$, whereas the circle shows the region with $\|\mathbf{m}\|_2 \leq c_2$. An arbitrary constraint equation in this 2-dimensional model space, $(\mathbf{Gm})_i - d_i = 0$, defines a line. The minimum 2-norm residual model satisfying the constraint, $\mathbf{m}_2$, will not generally be sparse. However, the minimum 1-norm model satisfying the constraint, $\mathbf{m}_1 = [0\ c_1]^T$, will tend to be sparse due to the presence of corners in the 1-norm contour.

reasonable conditions under which the solution to the 1-norm regularized problem is close to the solution of the 0-norm regularized problem (7.1) with very high probability [27,33–35].

The L_1 regularized least squares problem can be written as

$$\min \|\mathbf{m}\|_1$$
$$\|\mathbf{Gm} - \mathbf{d}\|_2 \leq \delta \ . \tag{7.2}$$

Using the standard approach of moving the constraint into the objective function, we can select a positive regularization parameter, α, so that this is equivalent to

$$\min \|\mathbf{Gm} - \mathbf{d}\|_2^2 + \alpha \|\mathbf{m}\|_1 \ . \tag{7.3}$$

This is a convex optimization problem that can be solved efficiently by a number of methods discussed in this chapter.

7.2. THE ITERATIVE SOFT THRESHHOLDING ALGORITHM (ISTA)

The objective function in (7.3) is the sum of two terms. An important family of methods for solving these kinds of problems is referred to as **proximal point** algorithms [9,10,44, 156]. In this section, we'll introduce the Iterative Soft Threshholding Algorithm (ISTA), which combines the gradient descent method for decreasing $\|\mathbf{Gm} - \mathbf{d}\|_2^2$ described in Chapter 6 with a proximal point step in which $\alpha\|\mathbf{m}\|_1$ is decreased.

In the kth iteration of this algorithm, we first move from $\mathbf{m}^{(k)}$ to $\mathbf{m}^{(k+1/2)}$ with a gradient descent step for the $\|\mathbf{Gm} - \mathbf{d}\|_2^2$ term

$$\mathbf{m}^{(k+1/2)} = \mathbf{m}^{(k)} - t_k\left(2\mathbf{G}^T(\mathbf{Gm}^{(k)} - \mathbf{d})\right). \tag{7.4}$$

As in the gradient descent method described in Chapter 6, the step length parameter t_k may be dynamically adjusted or a constant step length parameter, t, may be used. It can be shown that ISTA will converge if a constant step length $t \leq 1/(2\|\mathbf{G}^T\mathbf{G}\|_2)$ is used [10]. Since $\|\mathbf{G}^T\mathbf{G}\|_2 \leq \|\mathbf{G}\|_2\|\mathbf{G}^T\|_2$, we can use the MATLAB function **normest** (discussed in Exercise 6.3) on $\mathbf{G}$ and $\mathbf{G}^T$ to find a suitable step size.

Next, we reduce $\alpha\|\mathbf{m}\|_1$ while staying close to the gradient descent solution,

$$\mathbf{m}^{(k+1)} = \arg\min\left[\alpha\|\mathbf{m}\|_1 + \frac{1}{2}\|\mathbf{m} - \mathbf{m}^{(k+1/2)}\|_2^2\right], \tag{7.5}$$

where "arg min" simply refers to the argument ($\mathbf{m}$) that minimizes the accompanying expression. Here the second term insures that $\mathbf{m}^{(k+1)}$ will remain reasonably close to $\mathbf{m}^{(k+1/2)}$.

The minimization problem in (7.5) is particularly easy to solve because the objective function can be written as a sum of convex terms corresponding to the individual model parameters

$$\mathbf{m}^{(k+1)} = \arg\min\left[\sum_{j=1}^{n}\left(\alpha|m_j| + \frac{1}{2}(m_j - m_j^{(k+1/2)})^2\right)\right]. \tag{7.6}$$

We can minimize this sum by minimizing each of the n terms separately. First, suppose that $m_j^{(k+1/2)} \geq \alpha$. For $m_j > 0$, the derivative of the jth term is

$$\frac{d}{dm_j}\left[\alpha|m_j| + \frac{1}{2}(m_j - m_j^{(k+1/2)})^2\right] = \alpha + (m_j - m_j^{(k+1/2)}). \tag{7.7}$$

Setting this derivative equal to zero gives

$$m_j^{(k+1)} = m_j^{(k+1/2)} - \alpha. \tag{7.8}$$

Similarly, if $m_j^{(k+1/2)} \leq -\alpha$, then

$$m_j^{(k+1)} = m_j^{(k+1/2)} + \alpha. \tag{7.9}$$

The cases where $-\alpha < m_j^{(k+1/2)} < \alpha$ are more difficult. First consider the case where $0 < m_j^{(k+1/2)} < \alpha$. The solution $m_j^{(k+1)} = m_j^{(k+1/2)}$ is clearly better than any solution with

$m_j^{(k+1)} > m_j^{(k+1/2)}$. For $0 < m_j < m_j^{(k+1/2)}$, the derivative is positive, so these solutions cannot be optimal. Solutions with $m_j < 0$ are also clearly not optimal. Although the objective function is not differentiable at $m_j = 0$, this must be the optimal value. The case where $-\alpha < m_j < 0$ is similar. Since m_j is not greater than 0 and not less than 0, it must be 0. Thus the optimal solution to (7.5) is given by

$$m_j^{(k+1)} = \begin{cases} m_j^{(k+1/2)} - \alpha & m_j^{(k+1/2)} > \alpha \\ m_j^{(k+1/2)} + \alpha & m_j^{(k+1/2)} < -\alpha \\ 0 & \text{otherwise} . \end{cases} \qquad (7.10)$$

This operation of reducing the absolute value of $m_j^{(k+1/2)}$ by α is known as **soft thresholding**, which we denote by

$$\mathbf{m}^{(k+1)} = \text{sthresh}_\alpha(\mathbf{m}^{(k+1/2)}) . \qquad (7.11)$$

We can now summarize the iterative soft threshholding algorithm. We have implemented this algorithm in the library function **ista**.

Algorithm 7.1 Iterative Soft Threshholding Algorithm (ISTA)

Given the L_1 regularized least squares problem

$$\min \|\mathbf{Gm} - \mathbf{d}\|_2^2 + \alpha \|\mathbf{m}\|_1$$

and a step size $t < 1/(2\|\mathbf{G}^T\mathbf{G}\|_2)$. Let $\mathbf{m}^{(0)} = \mathbf{0}$ and $k = 0$.
1. Let $\mathbf{m}^{(k+1/2)} = \mathbf{m}^{(k)} - t\left(2\mathbf{G}^T(\mathbf{Gm}^{(k)} - \mathbf{d})\right)$.
2. Let $\mathbf{m}^{(k+1)} = \text{sthresh}_\alpha(\mathbf{m}^{(k+1/2)})$.
3. Let $k = k + 1$.
4. Repeat until convergence.

It was noted in Chapter 6 that the method of gradient descent can converge very slowly if $\mathbf{G}^T\mathbf{G}$ is badly conditioned. When the $\alpha \|\mathbf{m}\|_1$ term is small in comparison to $\|\mathbf{Gm} - \mathbf{d}\|_2^2$, the iterative soft threshholding algorithm performs similarly to the method of gradient descent and can also suffer from very slow convergence. In Chapter 6 we developed the method of conjugate gradients as a faster alternative to the method of gradient descent. Unfortunately, we can't combine proximal point steps with the method of conjugate gradients because the proximal point steps would disrupt the conjugacy of the search directions.

An alternative approach to accelerating the convergence of gradient descent is to use an extrapolation approach. After moving from $\mathbf{m}^{(k)}$ to $\mathbf{m}^{(k+1)}$, we extrapolate the step

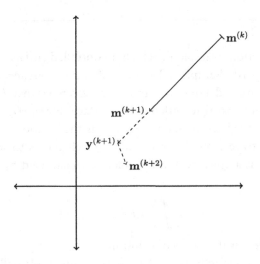

Figure 7.2 An accelerated gradient descent step.

to a point

$$\mathbf{y}^{(k+1)} = \mathbf{m}^{(k+1)} + \omega(\mathbf{m}^{(k+1)} - \mathbf{m}^{(k)}) \tag{7.12}$$

and then take a gradient descent step from the extrapolated point $\mathbf{y}^{(k+1)}$ to obtain $\mathbf{m}^{(k+2)}$. See Fig. 7.2. If the extrapolation parameter ω is chosen carefully, then convergence of the gradient descent method can be substantially improved. This approach was first studied by Polyak in the 1960's, and later by Nesterov in the 1980's [146,166].

The fast iterative soft threshholding algorithm (FISTA) of Beck and Teboulle is an accelerated version of ISTA [10]. We will not analyze FISTA in detail, but have implemented a FISTA algorithm in the library function **fista**.

Algorithm 7.2 Fast Iterative Soft Threshholding Algorithm (FISTA)

Given the L_1 regularized least squares problem

$$\min \|\mathbf{Gm} - \mathbf{d}\|_2^2 + \alpha \|\mathbf{m}\|_1$$

and a step size $t < 1/(2\|\mathbf{G}^T\mathbf{G}\|_2)$. Let $\mathbf{m}^{(0)} = \mathbf{0}$, $\mathbf{y}^{(0)} = \mathbf{0}$, $\theta_0 = 1$, and $k = 0$.
1. Let $\mathbf{m}^{(k+1/2)} = \mathbf{y}^{(k)} - t\left(2\mathbf{G}^T(\mathbf{Gy}^{(k)} - \mathbf{d})\right)$.
2. Let $\mathbf{m}^{(k+1)} = \text{sthresh}_\alpha(\mathbf{m}^{(k+1/2)})$.
3. Let $\theta_{k+1} = \dfrac{1 + \sqrt{1 + 4\theta_k^2}}{2}$.
4. Let $\mathbf{y}^{(k+1)} = \mathbf{m}^{(k+1)} + \dfrac{\theta_k - 1}{\theta_{k+1}}(\mathbf{m}^{(k+1)} - \mathbf{m}^{(k)})$.
5. Let $k = k + 1$.
6. Repeat until convergence.

Example 7.1

Consider a deconvolution example in which a controlled surface source sends seismic waves down into the earth. Seismic reflections will occur from abrupt material discontinuities at various depths, and a seismogram of the reflected signals (which will generally include multiple reflections) is recorded at the surface. Here, $g(t)$ is the known source signal, $d(t)$ is the recorded seismogram, and $m(t)$ is the unknown impulse response of the Earth. We'd like to recover the times of these reflections to study the responsible Earth structure. This is a linear forward problem characterized by a convolution (e.g., (1.11))

$$d(t) = \int_{-\infty}^{\infty} g(t - \xi)\, m(\xi)\, d\xi \tag{7.13}$$

and the inverse problem is thus a deconvolution.

In this highly simplified reflection seismology example, we will assume a 1-dimensional structure with depth and consider downward- and upward-traveling plane compressional seismic waves detected by a vertically-oriented seismometer. In this case, the presence of horizontally oriented seismic structural discontinuities in an otherwise smooth medium will ideally be manifested in the Earth impulse response $m(t)$ as a sequence of delta functions. We discretize $m(t)$ as a model vector $\mathbf{m}$ and will seek models that consist of an optimal sequence of spikes.

The source signal is a short pulse with a characteristic time of around 0.4 s

$$g(t) = e^{-5t} \sin(10t) . \tag{7.14}$$

We add independent normally distributed random noise to the corresponding data resulting from the convolution (7.13) and attempt to recover $m(t)$ in regularized inversions. All time series are sampled uniformly at 100 samples/s to create model and data vectors of 1000 points, or spanning a time interval of 10 s. The true model (Fig. 7.3) is the impulse response of a single subsurface reflector with a reflection coefficient of $r = 0.4$, located at a depth corresponding to a two-way seismic travel time from the surface of $\tau = 1.5$ s. The source signal begins at 0 s and the primary reflection and surface-reflected multiples appear in Fig. 7.3 as alternating-sign spikes at intervals of τ with amplitudes that decrease by a factor of r upon each reflection. Fig. 7.4 shows the corresponding data from (7.13) with $N(0, (0.002^2))$ noise added. The noise and smoothing from the convolution largely obscure the signal corresponding to the later impulses of Fig. 7.3.

Fig. 7.5 shows the L-curve for zeroth-order Tikhonov regularization. Fig. 7.6 shows the model corresponding to the corner value of $\alpha = 0.01$. The impulses are broadened

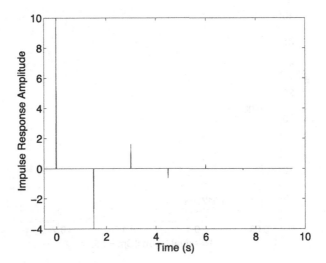

Figure 7.3 The true impulse response model $m(t)$.

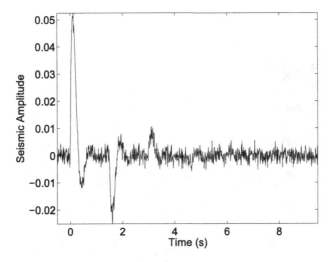

Figure 7.4 Data (7.13) with $N(0, (0.002)^2)$ noise added.

relative to the true model due to the associated regularization bias, and there are noise-generated secondary peaks. However, solving (7.2) for this problem using FISTA produces excellent model recovery (Figs. 7.7, 7.8, and 7.9). Note that the spikes are correctly placed in time and that all are resolved, except for the tiny pulse at 7.5 s, which has an amplitude of only $r^5 \approx 0.01$, or just 1% of the amplitude of the source

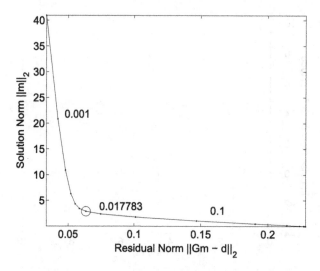

Figure 7.5 L-curve for zeroth-order Tikhonov solution with corner at $\alpha = 0.01$.

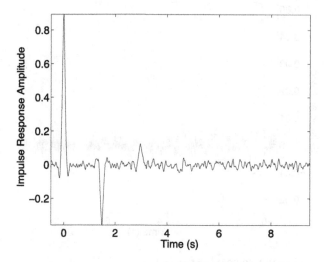

Figure 7.6 Zeroth-order Tikhonov solution.

spike. The amplitudes of the spikes are reduced and they are slightly broader than they should be, but the model is vastly better than that obtained with Tikhonov regularization.

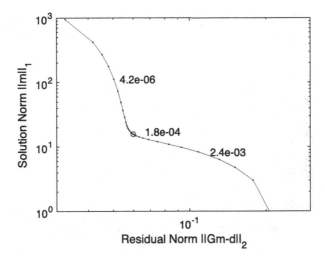

Figure 7.7 L-curve for the 1-norm regularized solution with corner near $\alpha = 1.8 \times 10^{-4}$.

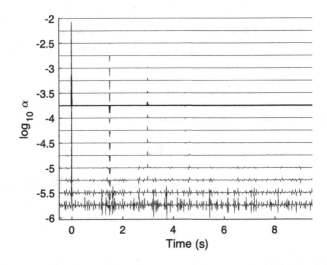

Figure 7.8 1-norm regularized solutions corresponding to the L-curve shown in Fig. 7.7. Model corresponding to the L-curve corner is shown in bold and is reproduced in Fig. 7.9.

7.3. SPARSE REPRESENTATION AND COMPRESSIVE SENSING

In situations where there is no reason to assume that the model is sparse, sparsity regularization will of course produce poor model recovery. However, it is sometimes possible to implement a change of variables so that the coefficients of a model with respect to the new basis will be sparse, or can at least be well approximated by a sparse model that can be stored and manipulated very efficiently.

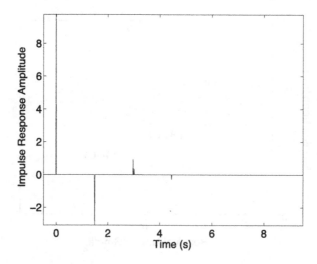

Figure 7.9 1-norm regularized solution for $\alpha = 1.8 \times 10^{-4}$ (Fig. 7.8).

Appendix A notes that if a set of vectors $\mathbf{w}_1$, $\mathbf{w}_2$, ..., $\mathbf{w}_n$ form a basis for R^n, then any vector $\mathbf{m}$ in R^n can be written as a linear combination of the basis vectors with n coefficients, x_i

$$\mathbf{m} = x_1\mathbf{w}_1 + x_2\mathbf{w}_2 + \cdots + x_n\mathbf{w}_n .\tag{7.15}$$

In vector–matrix form, this can be written as

$$\mathbf{m} = \mathbf{W}\mathbf{x}\tag{7.16}$$

where the basis vectors $\mathbf{w}_1$, $\mathbf{w}_2$, ..., $\mathbf{w}_n$ form the columns of $\mathbf{W}$. We can also perform the change of variables in the reverse direction as

$$\mathbf{x} = \mathbf{W}^{-1}\mathbf{m} .\tag{7.17}$$

In practice, linear transformation algorithms, such as the Fast Fourier Transform (FFT; Chapter 8), Discrete Cosine Transform (DCT), or Discrete Wavelet Transform (DWT) are often used in place of the matrix–vector multiplications of (7.16) and (7.17) to effect the corresponding forward and inverse operations.

Example 7.2

Fig. 7.10 shows the 40,000-pixel image from Example 6.3. If we were to zero out any significant fraction of these pixels, the effect on the image would be quite obvious. However, if instead we encode the image in terms of the discrete cosine

Figure 7.10 An image before discrete cosine transform threshholding.

Figure 7.11 Image after removing the smallest 40% of the DCT coefficients.

transform (DCT) of the pixel values, a sparser representation of the image is possible that closely approximates the original. Calculating the discrete cosine transform of the two-dimensional image using MATLAB's **dct2** command produces a set of 40,000 coefficients. Of these 40,000 coefficients, the 16,000 coefficients (40% of the original coefficients) that were smallest in absolute value were set to zero.

After inverting the DCT, we produced the image shown in Fig. 7.11. This second image appears to be very similar to the original image, yet 40% fewer coefficients were used to represent the image in terms of the DCT basis. This is because the coefficients associated with the 16,000 unused basis vectors in the original image were small.

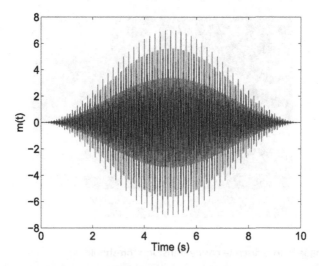

Figure 7.12 A 1001-point signal consisting of 25 and 35 Hz cosines (7.19), multiplied by a Hann taper.

The discrete cosine transform is the basis for the widely used JPEG standard for image compression. Wavelet bases are also widely used in image processing, and thresholding the wavelet transform produces similar results in compressing images. The newer JPEG 2000 standard for image compression uses wavelets rather than the discrete cosine transform.

In **compressive sensing** we apply sparsity regularization, using a basis chosen so that the model can be represented by a linear combination of the basis vectors with a sparse set of coefficients. The regularized problem can be written using (7.16) and (7.3) as

$$\min \|\mathbf{G}\mathbf{W}\mathbf{x} - \mathbf{d}\|_2^2 + \alpha \|\mathbf{x}\|_1 \tag{7.18}$$

where $\mathbf{G}$ is called the **measurement matrix**. The underlying idea is to use the measurement matrix $\mathbf{G}$ to extract the sparse solution without first computing the full solution and then discarding the "small" parts.

The recovery of a sparse solution using (7.18) depends on the details and appropriateness of the particular $\mathbf{G}$ and $\mathbf{W}$ that are used. The analysis and implementation of compressive schemes become relatively simple if the measurement matrix is random [36].

Example 7.3

Consider the recovery of a signal, $\mathbf{m}$, shown in Fig. 7.12. This 10 s-long time series of $n = 1001$ time points, t_i, is sampled at 100 samples/s and consists of two sine waves at $f_1 = 25$ and $f_2 = 35$ Hz.

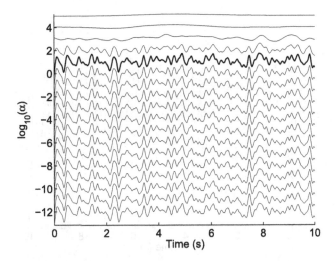

Figure 7.13 Signal recovery obtained using second-order Tikhonov regularization. Solution amplitudes are normalized to improve legibility.

$$m_i = h_i \cdot \left(5\cos(2\pi f_1 \cdot t_i) + 2\cos(2\pi f_2 \cdot t_i)\right) \quad 1 \le i \le n \qquad (7.19)$$

where the signal envelope has also been smoothed with term-by-term multiplication by a Hann taper function

$$h_i = \frac{1}{2}\left(1 - \cos(2\pi(i-1)/n)\right) \quad 1 \le i \le n . \qquad (7.20)$$

In the standard basis, the signal (7.19), shown in Fig. 7.12, is obviously not sparse. However, because it is constructed from two cosine components, this model has a very sparse representation in the DCT basis.

We use a random measurement matrix $\mathbf{G}$ that is $m = 100$ by $n = 1001$, with entries chosen independently from $N(0, 1)$ to generate a synthetic set of 100 data measurements, $\mathbf{d} = \mathbf{Gm}$, and then add independent $N(0, 25)$ noise. The matrix of basis vectors, $\mathbf{W}$, is constructed by applying the discrete cosine transform to standard basis vectors for the model space R^n.

Fig. 7.13 shows a range of solutions obtained by solving $\mathbf{Gm} = \mathbf{d}$ using second-order Tikhonov regularization. The highlighted solution ($\alpha = 100$; Fig. 7.14) fits the data with a residual norm that approximately satisfies the discrepancy principle, so that $\|\mathbf{Gm} - \mathbf{d}\| \approx 5 \cdot \sqrt{m} = 50$. However, this solution and others across a wide range of α values bear essentially no resemblance to the true signal shown in Fig. 7.12.

With the compressive sensing approach, we consistently achieve good recovery of the original signal for values of α greater than 1×10^{-4} up to 1000. For α values greater

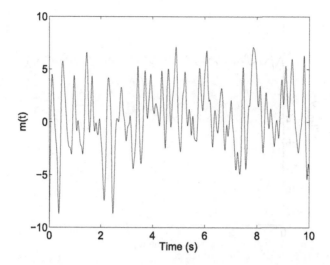

Figure 7.14 A representative solution (Fig. 7.13; $\alpha = 100$) obtained using second-order Tikhonov regularization that approximately satisfies the discrepancy principle.

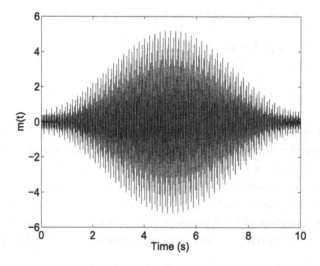

Figure 7.15 A representative solution obtained using compressive sensing with $\alpha = 1000$ that approximately satisfies the discrepancy principle.

than 1000, on the other hand, solutions rapidly approach the zero model. For a wide range of α values between these extreme values, the recovered signal $\mathbf{Wx}$ is stable and is quite close to the true signal. The signal corresponding to $\alpha = 1000$ roughly satisfies the discrepancy principle, with $\|\mathbf{GWx} - \mathbf{d}\|_2 \approx 150$. This solution is shown in Fig. 7.15.

It may seem paradoxical that we can recover a 1001–point signal so well from only 100 random linear combination measurements, when the Nyquist sampling theorem (8.36) suggests that at least $10 \text{ s} \times 35 \text{ Hz} \times 2 = 700$ equally-spaced samples should be required to accurately represent it. The reason that this is possible in this example is because the measured signal is sparse with respect to the DCT basis, and just a few cosine basis functions are therefore necessary to represent it well. The corresponding sparse set of coefficients is easily recovered when sparsity regularization is employed.

7.4. TOTAL VARIATION REGULARIZATION

Total Variation (TV) regularization is appropriate for problems where we expect there to be discontinuous jumps in the model. In the one-dimensional case, the TV regularization function is just

$$TV(\mathbf{m}) = \sum_{i=1}^{n-1} |m_{i+1} - m_i| \tag{7.21}$$

$$= \|\mathbf{Lm}\|_1 \tag{7.22}$$

where $\mathbf{L} = \mathbf{L}_1$ as in (4.27).

In first- and second-order Tikhonov regularization, discontinuities in the model are smoothed out and are not well recovered in the inverse solution. This is because smooth model differences are generally penalized less by the regularization term than sharp ones. A useful feature of (7.22) is that the regularization term imposes a constraint to keep the number of discontinuous transitions to a minimum (i.e., keeping $\mathbf{Lm}$ sparse). We can insert the TV regularization term (7.22) in place of $\|\mathbf{Lm}\|_2^2$ in the Tikhonov regularization optimization problem to obtain

$$\min \|\mathbf{Gm} - \mathbf{d}\|_2^2 + \alpha \|\mathbf{Lm}\|_1 \tag{7.23}$$

which is a convex optimization problem. Unfortunately, (7.23) is not in the form of (7.3) because the $\alpha \|\mathbf{m}\|_1$ term has been replaced with $\alpha \|\mathbf{Lm}\|_1$.

For two-dimensional models, it would be desirable to define the total variation as the norm of the discretization of the gradient operator rather than a discretization of the first derivative. If the 2-norm is used, for a model vector $\mathbf{m}$ mapped into an appropriately indexed two-dimensional matrix $\mathbf{M}$, we have

$$TV_2(\mathbf{m}) = \sum_{i=1}^{n-1} \sum_{j=1}^{n-1} \sqrt{|M_{i+1,j} - M_{i,j}|^2 + |M_{i,j+1} - M_{i,j}|^2} \; . \tag{7.24}$$

Unfortunately, the expression in (7.24) is not amenable to being cast in the form $\|\mathbf{Lm}\|_1$, so specialized algorithms are required to implement this type of regularization [172].

However, if the 1-norm is used, then we have

$$TV_1(\mathbf{m}) = \sum_{i=1}^{n-1}\sum_{j=1}^{n-1} |M_{i+1,j} - M_{i,j}| \; + \; \sum_{i=1}^{n-1}\sum_{j=1}^{n-1} |M_{i,j+1} - M_{i,j}| \; . \tag{7.25}$$

(7.25) can be put into the form $\|\mathbf{Lm}\|_1$, and algorithms for solving (7.23) can thus be used to solve associated TV problems.

7.5. USING IRLS TO SOLVE L_1 REGULARIZED PROBLEMS

To solve problems of the form (7.23) we can extend the IRLS algorithm introduced in Chapter 2 for L_1 regression. Let

$$f(\mathbf{m}) = \|\mathbf{Gm} - \mathbf{d}\|_2^2 + \alpha \|\mathbf{Lm}\|_1 \tag{7.26}$$

and let

$$\mathbf{y} = \mathbf{Lm} \; . \tag{7.27}$$

At points where any element of $\mathbf{Lm}$ is zero, $f(\mathbf{m})$ is not differentiable because of the non-differentiability of the absolute value functions in the L_1 norm. However, at other points we can easily compute the gradient of f.

$$\nabla f(\mathbf{m}) = 2\mathbf{G}^T\mathbf{Gm} - 2\mathbf{G}^T\mathbf{d} + \alpha \sum_{i=1}^{m} \nabla |y_i| \; . \tag{7.28}$$

For y_i nonzero,

$$\frac{\partial |y_i|}{\partial m_k} = L_{i,k}\,\mathrm{sgn}(y_i) \; . \tag{7.29}$$

Writing $\mathrm{sgn}(y_i)$ as $y_i/|y_i|$, and using $\|\mathbf{Lm}\|_1 = \sum_{i=1}^{m} |y_i|$, we obtain

$$\frac{\partial \|\mathbf{Lm}\|_1}{\partial m_k} = \sum_{i=1}^{m} L_{i,k}\frac{y_i}{|y_i|} \; . \tag{7.30}$$

Let $\mathbf{W}$ be the diagonal matrix with elements

$$W_{i,i} = \frac{1}{|y_i|} \; . \tag{7.31}$$

Then

$$\nabla\left(\|\mathbf{Lm}\|_1\right) = \mathbf{L}^T\mathbf{WLm} \tag{7.32}$$

and

$$\nabla f(\mathbf{m}) = 2\mathbf{G}^T\mathbf{Gm} - 2\mathbf{G}^T\mathbf{d} + \alpha \mathbf{L}^T\mathbf{WLm} \; . \tag{7.33}$$

Setting $\nabla f(\mathbf{m}) = \mathbf{0}$, we obtain

$$\left(2\mathbf{G}^T\mathbf{G} + \alpha\mathbf{L}^T\mathbf{W}\mathbf{L}\right)\mathbf{m} = 2\mathbf{G}^T\mathbf{d} .\tag{7.34}$$

Because $\mathbf{W}$ depends on $\mathbf{Lm}$, this is a nonlinear system of equations. Furthermore, $\mathbf{W}$ is not defined at any point where $\mathbf{Lm}$ has a zero element.

To accommodate the non-differentiability of the 1-norm at points where some components of $\mathbf{m}$ are 0, we set a tolerance ϵ, and let

$$W_{i,i} = \begin{cases} 1/|\gamma_i| & |\gamma_i| \geq \epsilon \\ 1/\epsilon & |\gamma_i| < \epsilon . \end{cases}\tag{7.35}$$

In the IRLS procedure, we solve (7.34), update $\mathbf{W}$ according to (7.35), and iterate until the solution has converged.

The system of equations (7.34) can be seen as the normal equations for the least squares problem

$$\min \left\| \begin{bmatrix} \mathbf{G} \\ \sqrt{\frac{\alpha}{2}}\sqrt{\mathbf{W}}\mathbf{L} \end{bmatrix} \mathbf{m} - \begin{bmatrix} \mathbf{d} \\ \mathbf{0} \end{bmatrix} \right\|_2 .\tag{7.36}$$

When $\mathbf{G}$ is large and sparse it can be extremely advantageous to apply LSQR to solve the least squares problem (7.36) rather than solving the system of equations (7.34).

We can now summarize the IRLS algorithm, which we have implemented in the library function **irlsl1reg**.

Algorithm 7.3 Iteratively Reweighted Least Squares (IRLS) for an L_1 Regularized Least Squares Problem

---

Given the L_1 regularized least squares problem

$$\min \|\mathbf{Gm} - \mathbf{d}\|_2^2 + \alpha\|\mathbf{Lm}\|_1 .$$

Let $\mathbf{m}^{(0)} = \mathbf{0}$ and $k = 0$.

1. Let $\mathbf{y}^{(k)} = \mathbf{Lm}^{(k)}$.
2. Let $\mathbf{W}$ be the diagonal matrix with elements

$$W_{i,i} = \begin{cases} 1/|\gamma_i^{(k)}| & |\gamma_i^{(k)}| \geq \epsilon \\ 1/\epsilon & |\gamma_i^{(k)}| < \epsilon . \end{cases}\tag{7.37}$$

3. Let

$$\mathbf{m}^{(k+1)} = \arg\min \left\| \begin{bmatrix} \mathbf{G} \\ \sqrt{\frac{\alpha}{2}}\sqrt{\mathbf{W}}\mathbf{L} \end{bmatrix} \mathbf{m} - \begin{bmatrix} \mathbf{d} \\ \mathbf{0} \end{bmatrix} \right\|_2 .\tag{7.38}$$

4. Let $k = k + 1$.
5. Repeat until convergence.

7.6. THE ALTERNATING DIRECTION METHOD OF MULTIPLIERS (ADMM)

Another popular algorithm for solving (7.23) is the Alternating Direction Method of Multipliers (ADMM) [9,18,25,70].

In ADMM, we begin by introducing an additional vector of variables, $\mathbf{z} = \mathbf{Lm}$, and write (7.23) as a constrained optimization problem

$$\min \|\mathbf{Gm} - \mathbf{d}\|_2^2 + \alpha\|\mathbf{z}\|_1$$
$$\text{subject to } \mathbf{Lm} = \mathbf{z} \, . \tag{7.39}$$

As discussed in Appendix C, we can write the Lagrangian of this constrained problem as

$$L(\mathbf{m}, \mathbf{z}, \boldsymbol{\lambda}) = \|\mathbf{Gm} - \mathbf{d}\|_2^2 + \alpha\|\mathbf{z}\|_1 + \boldsymbol{\lambda}^T(\mathbf{Lm} - \mathbf{z}) \tag{7.40}$$

where $\boldsymbol{\lambda}$ is a vector of Lagrange multipliers. The optimality condition requires the gradient of the Lagrangian with respect to $\mathbf{m}$, $\mathbf{z}$, and $\boldsymbol{\lambda}$ to be 0. At such a stationary point, the gradient of the Lagrangian with respect to $\boldsymbol{\lambda}$ is $\mathbf{Lm} - \mathbf{z}$ and since the gradient is zero, the constraints $\mathbf{Lm} = \mathbf{z}$ are satisfied. For fixed $\boldsymbol{\lambda}$, the Lagrangian is a convex function of $\mathbf{m}$ and $\mathbf{z}$. For fixed $\mathbf{m}$ and $\mathbf{z}$, the Lagrangian is a linear function of $\boldsymbol{\lambda}$. Any stationary point will have to be a saddle point of the Lagrangian rather than a minimum or maximum. By minimizing L with respect to $\mathbf{m}$ and $\mathbf{z}$ and alternately maximizing L with respect to $\boldsymbol{\lambda}$, we seek convergence to a stationary point of the Lagrangian. The corresponding values of $\mathbf{m}$ and $\mathbf{z}$ will be optimal for (7.39).

In practice, the convergence of this method can be accelerated significantly by augmenting the Lagrangian with a term that penalizes violations of the constraint $\mathbf{Lm} - \mathbf{z} = 0$. The **augmented Lagrangian** is

$$L_\rho(\mathbf{m}, \mathbf{z}, \boldsymbol{\lambda}) = \|\mathbf{Gm} - \mathbf{d}\|_2^2 + \alpha\|\mathbf{z}\|_1 + \boldsymbol{\lambda}^T(\mathbf{Lm} - \mathbf{z}) + \frac{\rho}{2}\|\mathbf{Lm} - \mathbf{z}\|_2^2 \tag{7.41}$$

where the positive penalty parameter ρ will be adjusted to improve the rate of convergence.

Each major iteration of the algorithm consists of three steps. First, we minimize $L_\rho(\mathbf{m}, \mathbf{z}^{(k)}, \boldsymbol{\lambda}^{(k)})$ with respect to $\mathbf{m}$ to obtain $\mathbf{m}^{(k+1)}$.

$$\mathbf{m}^{(k+1)} = \arg\min L_\rho(\mathbf{m}, \mathbf{z}^{(k)}, \boldsymbol{\lambda}^{(k)}) \, . \tag{7.42}$$

Since $\mathbf{z}^{(k)}$ and $\boldsymbol{\lambda}^{(k)}$ are held constant in (7.42), this minimization problem simplifies to

$$\mathbf{m}^{(k+1)} = \arg\min \left[\|\mathbf{Gm} - \mathbf{d}\|_2^2 + (\boldsymbol{\lambda}^{(k)})^T \mathbf{Lm} + \frac{\rho}{2}\|\mathbf{Lm} - \mathbf{z}^{(k)}\|_2^2 \right] \, . \tag{7.43}$$

It is convenient to let $\mathbf{u}^{(k)} = (1/\rho)\boldsymbol{\lambda}^{(k)}$ and write the minimization problem as

$$\mathbf{m}^{(k+1)} = \arg\min \left[\|\mathbf{Gm} - \mathbf{d}\|_2 + \rho(\mathbf{u}^{(k)})^T \mathbf{Lm} + \frac{\rho}{2}\|\mathbf{Lm} - \mathbf{z}^{(k)}\|_2^2 \right] \, . \tag{7.44}$$

Taking the derivative with respect to **m** and setting it equal to zero gives the system of equations

$$\left(2\mathbf{G}^T\mathbf{G} + \rho\mathbf{L}^T\mathbf{L}\right)\mathbf{m}^{(k+1)} = 2\mathbf{G}^T\mathbf{d} + \rho\mathbf{L}^T(\mathbf{z}^{(k)} - \mathbf{u}^{(k)}) . \tag{7.45}$$

These are the normal equations for the linear least squares problem

$$\mathbf{m}^{(k+1)} = \arg\min \left\| \begin{bmatrix} \mathbf{G} \\ \sqrt{\frac{\rho}{2}}\mathbf{L} \end{bmatrix} \mathbf{m} - \begin{bmatrix} \mathbf{d} \\ \sqrt{\frac{\rho}{2}}(\mathbf{z}^{(k)} - \mathbf{u}^{(k)}) \end{bmatrix} \right\|_2^2 , \tag{7.46}$$

which can be solved using LSQR or another iterative method.

Next, we minimize $L_\rho(\mathbf{m}^{(k+1)}, \mathbf{z}, \boldsymbol{\lambda}^{(k)})$ with respect to $\mathbf{z}$ to obtain

$$\mathbf{z}^{(k+1)} = \arg\min L_\rho(\mathbf{m}^{(k+1)}, \mathbf{z}, \boldsymbol{\lambda}^{(k)}) . \tag{7.47}$$

Since $\mathbf{m}^{(k+1)}$ and $\boldsymbol{\lambda}^{(k)}$ are constant in this minimization, we can write (7.47) as

$$\mathbf{z}^{(k+1)} = \arg\min \left[\alpha\|\mathbf{z}\|_1 + (\boldsymbol{\lambda}^{(k)})^T(\mathbf{L}\mathbf{m}^{(k+1)} - \mathbf{z}) + \frac{\rho}{2}\|\mathbf{L}\mathbf{m}^{(k+1)} - \mathbf{z}\|_2^2 \right] . \tag{7.48}$$

Expanding the 2-norm expression and rearranging terms, we can see that this is equivalent to

$$\mathbf{z}^{(k+1)} = \arg\min \left[\alpha\|\mathbf{z}\|_1 + \frac{\rho}{2}\|\mathbf{L}\mathbf{m}^{(k+1)} - \mathbf{z} + \frac{1}{\rho}\boldsymbol{\lambda}^{(k)}\|_2^2 - \frac{1}{2\rho}\|\boldsymbol{\lambda}^{(k)}\|_2^2 \right] \tag{7.49}$$

or

$$\mathbf{z}^{(k+1)} = \arg\min \left[\alpha\|\mathbf{z}\|_1 + \frac{\rho}{2}\|\mathbf{L}\mathbf{m}^{(k+1)} - \mathbf{z} + \frac{1}{\rho}\boldsymbol{\lambda}^{(k)}\|_2^2 \right] . \tag{7.50}$$

In terms of $\mathbf{u}^{(k)}$, this is

$$\mathbf{z}^{(k+1)} = \arg\min \left[\alpha\|\mathbf{z}\|_1 + \frac{\rho}{2}\|\mathbf{L}\mathbf{m}^{(k+1)} - \mathbf{z} + \mathbf{u}^{(k)}\|_2^2 \right] . \tag{7.51}$$

This is not quite in the form of (7.5), but (7.51) is equivalent to

$$\mathbf{z}^{(k+1)} = \arg\min \left[\frac{\alpha}{\rho}\|\mathbf{z}\|_1 + \frac{1}{2}\|\mathbf{L}\mathbf{m}^{(k+1)} - \mathbf{z} + \mathbf{u}^{(k)}\|_2^2 \right] . \tag{7.52}$$

We showed earlier in the chapter how soft threshholding can be used to quickly solve (7.52), i.e.,

$$\mathbf{z}^{(k+1)} = \text{sthresh}_{\alpha/\rho}(\mathbf{L}\mathbf{m}^{(k+1)} - \mathbf{z}^{(k)}). \tag{7.53}$$

Finally, we update the Lagrange multipliers $\boldsymbol{\lambda}$. The gradient of $L_\rho(\mathbf{m}^{(k+1)}, \mathbf{z}^{(k+1)}, \boldsymbol{\lambda})$ with respect to $\boldsymbol{\lambda}$ is

$$\nabla_\lambda L_\rho(\mathbf{m}^{(k+1)}, \mathbf{z}^{(k+1)}, \boldsymbol{\lambda}) = \mathbf{L}\mathbf{m}^{(k+1)} - \mathbf{z}^{(k+1)} . \tag{7.54}$$

To increase L_ρ, we'll step in this gradient direction. It can be shown that the method converges if we use a step length of ρ [25]. So, we let

$$\lambda^{(k+1)} = \lambda^{(k)} + \rho(\mathbf{Lm}^{(k+1)} - \mathbf{z}^{(k+1)}) . \tag{7.55}$$

In terms of the scaled Lagrange multipliers $\mathbf{u}$, the update is

$$\mathbf{u}^{(k+1)} = \mathbf{u}^{(k)} + \mathbf{Lm}^{(k+1)} - \mathbf{z}^{(k+1)} . \tag{7.56}$$

The ADMM algorithm enforces the constraint $\mathbf{Lm} = \mathbf{z}$ only in the limit as the iterates converge to an optimal solution. Thus, it is important to track the **primal residual** vector

$$\mathbf{r}^{(k+1)} = \mathbf{Lm}^{(k+1)} - \mathbf{z}^{(k+1)} \tag{7.57}$$

and not declare convergence until its 2-norm is smaller than a desired tolerance.

Even though we have exact solutions to (7.42) and (7.47), the combination of $\mathbf{m}^{(k)}$ and $\mathbf{z}^{(k)}$ must also be a minimizer of (7.41). After the solution of (7.42), we have

$$\nabla_{\mathbf{m}} L(\mathbf{m}^{(k+1)}, \mathbf{z}^{(k)}, \lambda^{(k)}) = 0 \tag{7.58}$$

and

$$2\mathbf{G}^T(\mathbf{Gm}^{(k+1)} - \mathbf{d}) + \mathbf{L}^T\lambda^{(k)} + \rho(\mathbf{L}^T\mathbf{Lm}^{(k+1)} - \mathbf{L}^T\mathbf{z}^{(k)}) = 0 . \tag{7.59}$$

In the solution to (7.47), we updated $\mathbf{z}^{(k)}$ to $\mathbf{z}^{(k+1)}$ without changing $\mathbf{m}^{(k+1)}$ or $\lambda^{(k)}$. To maintain the optimality of $\mathbf{m}^{(k+1)}$, the difference between $\rho\mathbf{L}^T\mathbf{z}^{(k)}$ and $\rho\mathbf{L}^T\mathbf{z}^{(k+1)}$ must be zero. Thus, in addition to the primal residual, we also track the **dual residual**

$$\mathbf{s}^{(k+1)} = \rho\mathbf{L}^T(\mathbf{z}^{(k+1)} - \mathbf{z}^{(k)}) \tag{7.60}$$

and do not declare convergence until its 2-norm is smaller than a desired tolerance.

In practice, with a fixed penalty parameter ρ, we often find that either the primal residual $\mathbf{r}^{(k+1)}$ decreases much faster than the dual residual $\mathbf{s}^{(k+1)}$ or vice versa. If the primal residual is much smaller than the dual residual, then we decrease the penalty parameter ρ to balance the terms. Conversely, if the dual residual is much smaller than the primal residual then we can increase the penalty parameter to put more weight on achieving $\mathbf{Lm} - \mathbf{z} = 0$.

Algorithm 7.4 Alternating Direction Method of Multipliers (ADMM)

Given

$$\min \|\mathbf{Gm} - \mathbf{d}\|_2^2 + \alpha \|\mathbf{Lm}\|_1 \,,$$

let $\mathbf{m}^{(0)} = \mathbf{0}$, $\mathbf{z}^{(0)} = \mathbf{0}$, $\mathbf{u}^0 = \mathbf{0}$, and $k = 0$.

1. Let

$$\mathbf{m}^{(k+1)} = \arg\min \left\| \begin{bmatrix} \mathbf{G} \\ \sqrt{\frac{\rho}{2}}\mathbf{L} \end{bmatrix} \mathbf{m} - \begin{bmatrix} \mathbf{d} \\ \sqrt{\frac{\rho}{2}}(\mathbf{z}^{(k)} - \mathbf{u}^{(k)}) \end{bmatrix} \right\|_2^2 . \tag{7.61}$$

2. Let

$$\mathbf{z}^{(k+1)} = \mathrm{sthresh}_{\alpha/\rho}(\mathbf{Lm}^{(k+1)} - \mathbf{z}^{(k)}) \,. \tag{7.62}$$

3. Let

$$\mathbf{u}^{(k+1)} = \mathbf{u}^{(k)} + \mathbf{Lm}^{(k+1)} - \mathbf{z}^{(k+1)} \,. \tag{7.63}$$

4. Calculate the primal residual vector $\mathbf{r}^{(k+1)} = \mathbf{Lm}^{(k+1)} - \mathbf{z}^{(k+1)}$.
5. Calculate the dual residual vector $\mathbf{s}^{(k+1)} = \rho\mathbf{L}^T(\mathbf{z}^{(k+1)} - \mathbf{z}^{(k)})$.
6. Adjust ρ if necessary to balance the 2-norms of the primal and dual residuals.
7. Let $k = k + 1$.
8. Repeat until both $\|\mathbf{r}^{(k+1)}\|_2$ and $\|\mathbf{s}^{(k+1)}\|_2$ are sufficiently small.

We have implemented ADMM for (7.23) in MATLAB using LSQR to solve the least squares problems in the library function **admml1reg**.

Example 7.4

Returning to the source history reconstruction problem of Example 4.9, we test the application of TV as an alternative regularization technique. Fig. 7.16 shows the L-curve, which has a distinct corner at $\alpha = 5.6 \times 10^{-6}$, and Fig. 7.17 shows a corresponding suite of solutions for a range of α values, with the model corresponding to the L-curve corner shown in bold. The L-curve corner solution is compared to the true source history in Fig. 7.18. Because the source history is smooth and not discontinuous, the TV solution cannot fit the gentle rising and falling of the contaminant history, but it performs much better than the NNLS or BVLS methods (Chapter 4).

Next, consider the solution to this problem when the true source history does have sharp changes (Fig. 7.19), with three adjoining periods of constant contaminant concentration. The TV L-curve (Fig. 7.20) is quite similar to the previous L-curve, and again has a clear corner for $\alpha = 5.6 \times 10^{-6}$, and Fig. 7.21 shows the suite of solutions for the same range of α values as before. The TV solution for the L-curve corner (Fig. 7.22)

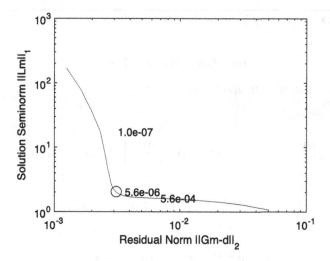

Figure 7.16 TV L-curve, with a corner at $\alpha = 5.6 \times 10^{-6}$.

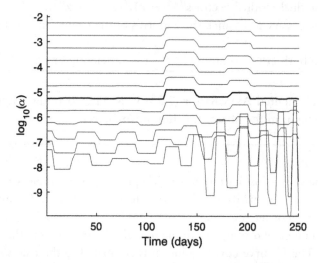

Figure 7.17 TV solutions for a range of α values, with the solution for the corner at $\alpha = 5.6 \times 10^{-6}$ (Fig. 7.16) shown in bold.

in this case fits the true model quite well, with only small spurious precursor features and a slight underestimate of contaminant concentration for the first two phases. The good fit of the TV solution can be attributed to the true model having sharp changes, so that the bias imposed by the TV solution is, in this case, highly appropriate.

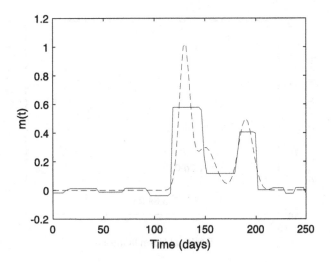

Figure 7.18 TV solution corresponding to the L-curve corner solution (Fig. 7.17) compared with the (smooth) true model (dashed curve).

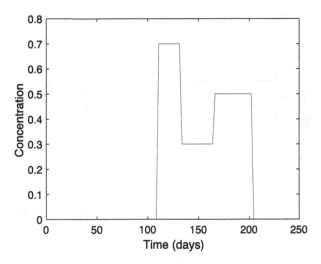

Figure 7.19 An alternative true source history for the source reconstruction problem with sharp features.

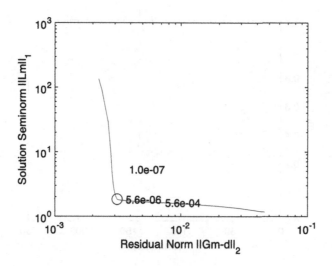

Figure 7.20 TV L-curve, with a corner at $\alpha = 5.6 \times 10^{-6}$.

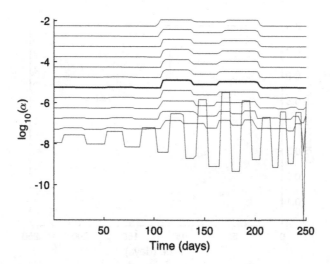

Figure 7.21 TV solutions for a range of α values, with the solution for the corner at $\alpha = 5.6 \times 10^{-6}$ (Fig. 7.20) shown in bold.

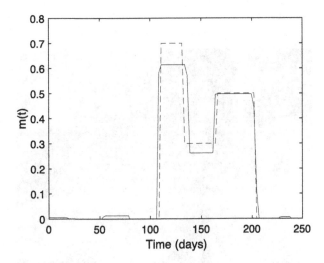

Figure 7.22 TV solution corresponding to the L-curve corner solution (Fig. 7.21) compared with the (sharp features) true model (dashed curve).

7.7. TOTAL VARIATION IMAGE DENOISING

Total variation regularization has seen wide use in the problem of "denoising" a model [153]. Denoising is a linear inverse problem in which $\mathbf{G} = \mathbf{I}$, where the general goal is to remove noise from a signal or image while still retaining the overall structure of the signal. In image processing, the quality of the resulting model is often then subjectively evaluated depending on specific motivations.

Example 7.5

Fig. 7.23 shows a 512 by 512-pixel black and white image in which the pixels have 8-bit brightness levels (from 0 to 255). In Fig. 7.24, $N(0, 50^2)$ noise has been added to the image. This produces general image degradation, including a speckled effect (similar to the noise that may be seen when a photograph is taken with an overly short exposure). Fig. 7.25 shows the image obtained using ADMM after applying total variation denoising with $\alpha = 50$, selected after examination of the broad L-curve between the 2-norm misfit $\|\mathbf{Gm} - \mathbf{d}\|_2 = \|\mathbf{m} - \mathbf{d}\|_2$, and the 1-norm regularization term $\|\mathbf{Lm}\|_1$. The noise is significantly reduced in areas of the original image that had nearly constant values, while the edges remain sharp. Fig. 7.26 shows the image obtained after applying total variation denoising with $\alpha = 300$, which is overly biased towards constant value regions.

Figure 7.23 Original image.

Figure 7.24 Image with $N(0, 50^2)$ noise added. Note in particular the speckled appearance of the white and light gray areas, which are nearly constant shade in the original image.

Figure 7.25 Image after TV denoising with $\alpha = 50$.

Figure 7.26 Image after TV denoising with $\alpha = 300$.

7.8. EXERCISES

1. Noisy seismic travel time data (units of s) from equally-spaced (every 20 m) seis-mometers deployed in a 1000-m deep vertical seismic profile experiment (the same geometry as in Example 4.4) are to be inverted to reveal the slowness of the Earth as a function of depth. Invert the data in **vsp.mat** using TV regularization with an L-curve analysis and assess the presence of any likely discontinuities. Compare this with the least squares solution and with a direct finite-difference solution suggested by (1.22).

2. Apply total variation to solve the (straight-ray) tomography problem described in Exercise 3.4. Construct an appropriate **L** matrix that implements (7.25), use ADMM, and explore a regularization parameter range of $\log_{10}\alpha = -8, -8.1, \ldots, -3$ to find a best L-curve determined solution (to obtain a smooth L-curve, you may have to tighten the default ADMM tolerances). Compare your total variation solution to the second-order Tikhonov regularized solution.

3. Apply total variation to solve the (straight-ray) cross-well tomography problem described in Exercise 4.3. Construct an appropriate **L** matrix using (7.25), use ADMM, and explore a regularization parameter range of $\log_{10}\alpha = -11.5, -10.5, \ldots, -1.5$ to find a best (L-curve determined) solution. Compare your total variation solution to the second-order Tikhonov regularized solution.

4. Returning to the problem in Example 5.2, solve for the density profile of the Earth using total variation regularization. How does your optimal solution compare to that obtained in Exercise 5.3?

5. The IRLS and ADMM methods can also be used to solve problems of the form of (7.2). Modify Examples 7.2 and 7.4 to use IRLS and ADMM. How do your results compare to the original results? Which methods are most efficient?

6. Total variation regularization can be used to repair images in which some pixels have been lost or overwritten. The file **inpaint.png** contains a black and white image in which portions of the image have been overwritten with "Test." The overwritten pixels have the value 255, while other pixels range in value from 0 to 254. Use total variation to restore the portions of the image that have been overwritten with text.

7. There are a number of free and open source codes available for solving (7.3) and (7.23) including GPSR [62], SPGL1 [214], and TFOCS [11]. Pick one or more of these packages and solve the examples from this chapter. How did these solutions compare with those obtained with ISTA, FISTA, ADMM, and IRLS? Which methods did you find to be fastest and most accurate, and what issues did you notice in using the various solvers?

7.9. NOTES AND FURTHER READING

There is a long history of sparsity regularization in geophysical inverse problems. Claerbout was an early advocate of this approach [42], however, there was little theoretical justification for this prior to the work of Candes, Romberg, and Tao, which brought about a resurgence of interest in this methodology [32–35]. In recent years there has been an explosion of interest in compressive (or compressed) sensing, with applications in many areas of signal and image processing and inverse problems. There are numerous surveys and tutorials on compressive sensing, including [27,36,132]. Methods for solution of the convex optimization problems arising in compressive sensing are discussed in [9,18,118]. Specialized methods for total variation regularization are discussed in [153, 172,216].

CHAPTER EIGHT

Fourier Techniques

Synopsis

The formulation of a general linear forward problem as a convolution is derived. The Fourier transform, Fourier basis functions, and the convolution theorem are introduced for continuous- and discrete-time systems. The inverse problem of deconvolution is explored in the context of the convolution theorem. Water level and Tikhonov regularization in the frequency domain are employed to solve discrete deconvolution problems.

8.1. LINEAR SYSTEMS IN THE TIME AND FREQUENCY DOMAINS

Many linear forward problems can be naturally described by a **convolution equation** (1.11),

$$d(t) = \int_{-\infty}^{\infty} m(\tau)g(t - \tau)\,d\tau \;. \tag{8.1}$$

Here, the independent variable t is time and the data d, model m, and kernel g are all time functions. However, the results here are equally applicable to spatial problems (e.g. Example 8.1) and to problems in higher dimensions. Inverse solutions to problems of this form can be obtained via the process of **deconvolution**.

Linear inverse problems that incorporate convolution and deconvolution operations can often be valuably analyzed in the context of Fourier theory. To see why this is the case, consider a linear forward operator G that maps an unknown model function $m(t)$ into a data function $d(t)$,

$$d(t) = G[m(t)] \;. \tag{8.2}$$

Because this mapping is linear, and we assume that the response of the system is **time-invariant**, so that $g(t)$ is functionally identical to $g(t - t_0)$, for any t_0, except for being shifted in time, (8.2) obeys the principles of superposition (1.5) and scaling (1.6).

We now show that any linear system can be cast in the form of (8.1), by utilizing the **sifting property** of the **impulse** or **delta function** $\delta(t)$. The delta function can be conceptualized as the limiting case of a pulse as its width goes to zero, its height goes to infinity, and its area stays constant and equal to one, for example

$$\delta(t) = \lim_{\sigma \to 0} \frac{1}{\sigma\sqrt{2\pi}} e^{-\frac{1}{2}t^2/\sigma^2} \;. \tag{8.3}$$

Parameter Estimation and Inverse Problems
https://doi.org/10.1016/B978-0-12-804651-7.00013-4
211

The sifting property of the delta function extracts the functional value at a particular point by an integral operation

$$\int_a^b f(t)\delta(t - t_0)\ dt = \begin{cases} f(t_0) & a \le t_0 \le b \\ 0 & \text{elsewhere} \end{cases} \tag{8.4}$$

for any $f(t)$ that is continuous at finite $t = t_0$.

The **impulse response**, or **Green's function**, $g(t)$, for a system where the model and data are related by an operator G, as in (8.2), is defined as the system output (data) for a delta function input (model), or

$$g(t) = G[\delta(t)]\ . \tag{8.5}$$

The concept of the impulse response can be used to demonstrate our assertion that linear time–invariant forward problems can generally be expressed as convolutions.

Theorem 8.1. We begin by using (8.4) to expand $m(t)$

$$m(t) = \int_{-\infty}^{\infty} m(\tau)\delta(t - \tau)\ d\tau\ . \tag{8.6}$$

From (8.2),

$$d(t) = G\left[\int_{-\infty}^{\infty} m(\tau)\delta(t - \tau)\ d\tau\right]\ . \tag{8.7}$$

Applying the definition of the integral as a limit of a sum of Δt-width rectangular areas as Δt goes to zero, we obtain the Riemann sum

$$d(t) = G\left[\lim_{\Delta\tau \to 0} \sum_{n=-\infty}^{\infty} m(\tau_n)\delta(t - \tau_n)\Delta\tau\right]\ . \tag{8.8}$$

Because G is a linear and time-invariant operator, superposition allows us to move the operator inside of the summation in (8.8) and apply it to each term. Furthermore we can factor out the coefficients $m(\tau_n)$ from inside of the operator to obtain

$$d(t) = \lim_{\Delta\tau \to 0} \sum_{n=-\infty}^{\infty} m(\tau_n)G[\delta(t - \tau_n)]\Delta\tau\ . \tag{8.9}$$

In the limit as $\Delta t \to 0$ and substituting (8.5), (8.9) thus defines the integral

$$d(t) = \int_{-\infty}^{\infty} m(\tau)g(t - \tau)\ d\tau \tag{8.10}$$

which is the convolution of $m(t)$ and $g(t)$ (8.1).

It is common to denote the convolution of two functions with the shorthand nomenclature $g_1(t) * g_2(t) = g_2(t) * g_1(t)$. The indicated reciprocity is easily demonstrated from (8.10) with a change of variables.

We define the **Fourier transform** of a function $g(t)$ as

$$\mathcal{G}(f) = \mathcal{F}[g(t)] \tag{8.11}$$

$$= \int_{-\infty}^{\infty} g(t)e^{-\imath 2\pi ft}\,dt \tag{8.12}$$

and the corresponding **inverse Fourier transform** as

$$g(t) = \mathcal{F}^{-1}[\mathcal{G}(f)] \tag{8.13}$$

$$= \int_{-\infty}^{\infty} \mathcal{G}(f)e^{\imath 2\pi ft}\,df\,, \tag{8.14}$$

where the frequency f has units of reciprocal time.

The Fourier transform or **spectrum** of the impulse response (8.5) when the independent variable is time is commonly referred to as the **frequency response**, or **transfer function**. Note, however, that the impulse response concept and Fourier theory are fully generalizable to functions of space or other variables.

The Fourier transform (8.12) provides a formula for evaluating the spectrum. The inverse Fourier transform (8.14) states that the time domain function $g(t)$ can be perfectly reconstructed via a weighted integration of functions of the form $e^{\imath 2\pi ft}$, where the weighting is provided by the spectrum $\mathcal{G}(f)$. A key implication expressed in (8.12) and (8.14) is that general functions $g(t)$ can be expressed and analyzed as a continuous weighted superposition of **Fourier basis functions** of the form $e^{\imath 2\pi ft}$. In the common case where t is time, operations or analysis involving $g(t)$ and t are referred to as occurring in the **time domain** whereas those involving $\mathcal{G}(f)$ and f are referred to as occurring in the **frequency domain**. Note that this representation differs from the finite basis function sets considered in Chapter 5 in that the sets of Fourier basis functions in (8.12) and (8.14) are infinite.

Because the Fourier basis functions, $e^{\imath 2\pi ft} = \cos(2\pi ft) + \imath \sin(2\pi ft)$, are complex, a real-valued function $g(t)$ will have a corresponding, generally complex-valued, spectrum, $\mathcal{G}(f)$. It is commonly convenient to express spectra in polar form

$$\mathcal{G}(f) = |\mathcal{G}(f)|e^{\imath\theta(f)} \tag{8.15}$$

where $|\mathcal{G}(f)|$ is the **spectral amplitude**, and the angle that $\mathcal{G}(f)$ makes in the complex plane with the real axis

$$\theta(f) = \tan^{-1}\left(\frac{\text{imag}(\mathcal{G}(f))}{\text{real}(\mathcal{G}(f))}\right) \tag{8.16}$$

is the **spectral phase**.

A useful feature of the Fourier transform is that it is **length preserving** for the 2-norm measure, in the sense that the 2-norms of a function and its Fourier transform are identical. The following theorem demonstrates this.

Theorem 8.2. Consider a time domain norm of the form of (5.3) for a general complex function $g(t)$

$$\|g(t)\|_2^2 = \int_{-\infty}^{\infty} g(t)g^*(t)\ dt\ . \tag{8.17}$$

Expressing $g(t)$ using the inverse Fourier transform (8.14) and applying complex conjugation (denoted by a superscript asterisk) gives

$$\|g(t)\|_2^2 = \int_{-\infty}^{\infty} g(t)\left(\int_{-\infty}^{\infty} \mathcal{G}^*(f)e^{-i2\pi ft}\ df\right)\ dt\ . \tag{8.18}$$

Interchanging the order of integration and utilizing (8.12), we have

$$\begin{aligned}\|g(t)\|_2^2 &= \int_{-\infty}^{\infty} \mathcal{G}^*(f)\left(\int_{-\infty}^{\infty} g(t)e^{-i2\pi ft}\ dt\right)\ df \\ &= \int_{-\infty}^{\infty} \mathcal{G}^*(f)\mathcal{G}(f)df \\ &= \|\mathcal{G}(f)\|_2^2\end{aligned} \tag{8.19}$$

where the right hand side is the 2-norm of the Fourier transform of $g(t)$.

The result (8.19) is most commonly referred to as **Parseval's theorem**. Parseval's theorem implies that the minimization of the 2-norm of a function can be equivalently performed in the time or frequency domains.

An especially important result is that convolution in the time domain is equivalent to multiplication in the frequency domain, as shown in the following theorem.

Theorem 8.3.

$$\mathcal{F}[m(t) * g(t)] = \int_{-\infty}^{\infty}\left(\int_{-\infty}^{\infty} m(\tau)g(t-\tau)\ d\tau\right)e^{-i2\pi ft}\ dt\ . \tag{8.20}$$

Reversing the order of integration gives

$$\mathcal{F}[m(t) * g(t)] = \int_{-\infty}^{\infty} m(\tau)\left(\int_{-\infty}^{\infty} g(t-\tau)e^{-i2\pi ft}dt\right)\ d\tau\ . \tag{8.21}$$

Introducing a change of variables, $\xi = t - \tau$, we obtain

$$\mathcal{F}[m(t) * g(t)] = \int_{-\infty}^{\infty} m(\tau) \left(\int_{-\infty}^{\infty} g(\xi) e^{-i2\pi f(\xi + \tau)} \, d\xi \right) \, d\tau$$

$$= \left(\int_{-\infty}^{\infty} m(\tau) e^{-i2\pi f\tau} \, d\tau \right) \left(\int_{-\infty}^{\infty} g(\xi) e^{-i2\pi f\xi} \, d\xi \right) \qquad (8.22)$$

$$= \mathcal{M}(f)\mathcal{G}(f) \, .$$

Eq. (8.22) is the **convolution theorem**, which shows that convolution in the time domain corresponds to the multiplication of the respective function Fourier transforms in the frequency domain. When the two functions are a model $m(t)$ and a forward problem impulse response $g(t)$, as in (8.1), (8.22) indicates that the corresponding spectrum of the data is the product of the model and impulse response spectra.

Consider the data produced by a linear time–invariant system, characterized by the spectrum of the impulse response $\mathcal{G}(f)$ for a model $m_0(t)$ that is a Fourier basis function of frequency f_0,

$$m_0(t) = e^{i2\pi f_0 t} \, . \qquad (8.23)$$

The spectrum of (8.23) is a delta function located at $f = f_0$, $\delta(f - f_0)$ [26]. This can be demonstrated by constructing an inverse Fourier transform (8.14) and invoking the sifting property of the delta function (8.4)

$$e^{i2\pi f_0} = \int_{-\infty}^{\infty} \mathcal{F}(e^{i2\pi f_0}) = \int_{-\infty}^{\infty} \delta(f - f_0) e^{i2\pi ft} \, df \, . \qquad (8.24)$$

The spectrum of the corresponding data is thus, using the convolution theorem (8.22), a delta function at $f = f_0$, scaled by the spectrum of $G(t)$

$$\mathcal{F}[G[e^{i2\pi f_0 t}]] = \mathcal{F}[\mathcal{G}(f)e^{i2\pi f_0 t}] = \delta(f - f_0)\mathcal{G}(f_0) \, . \qquad (8.25)$$

The corresponding time domain response is, by (8.14),

$$\int_{-\infty}^{\infty} \mathcal{G}(f_0)\delta(f - f_0)e^{i2\pi ft} \, df = \mathcal{G}(f_0)e^{i2\pi f_0 t}$$

$$= |\mathcal{G}(f_0)|e^{i\theta(f_0)} e^{i2\pi f_0 t} \qquad (8.26)$$

$$= |\mathcal{G}(f_0)|e^{i\theta(f_0)} m_0(t)$$

where $\theta(f_0)$ is the spectral phase (8.16) at $f = f_0$.

Linear time–invariant systems thus transform Fourier basis functions (8.23) to identical functions of the same frequency, f_0, but altered in amplitude and phase by the factor $|\mathcal{G}(f_0)|e^{i\theta(f_0)}$. For a linear system, the transformation for a general input function is just the superposition of all such components at all frequencies, where the appropriate functional weighting is $\mathcal{G}(f)$.

Of particular relevance for inverse methods is that model basis function amplitudes at frequencies that are weakly mapped to the data (i.e., frequencies for which $|\mathcal{G}(f)|$ is small) in a forward problem, and/or are obscured by noise, may be difficult or impossible to recover in an inverse problem.

The spectrum of the impulse response (transfer function) can be evaluated in a particularly useful analytical manner if the forward problem $d(t) = G[m(t)]$ can be expressed as a linear differential equation

$$a_r m^{(r)} + a_{r-1} m^{(r-1)} + \cdots + a_1 m^{(1)} + a_0 m = b_q d^{(q)} + b_{q-1} d^{(q-1)} + \cdots + b_1 d^{(1)} + b_0 d \quad (8.27)$$

where the superscripts denote the order of time differentiation of m and d, and the a_i and b_i are constant coefficients. Because each term in (8.27) is linear (there are no powers or other nonlinear functions of m, d, or their derivatives), and because differentiation is itself a linear operation, (8.27) expresses a linear time-invariant system obeying superposition and scaling.

If (8.27) operates on a model of the form $m(t) = e^{i2\pi ft}$, (8.26) indicates that the corresponding output will be $d(t) = \mathcal{G}(f)e^{i2\pi ft}$. Substituting this form of $m(t)$ and differentiating each term, each time derivative, $m^{(k)}$ and $d^{(k)}$, will generate corresponding multipliers $(2\pi if)^k$. Finally, dividing the resulting equation on both sides by $e^{i2\pi ft}$, and solving for $\mathcal{G}(f)$ gives the system transfer function

$$\mathcal{G}(f) = \frac{\mathcal{D}(f)}{\mathcal{M}(f)} \quad (8.28)$$

$$= \frac{\sum_{j=0}^{q} b_j (2\pi if)^j}{\sum_{k=0}^{r} a_k (2\pi if)^k}. \quad (8.29)$$

The transfer function can thus be expressed as a ratio of complex polynomials in f for any system having the form of (8.27). The $q+1$, generally complex, frequencies f_z, for which the numerator of (8.29) is equal to zero are referred to as **zeros** of the transfer function (8.29). The predicted data for the forward problem will thus be zero for inputs of the form $e^{i2\pi f_z t}$, regardless of their amplitude. Any real-valued frequency, f_z corresponding to the Fourier model basis function $e^{i2\pi f_z t}$ will thus lie in the model null space and be unrecoverable by any inverse methodology. The $r+1$, generally complex, frequencies f_p for which the denominator of (8.29) is equal to zero are called **poles** of the transfer function. The system will be unstable when excited by model basis functions $e^{i2\pi f_p t}$. A transfer function can be completely characterized by tabulating the pole and zero frequencies, along with a scalar gain factor. This practice is, for example, commonly employed in specifying instrument responses.

8.2. LINEAR SYSTEMS IN DISCRETE TIME

To implement Fourier methods numerically, a discrete Fourier theory and transforms with properties that are analogous to the continuous transforms (8.12) and (8.14) are required. This is achieved by the **discrete Fourier transform**, or **DFT**. In its most basic formulation, the DFT operates on a uniformly sampled (e.g., in space or time) sequence, stored as a vector, with n elements. The frequency, f_s, at which sampling occurs is called the **sampling rate**. The forward discrete Fourier transform of an n-point sequence, $m_j, j = 0, 1, \ldots, n-1$, is

$$\mathcal{M}_k = (\mathrm{DFT}[\mathbf{m}])_k \tag{8.30}$$

$$= \sum_{j=0}^{n-1} m_j e^{-\iota 2\pi jk/n} \tag{8.31}$$

and its inverse is

$$m_j = \left(\mathrm{DFT}^{-1}[\mathcal{M}]\right)_j \tag{8.32}$$

$$= \frac{1}{n} \sum_{k=0}^{n-1} \mathcal{M}_k e^{\iota 2\pi jk/n} . \tag{8.33}$$

Eq. (8.33) states that the sequence m_j can be expressed as a linear combination of the n basis functions $e^{\iota 2\pi jk/n}$, where the complex weights in the linear combination are the discrete spectral elements $\mathcal{M}_k$ (with an additional scaling factor of $1/n$ in (8.33)). The DFT basis functions are orthonormal on an n-point interval in that, for any integers l and k:

$$\frac{1}{n} \sum_{j=0}^{n-1} e^{\iota 2\pi jl} e^{-\iota 2\pi jk/n} = \begin{cases} 1 & l = k \\ 0 & l \neq k . \end{cases} \tag{8.34}$$

The DFT operations (8.31) and (8.33) are also widely referred to as the **FFT** and **IFFT** because a particularly efficient algorithm, the **Fast Fourier transform**, is nearly ubiquitously used for their evaluation. The transforms can be calculated in MATLAB using the **fft** and **ifft** commands (note that FFT vectors and arrays in MATLAB, as is standard, are indexed beginning with one instead of zero). The corresponding discrete formulation of Parseval's Theorem (8.19) is

$$\sum_{j=0}^{n-1} |m_j|^2 = \frac{1}{n} \sum_{k=0}^{n-1} |\mathcal{M}_k|^2 . \tag{8.35}$$

DFT spectra, $\mathcal{M}_k$, are generally complex, discrete, and periodic, where the period is n. There is an implicit assumption in DFT theory that the associated time domain sequence, m_j, is also periodic with period n. Because of these periodicities, DFT results

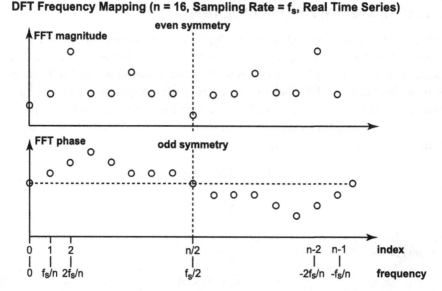

Figure 8.1 Frequency and index mapping for the DFT of a real-valued sequence ($n = 16$) sampled at f_s. For DFT theory to accurately represent Fourier operations on a continuous time signal, f_s must be greater than or equal to the Nyquist frequency (8.36).

can be stored in complex vectors of length n without loss of information, although (8.31) and (8.33) are valid for any integer index k. The DFT of a real-valued sequence has **Hermitian symmetry** about $k = 0$ and $k = n/2$, as defined by $\mathcal{M}_k = \mathcal{M}^*_{n-k}$, where the asterisk implies complex conjugation (see Exercise 8.2).

The mapping of the discrete spectrum index to specific frequencies is proportional to the sampling rate f_s. For n even, the positive frequencies, lf_s/n, where $l = 1, \ldots, n/2 - 1$, correspond to indices $k = 1, \ldots, n/2 - 1$, and the negative frequencies, $-lf_s/n$, correspond to indices $k = n/2 + 1, \ldots, n - 1$. The frequencies $\pm f_s/2$ have identical DFT values and correspond to index $k = n/2$. For n odd, there is no integer k corresponding to exactly half of the sampling rate. In this case positive frequencies correspond to indices 1 through $(n-1)/2$ and negative frequencies correspond to indices $(n+1)/2$ through $n - 1$. Fig. 8.1 displays the discrete spectrum index-frequency mapping with respect to k for an $n = 16$-length DFT.

The Hermitian symmetry of the DFT implies that, for a real-valued sequence, the spectral amplitude, $|\mathcal{M}|$, is symmetric and the spectral phase is antisymmetric with respect to $k = 0$ and $k = n/2$. See Fig. 8.1. Because of this symmetry it is customary to plot the spectral amplitude and phase for just positive frequencies in depicting the spectrum of a real signal.

For a uniformly sampled sequence to accurately represent a continuous function that contains nonnegligible spectral energy up to some maximum frequency f_{max}, the sampling rate, f_s, must be at least as large as the **Nyquist frequency**, f_N, so that

$$f_s \geq f_N = 2f_{max} \,. \tag{8.36}$$

Should (8.36) not be met, a nonlinear and generally undesirable and irreversible distortion called **aliasing** occurs. Generally speaking, aliasing causes spectral energy at frequencies $f > f_s/2$ to be "folded" and superimposed onto the DFT spectrum within the frequency range $-f_s/2 \leq f \leq f_s/2$.

Consider a model sequence m_j of length n and an impulse response sequence g_j of length p, where both sequences are synchronously and uniformly sampled at $f_s = 1/\Delta t$. The discrete convolution of the two sequences can be performed in two ways.

The first method is a **serial** convolution, which produces a sequence of length $n + p - 1$

$$d_j = \sum_{k=0}^{n-1} m_k g_{j-k} \, \Delta t \quad j = 0, \ 1, \ \dots, \ n+p-2 \,. \tag{8.37}$$

Serial convolution implements the discrete approximation to the convolution integral, and is implemented in MATLAB by the **conv** command.

The second type of discrete convolution is a **circular convolution**. Circular convolution is applicable to two sequences of equal length. If the lengths initially differ, they may be equalized by padding the shorter of the two sequences with zeros. The result of a circular convolution is as if each sequence is first expanded to be periodic (with period n), and then serially convolved solely on the index interval $j = 0, \ 1, \ \dots, \ n-1$. A circular convolution can be implemented using the discrete counterpart of the convolution theorem

$$\begin{aligned} \mathbf{d} &= \mathrm{DFT}^{-1}[\mathrm{DFT}[\mathbf{m}] \odot \mathrm{DFT}[\mathbf{g}]]\Delta t \\ &= \mathrm{DFT}^{-1}[\mathcal{M} \odot \mathcal{G}]\Delta t \end{aligned} \tag{8.38}$$

where $\mathcal{M} \odot \mathcal{G}$ indicates element-by-element multiplication of the vectors $\mathcal{M}$ and $\mathcal{G}$.

A very important practical consideration is that circular convolution implemented using the discrete convolution theorem (8.38) is much more efficient than serial convolution invoked with simple summation as in (8.37). To avoid **wrap-around** effects arising due to the implied n-length periodicity of $\mathbf{m}$ and $\mathbf{g}$ in the circular convolution, and thus obtain a result that is indistinguishable from the serial convolution (8.37), it may be necessary to pad one or both series with up to n zeros and apply (8.38) on the extended sequences. Because of the factoring strategy used in the FFT algorithm, it is also desirable for computational efficiency to pad $\mathbf{m}$ and $\mathbf{g}$ to lengths that are powers of two, or at least have many small prime factors.

Consider the case where we have a theoretically known, or accurately estimated, system impulse response, $g(t)$, convolved with an unknown model, $m(t)$. The continuous forward problem is

$$d(t) = \int_a^b g(t - \tau)m(\tau) \, d\tau \, . \tag{8.39}$$

Uniformly discretizing this expression using simple collocation with a sampling rate, $f_s = 1/\Delta t$, that is rapid enough to satisfy (8.36) and thus avoid aliasing, gives

$$\mathbf{d} = \mathbf{Gm} \tag{8.40}$$

where $\mathbf{d}$ and $\mathbf{m}$ are m and n length sequences, respectively, and $\mathbf{G}$ is a matrix with m rows of length n. Each row of $\mathbf{G}$ is a time-reversed and time-shifted representation of the impulse response, scaled by Δt to approximate the convolution integral.

$$G_{j,k} = g_{j-k}\Delta t \, . \tag{8.41}$$

This time domain representation of a forward problem convolution was previously examined in Example 3.2.

An inverse solution using Fourier methods can be obtained by first padding $\mathbf{d}$ and $\mathbf{g}$ appropriately with zeros so that they are of some equal and sufficient length n to obviate any wrap-around artifacts associated with circular convolution. $\mathbf{G}$ then becomes an n by n matrix. Applying the DFT and (8.22) allows us to cast the forward problem as a complex-valued linear system

$$\mathcal{D} = \tilde{\mathbf{G}}\mathcal{M} \, . \tag{8.42}$$

$\tilde{\mathbf{G}}$ in (8.42) is a complex-valued diagonal matrix with

$$\tilde{G}_{k,k} = \mathcal{G}_k \tag{8.43}$$

where $\mathcal{G}$ is the discrete Fourier transform of the sampled impulse response $\mathbf{g}$, $\mathcal{D}$ is the discrete Fourier transform of the data vector $\mathbf{d}$, and $\mathcal{M}$ is the discrete Fourier transform of the model vector $\mathbf{m}$. We can write (8.42) more simply using the element-by-element multiplication operator as

$$\mathcal{D} = \mathcal{G} \odot \mathcal{M} \, . \tag{8.44}$$

Eq. (8.42) suggests a natural deconvolution solution by **spectral division**, where we first solve for the Fourier transform of the model using the element-by-element division operation (or equivalently, by inverting the diagonal matrix $\tilde{\mathbf{G}}$ in (8.42)),

$$\mathcal{M} = \mathcal{D} \oslash \mathcal{G} \tag{8.45}$$

and then obtain the model $\mathbf{m} = DFT^{-1}[\mathcal{M}]$.

Eq. (8.45) is appealing in its simplicity and efficiency. The application of (8.22), combined with the efficient FFT implementation of the DFT, reduces the necessary computational effort from solving a potentially very large linear system of time domain equations (8.40) to just three n-length DFT operations (taking the DFT of the data and impulse response vectors, and then the inverse DFT of the element-by-element quotient). If $\mathbf{d}$ and $\mathbf{g}$ are real-valued, packing/unpacking algorithms also exist that allow DFT operations to be further reduced to complex vectors of length $n/2$.

However, (8.45) does not avoid the instability that is potentially associated with deconvolution if any elements in $\mathcal{G}$ are small (i.e., at frequencies that are at or close to the zeros of the transfer function). Application of (8.45) thus typically requires regularization.

8.3. WATER LEVEL REGULARIZATION

A straightforward method of regularizing spectral division is **water level regularization**. The water level strategy employs a modified impulse response spectrum $\mathcal{G}_w$ in (8.45), where

$$(\mathcal{G}_w)_i = \begin{cases} \mathcal{G}_i & (|\mathcal{G}_i| > w) \\ w(\mathcal{G}_i/|\mathcal{G}_i|) & (0 < |\mathcal{G}_i| \leq w) \\ w & (\mathcal{G}_i = 0) . \end{cases} \tag{8.46}$$

The water level regularized model estimate is then

$$\mathbf{m}_w = \mathrm{DFT}^{-1}\left[\mathcal{D} \oslash \mathcal{G}_w\right] \Delta t . \tag{8.47}$$

The colorful name for this technique arises from the construction of $\mathcal{G}_w$, applying the analogy of pouring water into the low amplitude "holes" of $\mathcal{G}$ until the spectral amplitude levels there reach w. The effect in (8.47) is to prevent undesirable noise amplification from occurring at frequencies where $\|\mathcal{G}\|$ is small.

An optimal water level value w will reduce the sensitivity to noise in the inverse solution while still recovering important model features. As is typical of the regularization process, it is possible to choose a "best" solution by assessing the tradeoff between the norm of the residuals and the model norm as the regularization parameter w is varied. In calculating a tradeoff curve, Parseval's theorem (8.35) usefully facilitates calculations of the model and residual norms from spectra without calculating inverse Fourier transforms. Note that the 2-norm of the water level-regularized solution $\mathbf{m}_w$ will be nonincreasing with increasing w because $|(\mathcal{G}_w)_i| \geq |\mathcal{G}_i|$.

Example 8.1

In Example 3.2, we investigated time domain deconvolution for uniformly sampled data with a sampling rate of $f_s = 2$ Hz using the TSVD. Here, we solve this problem using frequency domain deconvolution regularized via the water level technique. The impulse

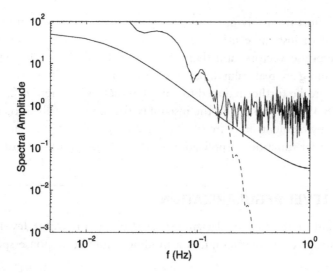

Figure 8.2 Amplitude spectra for the system impulse response (solid, smooth curve), noise-free data (dashed curve), and noisy data (solid, rough curve).

response, true model, and noisy data for this example are plotted in Figs. 3.9, 3.11, and 3.12, respectively. We first pad the 210-point data and impulse response vectors with 210 additional zeros to eliminate wrap-around artifacts, and apply the Fast Fourier transform to both vectors to obtain corresponding discrete spectra. The spectral amplitudes of the impulse response, data, and noise are critical in assessing the stability of the spectral division solution. See Fig. 8.2. The frequencies range from 0 to $f_s/2 = 1$ Hz. Because spectral amplitudes for real-valued sequences are symmetric about $k = 0$ and $k = n/2$ (Fig. 8.1), only positive frequencies are shown.

Examining the impulse response spectral amplitude $|\mathcal{G}_k|$ in Fig. 8.2, we note that it decreases by approximately three orders of magnitude between very low frequencies and half of the sampling frequency ($f_s/2 = 1$ Hz). The convolution theorem (8.22) shows that the forward problem convolution multiplies the spectrum of the model by $\mathcal{G}(f)$ in mapping it to the data. Thus, the convolution of a general signal with broad frequency content with this impulse response will strongly attenuate higher frequencies. Fig. 8.2 also shows that the spectral amplitudes of the noise-free data fall off more quickly than the impulse response. This indicates that spectral division will be a stable process for noise-free data in this problem. Fig. 8.2 also shows that the spectral amplitudes of the noisy data dominate the signal at frequencies higher than $f \approx 0.1$ Hz. Because of the small values of $\mathcal{G}_k$ at these frequencies, the spectral division solution using the noisy data will be dominated by noise (as was the case in the time domain solution of Example 3.2;

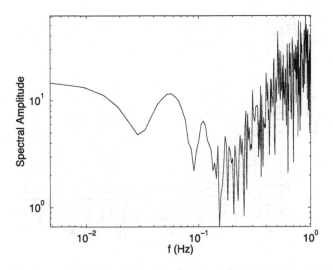

Figure 8.3 Spectral amplitudes resulting from the Fourier transform of the noisy data divided by the Fourier transform of the impulse response (the transfer function).

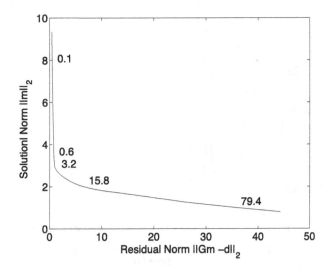

Figure 8.4 L-curve for a logarithmically distributed range of water level values, as indicated.

see Fig. 3.14). Fig. 8.3 shows the amplitude spectrum of the noisy data (Fig. 3.14) divided by the spectrum of the impulse response. The resulting model spectrum is dominated by noise at frequencies above about 0.1 Hz.

To regularize the spectral division solution, an optimal water level, w, is sought, where w should be large enough to avoid the undesirable amplification of noise. Figs. 8.2

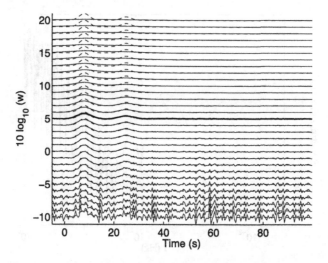

Figure 8.5 Models corresponding to the range of water level values used to construct Fig. 8.4. Dashed curves show the true model and bold trace shows the L-curve determined model with $w = 3.16$ (Fig. 8.4).

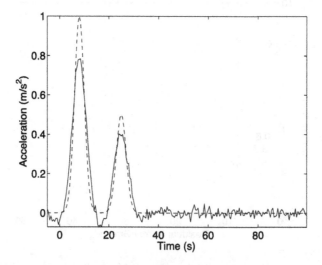

Figure 8.6 Water level regularized model corresponding to $w = 3.16$. Dashed curve shows the true model.

and 8.3 suggest that the optimal value of w is near the crossover spectral amplitude where the data spectrum is surpassed by noise, or near w somewhat greater than 1. However, such a determination might be more difficult for data with a more complex spectrum, and/or where the distinction between signal and noise spectra is generally less clear. Fig. 8.4 shows an L-curve for this example, which suggests an optimal w close to 3. Fig. 8.5 shows a corresponding range of solutions, and Fig. 8.6 shows the solution for $w = 3.16$.

The solution shown in Fig. 8.6, chosen from the corner of the tradeoff curve of Fig. 8.4, shows features of limited resolution that are typical of regularized solutions. In this case, imperfect resolution induced by regularization is manifested by reduced amplitude, oscillatory side lobes, and model broadening relative to the true model.

8.4. TIKHONOV REGULARIZATION IN THE FREQUENCY DOMAIN

An alternative regularized approach to solving the frequency domain forward problem (8.42) is to consider the system as a least squares minimization problem, as we did in Chapter 4 (e.g., (4.4))

$$\min \|\mathbf{G}\mathcal{M} - \mathcal{D}\|_2^2 + \alpha^2 \|\mathcal{M}\|_2^2 . \tag{8.48}$$

Expressing (8.48) as an augmented forward problem gives

$$\min \left\| \begin{bmatrix} \mathbf{G} \\ \alpha\mathbf{I} \end{bmatrix} \mathcal{M} - \begin{bmatrix} \mathcal{D} \\ \mathbf{0} \end{bmatrix} \right\|_2^2 . \tag{8.49}$$

We solve (8.49) by applying the normal equations using the conjugate transpose of the diagonal matrix $\mathbf{G}$, $\mathbf{G}^H$, which gives

$$\begin{bmatrix} \mathbf{G}^H & \alpha\mathbf{I} \end{bmatrix} \begin{bmatrix} \mathbf{G} \\ \alpha\mathbf{I} \end{bmatrix} \mathcal{M} = \begin{bmatrix} \mathbf{G}^H & \alpha\mathbf{I} \end{bmatrix} \begin{bmatrix} \mathcal{D} \\ \mathbf{0} \end{bmatrix} . \tag{8.50}$$

Eq. (8.50) corresponds to the zeroth-order Tikhonov-regularized formulation

$$(\mathbf{G}^H\mathbf{G} + \alpha^2\mathbf{I})\mathcal{M} = \mathbf{G}^H\mathcal{D} \tag{8.51}$$

with the frequency domain solution

$$\mathcal{M}_\alpha = (\mathbf{G}^H\mathbf{G} + \alpha^2\mathbf{I})^{-1}\mathbf{G}^H\mathcal{D} . \tag{8.52}$$

Because $(\mathbf{G}^H\mathbf{G} + \alpha^2\mathbf{I})$ is diagonal, (8.52) can be expressed solely, and efficiently evaluated, using n-length vector element-by-element operations as

$$\mathcal{M}_\alpha = (\mathcal{G}^H \odot \mathcal{D}) \oslash \mathrm{diag}(\mathbf{G}^H\mathbf{G} + \alpha^2\mathbf{I}) . \tag{8.53}$$

Applying the IDFT to $\mathcal{M}_\alpha$ then produces the corresponding regularized time domain solution $\mathbf{m}$.

Although we have implemented $\|\mathcal{M}\|_2 = \|\mathrm{DFT}[\mathbf{m}]\|_2$ rather than $\|\mathbf{m}\|_2$ to regularize the problem, by Parseval's theorem (8.35), the norm of the model spectrum is proportional to the norm of the time domain model. Thus, (8.53) is equivalent to that obtained if we were to employ zeroth-order Tikhonov regularization using the norm of the model in the time domain (although the specific values of α will be different).

An important consideration is that the inverse operation (8.53) is applied by element-by-element spectral division of n-length vectors, so the solution can be calculated extremely efficiently relative to inverting a nondiagonal n by n matrix. Even with the additional calculations associated with the FFT, this methodology is typically much more efficient than a time domain approach, such as is described in Example 3.2.

To implement higher–order Tikhonov regularization, we note that

$$
\begin{aligned}
\frac{d}{dt} m(t) &= \frac{d}{dt} \int_{-\infty}^{\infty} \mathcal{M}(f) e^{\imath 2\pi f t} \, df \\
&= \int_{-\infty}^{\infty} \frac{\partial}{\partial t} [\mathcal{M}(f) e^{\imath 2\pi f t}] \, df \\
&= \int_{-\infty}^{\infty} 2\pi \imath f \mathcal{M}(f) e^{\imath 2\pi f t} \, df \\
&= \mathcal{F}^{-1}[2\pi \imath f \mathcal{M}(f)] \, .
\end{aligned}
\tag{8.54}
$$

Taking the Fourier transform of both sides gives

$$
\mathcal{F}\left[\frac{d}{dt} m(t)\right] = 2\pi \imath f \mathcal{M}(f) \, .
\tag{8.55}
$$

Eq. (8.55) enables us to effect a first derivative seminorm by multiplying each element of $\mathcal{M}$ by $2\pi f_j$, choosing the f_j to be proportional to the spectral frequency of the jth element. Proportionality is sufficient, because the exact constant frequency scaling factor for converting the DFT index to frequency can be absorbed into the regularization parameter α. Thus, we can effect pth-order Tikhonov regularization by solving

$$
\min \left\| \begin{bmatrix} \mathbf{G} \\ \alpha \mathbf{K}^p \end{bmatrix} \mathcal{M} - \begin{bmatrix} \mathcal{D} \\ \mathbf{0} \end{bmatrix} \right\|_2^2
\tag{8.56}
$$

where $\mathbf{K}$ is an n by n diagonal matrix with diagonal elements (e.g., for n even)

$$
K_{j,j} = \begin{cases} (j-1)/n & j = 1, 2, \ldots, n/2 \\ (j-1)/n - 1 & j = n/2 + 1, n/2 + 2, \ldots, n \end{cases}
\tag{8.57}
$$

that are proportional to the frequency represented by the jth element of $\mathcal{M}$. The least squares solution, obtained utilizing the normal equations, is

$$
\mathcal{M}_\alpha = (\mathbf{G}^H \mathbf{G} + \alpha^2 \mathbf{K}^{2p})^{-1} \mathbf{G}^H \mathcal{D}
\tag{8.58}
$$

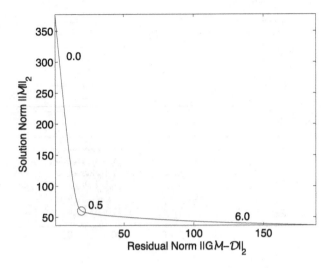

Figure 8.7 L-curve for a logarithmically distributed range of regularization parameter values, zeroth-order Tikhonov regularization.

or

$$\mathcal{M}_\alpha = (\mathcal{G}^H \odot \mathcal{D}) \oslash \text{diag}(\mathbf{G}^H \mathbf{G} + \alpha^2 \mathbf{K}^{2p}) \, . \tag{8.59}$$

As with the zeroth-order solution (8.53), (8.59) can be evaluated very efficiently with element-by-element n-length vector operations. Note that the implementation of the (non-identity) matrix $\mathbf{K}$ in (8.58) weights the higher frequency components of the model more in the regularization, and thus tends to make them small. It is thus straightforward to consider and implement more general regularization matrices that will penalize particular Fourier components (e.g., within a particular frequency range).

Example 8.2

Let us reconsider Example 8.1 in a Tikhonov regularization framework.

We first implement zeroth-order regularization using (8.52) and examine the trade-off curve (Fig. 8.7). The suite of solutions is shown in Fig. 8.8, and a solution selected from the tradeoff curve is shown in Fig. 8.9 compared with the true model. Note that, compared to the water level solution shown in Fig. 8.6, the Tikhonov regularized solution has better amplitude recovery and is somewhat smoother. On the other hand, the recovered zeroth-order Tikhonov regularized model shows higher amplitude structure later in the time series. Applying second-order Tikhonov regularization, we obtain a corresponding tradeoff curve (Fig. 8.10) and recover a still smoother model that is generally closer to the true model than either water level or zeroth-order Tikhonov

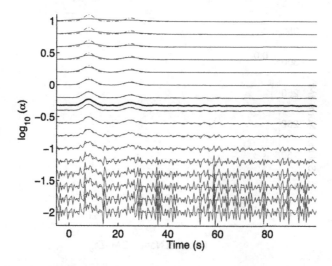

Figure 8.8 Zeroth-order Tikhonov regularized models corresponding to the range of regularization parameters used to construct Fig. 8.7. Dashed curves show the true model and bold trace shows the L-curve determined model with $\alpha = 0.48$.

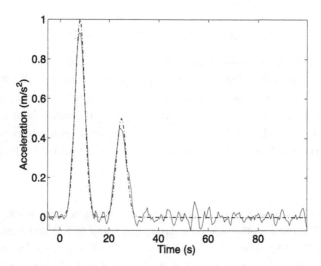

Figure 8.9 Zeroth-order Tikhonov regularized model corresponding to $\alpha = 0.48$. Dashed curve shows the true model.

regularization (Figs. 8.11 and 8.12). The final accuracy of any regularized solution will, of course, depend on properties of the true solution. In this particular case the true model is smooth, and the second-order regularization solution is closest.

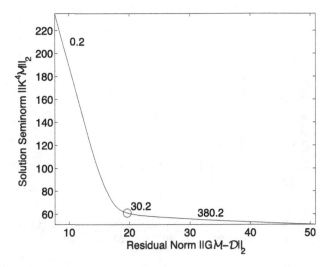

Figure 8.10 L-curve for a logarithmically distributed range of regularization parameter values, second-order Tikhonov regularization.

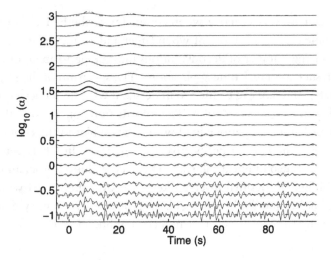

Figure 8.11 Second-order Tikhonov regularized models corresponding to the range of regularization parameters used to construct Fig. 8.10. Dashed curves show the true model and bold trace shows the L-curve determined model with $\alpha = 30.2$.

A significant new idea introduced by the Fourier methodology is that it provides a set of orthonormal model and data basis functions of the form of (8.23), the complex exponentials, that have the property of passing through a linear system altered in phase and amplitude, but not in frequency or functional character (8.26). This remarkable fact is the essence of the convolution theorem (8.22), which leads to frequency domain inverse

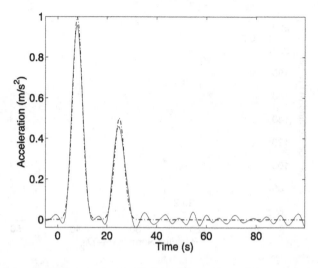

Figure 8.12 Second-order Tikhonov regularized model corresponding to $\alpha = 30.2$. Dashed curve shows the true model.

methodologies that are very efficient when coupled with the FFT algorithm. This efficiency can become critically important when larger and/or higher-dimensional models are of interest, a large number of deconvolutions must be performed, or computational speed is critical, such as in real-time applications. The spectrum of the impulse response (such as in Figs. 8.2 and 8.3) can be used to understand what frequency components may exhibit instability in an inverse solution. The information contained in the spectrum of Fig. 8.2 is thus analogous to that obtained with a Picard plot in the context of the SVD (Chapters 3 and 4). The Fourier perspective also provides a link between linear inverse theory and the (vast) field of linear filtering. The deconvolution problem in this context is identical to finding an optimal inverse filter to recover the model while suppressing the influence of noise, and Tikhonov regularization in the frequency domain applies a preferential filtering to the solution that reduces amplitudes of high-frequency Fourier components.

8.5. EXERCISES

1. Given that the Fourier transform of a real-valued linear system $g(t)$

$$\mathcal{F}[g(t)] = \mathcal{G}(f) = \text{real}(\mathcal{G}(f)) + \text{imag}(\mathcal{G}(f)) = \alpha(f) + \imath\beta(f) \qquad (8.60)$$

is Hermitian

$$\mathcal{G}(f) = \mathcal{G}^*(-f) \qquad (8.61)$$

show that convolving $g(t)$ with $\sin(2\pi f_0 t)$ and $\cos(2\pi f_0 t)$ produces the scaled and phase-shifted sinusoids

$$g(t) * \sin(2\pi f_0 t) = |\mathcal{G}(f_0)| \sin(2\pi f_0 t + \theta(f_0)) \tag{8.62}$$

$$g(t) * \cos(2\pi f_0 t) = |\mathcal{G}(f_0)| \cos(2\pi f_0 t + \theta(f_0)) \tag{8.63}$$

where the scale factor is the spectral amplitude

$$|\mathcal{G}(f_0)| = (\alpha^2(f_0) + \beta^2(f_0))^{\frac{1}{2}} \tag{8.64}$$

and the phase-shift factor is the spectral phase

$$\theta(f_0) = \tan^{-1}\left(\frac{\beta(f_0)}{\alpha(f_0)}\right) . \tag{8.65}$$

2. **a.** Demonstrate using (8.31) that the DFT of an n-point, real-valued sequence, **x**, is Hermitian, i.e.,

$$\mathcal{X}_{n-k} = \mathcal{X}_k^* . \tag{8.66}$$

  b. Demonstrate that the Hermitian symmetry shown in part (a) implies that the n independent elements in a real-valued time series **x** produce $n/2 + 1$ independent elements (n even) or $(n-1)/2 + 1$ independent elements (n odd) in the complex-valued DFT $\mathcal{X}$. As the DFT has an inverse (8.33) that reproduces **x** from $\mathcal{X}$, clearly information has not been lost in transforming to the DFT, yet the number of independent elements in **x** and $\mathcal{X}$ seems to differ. Reconcile this.

3. A linear damped vertical harmonic oscillator consisting of a mass suspended on a lossy spring is affixed to the surface of a terrestrial planet to function as a seismometer, where the recorded displacement $z(t)$ of the mass relative to its equilibrium position will depend on ground motion (note that when the surface of the planet moves upward, the inertia of the mass will tend to make it remain at rest, and the corresponding motion of the mass relative to its suspension system will be downwards). For an upward ground displacement $u(t)$ the system can be mathematically modeled as the linear differential equation

$$\frac{d^2 z}{dt^2} + \frac{D}{M}\frac{dz}{dt} + \frac{K}{M}z = \frac{d^2 u}{dt^2} \tag{8.67}$$

where the physical properties of the oscillator are defined by the mass M, the spring constant K, and the damping constant D.

  a. By taking the Fourier transform of (8.67), obtain the transfer function $\mathcal{G}(f) = \mathcal{Z}(f)/\mathcal{U}(f)$, where $\mathcal{Z}(f)$ and $\mathcal{U}(f)$ are the Fourier transforms of $z(t)$ and

$u(t)$, respectively. Express the transfer function in terms of M, K, D, and the undamped ($D = 0$) resonant frequency, $f_s = 1/(2\pi)\sqrt{K/M}$.

   b. For what general frequency range will the output of this instrument accurately reflect true ground motion in the absence of noise?

   c. For what general frequency range will the response of this instrument be increasingly difficult to deconvolve to recover true ground motion when noise is present?

4. A displacement seismogram is observed from a large earthquake at a far-field seismic station, from which the source region can be approximated as a point. A much smaller aftershock from the main shock region is used as an empirical Green's function for this event. It is supposed that the observed signal from the large event should be approximately equal to the convolution of the main shock's rupture history with this empirical Green's function. The 256-point seismogram is in the file **seis.mat**. The impulse response of the seismometer is in the file **impresp.mat**.

   a. Deconvolve the impulse response from the observed main shock seismogram using frequency domain Tikhonov zeroth-order deconvolution to solve for the source time function of the large earthquake. Note that the source time function is expected to consist of a nonnegative pulse or set of pulses. Estimate the source duration in samples and assess any evidence for subevents and their relative durations and amplitudes. Approximately what value of α do you believe is best for this data set? Why?

   b. Perform second-order ($p = 2$) frequency domain Tikhonov deconvolution to solve this problem.

   c. Recast the problem as a discrete linear inverse problem, as described in the example for Chapter 3, and solve the system using second-order Tikhonov regularization.

   d. Are the results in (c) better or worse than in (a) or (b)? How and why? Compare the amount of time necessary to find the solution in each case on your computing platform.

5. Many regularization matrices (e.g., the roughening matrices **L** for effecting one-dimensional Tikhonov regularization in (4.27) and (4.28)) are readily recognized as convolutions, where the rows are time-reversed sampled representations of the impulse response of the convolution effected by the regularization term **Lm**. Fourier analysis of the frequency response of regularization matrix convolutions can thus facilitate more sophisticated regularization matrix design in terms of penalizing the frequency content of the model.

Assuming a unit sampling rate, use the discrete Fourier transform to:

   a. Evaluate the frequency response of the first difference operator $[\ldots, 0, -1, 1, 0, \ldots]$ used in first-order Tikhonov regularization. How does this frequency response on the unit sampling rate Nyquist interval $-1/2 \le f \le 1/2$ differ from that of a perfect first derivative operation, which has the response $\Phi(f) = \imath 2\pi f$?

    b. Evaluate the frequency response of the second difference operator $[\ldots, 0, -1, 2, -1, 0, \ldots]$ used in second-order Tikhonov regularization. How does this frequency response on the unit sampling rate Nyquist interval $-1/2 \le f \le 1/2$ differ from that of a perfect second derivative operation, which has the response $\Phi(f) = -4\pi^2 f^2$?

8.6. NOTES AND FURTHER READING

Although we examine one-dimensional convolution and deconvolution problems here for conceptual simplicity, these methods are readily generalizable to higher dimensions, and higher-dimensional formulations of the DFT are widely utilized. MATLAB has a general n-dimensional set of FFT algorithms, **fftn** and **ifftn**.

In some physics and geophysics treatments the sign convention chosen for the complex exponentials in the Fourier transform and its inverse may be reversed, so that the forward transform (8.12) has a plus sign in the exponent and the inverse transform (8.14) has a minus sign in the exponent. This alternative sign convention merely induces a complex conjugation in the spectrum that is reversed when the corresponding inverse transform is applied. An additional convention issue arises as to whether to express frequency in Hz (f) or radians per second ($\omega = 2\pi f$). Alternative Fourier transform formulations using ω differ from (8.12) and (8.14) by a simple change of variables, and introduce scaling factors of 2π in the forward, reverse, or both transforms.

Gubbins [78] also explores connections between Fourier and inverse theory in a geophysical context. Kak and Slaney [106] give an extensive treatment of Fourier-based methods for tomographic imaging. Vogel [216] discusses Fourier methods for image deblurring. Because of the tremendous utility of Fourier techniques, there are numerous resources on their use in the physical sciences, engineering, and pure mathematics. A basic text covering theory and some applications at the approximate level of this text is [26], and a recommended advanced text on the topic is [168].

CHAPTER NINE

Nonlinear Regression

Synopsis

Common approaches to solving nonlinear regression problems are introduced, extending the development of linear regression in Chapter 2. We begin with a discussion of Newton's method, which provides a general framework for solving nonlinear systems of equations and nonlinear optimization problems. Then we discuss the Gauss–Newton (GN) and Levenberg–Marquardt (LM) methods, which are versions of Newton's method specialized for nonlinear regression problems. The distinction between LM and Tikhonov regularization is also made. Statistical aspects and implementation issues are addressed, and examples of nonlinear regression are presented.

9.1. INTRODUCTION TO NONLINEAR REGRESSION

In previous chapters we have concentrated on linear forward and inverse problems, and have seen that such problems are uniformly approachable using a variety of solution methods. We will next consider problems that are nonlinear, i.e., forward and inverse problems that do not obey the rules of superposition (1.5) and scaling (1.6). There is no general theory for the solution of nonlinear parameter estimation and inverse problems. However, we will see that iterative strategies incorporating linear concepts can be applied to usefully solve them in many circumstances.

9.2. NEWTON'S METHOD FOR SOLVING NONLINEAR EQUATIONS

Consider a nonlinear system of m equations in m unknowns

$$\mathbf{F}(\mathbf{x}) = \mathbf{0} . \tag{9.1}$$

We will construct a sequence of vectors, $\mathbf{x}^{(0)}$, $\mathbf{x}^{(1)}$, ..., that will converge to a solution $\mathbf{x}^*$. If the nonlinear vector function $\mathbf{F}$ is continuously differentiable, we can construct a Taylor series approximation about some starting solution estimate $\mathbf{x}^{(0)}$

$$\mathbf{F}(\mathbf{x}^{(0)} + \Delta\mathbf{x}) \approx \mathbf{F}(\mathbf{x}^{(0)}) + \mathbf{J}(\mathbf{x}^{(0)})\Delta\mathbf{x} \tag{9.2}$$

where $\mathbf{J}(\mathbf{x}^{(0)})$ is the Jacobian

$$\mathbf{J}(\mathbf{x}^{(0)}) = \left[\begin{array}{ccc} \frac{\partial F_1(\mathbf{x})}{\partial x_1} & \cdots & \frac{\partial F_1(\mathbf{x})}{\partial x_m} \\ \vdots & \ddots & \vdots \\ \frac{\partial F_m(\mathbf{x})}{\partial x_1} & \cdots & \frac{\partial F_m(\mathbf{x})}{\partial x_m} \end{array} \right]\Bigg|_{\mathbf{x}=\mathbf{x}^{(0)}} . \tag{9.3}$$

Parameter Estimation and Inverse Problems
https://doi.org/10.1016/B978-0-12-804651-7.00014-6

235

Using (9.2) and expressing the difference between the desired solution $\mathbf{x}^*$ and $\mathbf{x}^{(0)}$ as

$$\Delta \mathbf{x} = \mathbf{x}^* - \mathbf{x}^{(0)} \tag{9.4}$$

gives

$$\mathbf{F}(\mathbf{x}^*) = \mathbf{0} \approx \mathbf{F}(\mathbf{x}^{(0)}) + \mathbf{J}(\mathbf{x}^{(0)}) \Delta \mathbf{x} \tag{9.5}$$

which produces a linear system of equations

$$\mathbf{J}(\mathbf{x}^{(0)}) \Delta \mathbf{x} \approx -\mathbf{F}(\mathbf{x}^{(0)}) \tag{9.6}$$

that can be solved for $\Delta \mathbf{x}$ to improve the solution estimate.

Algorithm 9.1 Newton's Method

Given a system of equations $\mathbf{F}(\mathbf{x}) = \mathbf{0}$ and an initial solution $\mathbf{x}^{(0)}$, repeat the following steps to compute a sequence of solutions $\mathbf{x}^{(k)}$. Stop if and when the sequence adequately converges to a solution with $\mathbf{F}(\mathbf{x}) \approx \mathbf{0}$.
1. Calculate the Jacobian $\mathbf{J}(\mathbf{x}^{(k)})$ and $\mathbf{F}(\mathbf{x}^{(k)})$.
2. Solve $\mathbf{J}(\mathbf{x}^{(k)}) \Delta \mathbf{x} = -\mathbf{F}(\mathbf{x}^{(k)})$.
3. Let $\mathbf{x}^{(k+1)} = \mathbf{x}^{(k)} + \Delta \mathbf{x}$.
4. Let $k = k + 1$.

The theoretical properties of Newton's method are summarized in the following theorem. For a proof, see [51].

Theorem 9.1. If $\mathbf{x}^{(0)}$ is close enough to $\mathbf{x}^*$, $\mathbf{F}(\mathbf{x})$ is continuously differentiable in a neighborhood of $\mathbf{x}^*$, and $\mathbf{J}(\mathbf{x}^*)$ is nonsingular, then Newton's method will converge to $\mathbf{x}^*$. The convergence rate is quadratic in the sense that there is a positive constant c such that, for large k, the error at iteration $k + 1$ is less than or equal to c times the square of the error at iteration k, i.e.,

$$\|\mathbf{x}^{(k+1)} - \mathbf{x}^*\|_2 \leq c \|\mathbf{x}^{(k)} - \mathbf{x}^*\|_2^2 . \tag{9.7}$$

In practical terms, quadratic convergence means that as we approach $\mathbf{x}^*$, the number of accurate digits in the solution doubles at each iteration. Unfortunately, if the hypotheses in the above theorem are not satisfied, then Newton's method can converge very slowly or even fail altogether.

A simple modification to Newton's method often helps with convergence problems. In the **damped Newton's method**, we use the Newton's method equations at each iteration to compute an iterative correction to the working solution. However, instead of taking the full step $\mathbf{x}^{(k+1)} = \mathbf{x}^{(k)} + \Delta \mathbf{x}$, we perform a **line search** across solutions lying along the line $\mathbf{x}^{(k+1)} = \mathbf{x}^{(k)} + \alpha \Delta \mathbf{x}$ for a range of positive α values and take the one with minimum $\|\mathbf{F}(\mathbf{x}^{(k+1)})\|_2$.

Now suppose that we wish to minimize a scalar-valued function $f(\mathbf{x})$. If we assume that $f(\mathbf{x})$ is twice continuously differentiable, we can construct a Taylor series approximation

$$f(\mathbf{x}^{(0)} + \Delta \mathbf{x}) \approx f(\mathbf{x}^{(0)}) + \nabla f(\mathbf{x}^{(0)})^T \Delta \mathbf{x} + \frac{1}{2} \Delta \mathbf{x}^T \mathbf{H}(f(\mathbf{x}^{(0)})) \Delta \mathbf{x} \qquad (9.8)$$

where $\nabla f(\mathbf{x}^{(0)})$ is the gradient

$$\nabla f(\mathbf{x}^{(0)}) = \left[\begin{array}{c} \frac{\partial f(\mathbf{x})}{\partial x_1} \\ \vdots \\ \frac{\partial f(\mathbf{x})}{\partial x_m} \end{array} \right]_{\mathbf{x}=\mathbf{x}^{(0)}} \qquad (9.9)$$

and $\mathbf{H}(f(\mathbf{x}^{(0)}))$ is the Hessian

$$\mathbf{H}(f(\mathbf{x}^{(0)})) = \left[\begin{array}{ccc} \frac{\partial^2 f(\mathbf{x})}{\partial x_1^2} & \cdots & \frac{\partial^2 f(\mathbf{x})}{\partial x_1 \partial x_m} \\ \vdots & \ddots & \vdots \\ \frac{\partial^2 f(\mathbf{x})}{\partial x_m \partial x_1} & \cdots & \frac{\partial^2 f(\mathbf{x})}{\partial x_m^2} \end{array} \right]_{\mathbf{x}=\mathbf{x}^{(0)}} . \qquad (9.10)$$

A necessary condition for $\mathbf{x}^*$ to be a minimum of $f(\mathbf{x})$ is that $\nabla f(\mathbf{x}^*) = \mathbf{0}$. We can approximate the gradient in the vicinity of $\mathbf{x}^{(0)}$ by

$$\nabla f(\mathbf{x}^{(0)} + \Delta \mathbf{x}) \approx \nabla f(\mathbf{x}^{(0)}) + \mathbf{H}(f(\mathbf{x}^{(0)})) \Delta \mathbf{x} . \qquad (9.11)$$

Setting the approximate gradient (9.11) equal to zero gives

$$\mathbf{H}(f(\mathbf{x}^{(0)})) \Delta \mathbf{x} = -\nabla f(\mathbf{x}^{(0)}) . \qquad (9.12)$$

Solving the linear system of equations (9.12) for successive solution steps leads to **Newton's method for minimizing** $f(\mathbf{x})$.

Algorithm 9.2 Newton's Method for Minimizing $f(\mathbf{x})$

Given a twice continuously differentiable function $f(\mathbf{x})$, and an initial solution $\mathbf{x}^{(0)}$, repeat the following steps to compute a sequence of solutions $\mathbf{x}^{(k)}$. Stop if and when the sequence adequately converges to a solution with $\nabla f(\mathbf{x}^*) \approx \mathbf{0}$.

1. Calculate the gradient $\nabla f(\mathbf{x}^{(k)})$ and Hessian $\mathbf{H}(f(\mathbf{x}^{(k)}))$.
2. Solve $\mathbf{H}(f(\mathbf{x}^{(k)})) \Delta \mathbf{x} = -\nabla f(\mathbf{x}^{(k)})$.
3. Let $\mathbf{x}^{(k+1)} = \mathbf{x}^{(k)} + \Delta \mathbf{x}$.
4. Let $k = k+1$.

Since Newton's method for minimizing $f(\mathbf{x})$ is exactly Newton's method for solving a nonlinear system of equations applied to $\mathbf{F} = \nabla f(\mathbf{x}) = \mathbf{0}$, the convergence proof follows immediately from the proof of Theorem 9.1.

Theorem 9.2. If $f(\mathbf{x})$ is twice continuously differentiable in a neighborhood of a local minimizer $\mathbf{x}^*$, there is a positive constant λ such that $\|\mathbf{H}(f(\mathbf{x})) - \mathbf{H}(f(\mathbf{y}))\|_2 \leq \lambda \|\mathbf{x} - \mathbf{y}\|_2$ for every vector $\mathbf{y}$ in the neighborhood, $\mathbf{H}(f(\mathbf{x}^*))$ is positive definite, and $\mathbf{x}^{(0)}$ is close enough to $\mathbf{x}^*$, then Newton's method will converge quadratically to $\mathbf{x}^*$.

Newton's method for minimizing $f(\mathbf{x})$ can be very efficient, but the method can also fail to converge. As with Newton's method for solving a system of equations, the convergence properties of the algorithm can be improved in practice by modifying the model update step with a line search.

9.3. THE GAUSS–NEWTON AND LEVENBERG–MARQUARDT METHODS FOR SOLVING NONLINEAR LEAST SQUARES PROBLEMS

Newton's method is not directly applicable to most nonlinear regression and inverse problems. We may not have equal numbers of data points and model parameters, there may not be an exact solution to $\mathbf{G}(\mathbf{m}) = \mathbf{d}$, or $\mathbf{G}(\mathbf{m}) = \mathbf{d}$ may have multiple solutions. Here, we will use a specialized version of Newton's method to minimize a nonlinear least squares problem.

Given a nonlinear system of equations $\mathbf{G}(\mathbf{m}) = \mathbf{d}$, consider the problem of finding an n-length parameter vector $\mathbf{m}$ constrained by an m-length data vector $\mathbf{d}$ with associated specified data standard deviations. Our goal is to find a set of parameters that best fits the data in the sense of minimizing the 2-norm of the residuals.

As with linear regression, if we assume that the measurement errors are normally distributed, then the maximum likelihood principle leads us to minimizing the sum of squared residuals normalized by their respective standard deviations (2.13). We seek to minimize the weighted residual norm

$$\chi^2(\mathbf{m}) = \sum_{i=1}^{m} \left(\frac{\mathbf{G}(\mathbf{m})_i - d_i}{\sigma_i} \right)^2 . \tag{9.13}$$

We define the scalar-valued residual functions

$$f_i(\mathbf{m}) = \frac{\mathbf{G}(\mathbf{m})_i - d_i}{\sigma_i} \quad i = 1, 2, \ldots, m \tag{9.14}$$

and the vector-valued function

$$\mathbf{F}(\mathbf{m}) = \begin{bmatrix} f_1(\mathbf{m}) \\ \vdots \\ f_m(\mathbf{m}) \end{bmatrix} . \tag{9.15}$$

Let the squared 2-norm of the residuals be denoted by

$$\chi^2(\mathbf{m}) = \sum_{i=1}^{m} f_i(\mathbf{m})^2 = \|\mathbf{F}(\mathbf{m})\|_2^2 . \tag{9.16}$$

The gradient of $\chi^2(\mathbf{m})$ can be written as the sum of the gradients of the individual terms

$$\nabla \chi^2(\mathbf{m}) = \sum_{i=1}^{m} \nabla \left(f_i(\mathbf{m})^2 \right) . \tag{9.17}$$

The elements of the gradient are

$$(\nabla \chi^2(\mathbf{m}))_j = \sum_{i=1}^{m} 2 f_i(\mathbf{m})(\nabla f_i(\mathbf{m}))_j \tag{9.18}$$

and can thus be written in matrix notation as

$$\nabla \chi^2(\mathbf{m}) = 2 \mathbf{J}(\mathbf{m})^T \mathbf{F}(\mathbf{m}) \tag{9.19}$$

where $\mathbf{J}(\mathbf{m})$ is the Jacobian

$$\mathbf{J}(\mathbf{m}) = \begin{bmatrix} \frac{\partial f_1(\mathbf{m})}{\partial m_1} & \cdots & \frac{\partial f_1(\mathbf{m})}{\partial m_n} \\ \vdots & \ddots & \vdots \\ \frac{\partial f_m(\mathbf{m})}{\partial m_1} & \cdots & \frac{\partial f_m(\mathbf{m})}{\partial m_n} \end{bmatrix} . \tag{9.20}$$

Similarly, we can express the Hessian of $\chi^2(\mathbf{m})$ using the $f_i(\mathbf{m})$ terms

$$\mathbf{H}(\chi^2(\mathbf{m})) = \sum_{i=1}^{m} \mathbf{H}(f_i(\mathbf{m})^2) \tag{9.21}$$

$$= \sum_{i=1}^{m} \mathbf{H}^i(\mathbf{m}) \tag{9.22}$$

where $\mathbf{H}^i(\mathbf{m})$ is the Hessian of $f_i(\mathbf{m})^2$.
   The j, k element of $\mathbf{H}^i(\mathbf{m})$ is

$$H_{j,k}^i(\mathbf{m}) = \frac{\partial^2 (f_i(\mathbf{m})^2)}{\partial m_j \partial m_k} \tag{9.23}$$

$$= \frac{\partial}{\partial m_j} \left(2 f_i(\mathbf{m}) \frac{\partial f_i(\mathbf{m})}{\partial m_k} \right) \tag{9.24}$$

$$= 2 \left(\frac{\partial f_i(\mathbf{m})}{\partial m_j} \frac{\partial f_i(\mathbf{m})}{\partial m_k} + f_i(\mathbf{m}) \frac{\partial^2 f_i(\mathbf{m})}{\partial m_j \partial m_k} \right) . \tag{9.25}$$

Thus

$$\mathbf{H}(\chi^2(\mathbf{m})) = 2\,\mathbf{J}(\mathbf{m})^T\mathbf{J}(\mathbf{m}) + \mathbf{Q}(\mathbf{m}) \tag{9.26}$$

where

$$\mathbf{Q}(\mathbf{m}) = 2\sum_{i=1}^{m} f_i(\mathbf{m})\mathbf{H}(f_i(\mathbf{m})) \ . \tag{9.27}$$

In the **Gauss–Newton (GN) method**, we ignore the $\mathbf{Q}(\mathbf{m})$ term in (9.26) to approximate the Hessian as

$$\mathbf{H}(\chi^2(\mathbf{m})) \approx 2\,\mathbf{J}(\mathbf{m})^T\mathbf{J}(\mathbf{m}) \ . \tag{9.28}$$

In the context of nonlinear regression, we expect that the $f_i(\mathbf{m})$ terms will be small as we approach the optimal parameters $\mathbf{m}^*$, so that this should a reasonable approximation in the vicinity of the solution. Conversely, this is not a reasonable approximation for nonlinear least squares problems in which the values of $f_i(\mathbf{m})$ can be large.

Implementing Newton's method for minimizing $\chi^2(\mathbf{m})$ (9.12) using the gradient (9.19) and the approximate Hessian (9.28), and dividing both sides by two, we obtain

$$\mathbf{J}(\mathbf{m}^{(k)})^T\mathbf{J}(\mathbf{m}^{(k)})\Delta\mathbf{m} = -\mathbf{J}(\mathbf{m}^{(k)})^T\mathbf{F}(\mathbf{m}^{(k)}) \ , \tag{9.29}$$

which provides a formula for solving for successive update steps $\Delta\mathbf{m}$. The n by n matrix $\mathbf{J}(\mathbf{m}^{(k)})^T\mathbf{J}(\mathbf{m}^{(k)})$ is symmetric and positive semidefinite. If the matrix is actually positive definite then we can use the Cholesky factorization or another full-rank method to solve the system of equations for $\Delta\mathbf{m}$. However, if the matrix is singular then such straightforward approaches will fail. Although the GN method often works well in practice, it is based on Newton's method, and can thus fail by converging to a local maximum or saddle point where $\nabla\chi^2(\mathbf{m}) \approx 0$, or by not converging at all.

In the **Levenberg–Marquardt (LM) method**, the GN method model update equations (9.29) are modified to

$$(\mathbf{J}(\mathbf{m}^{(k)})^T\mathbf{J}(\mathbf{m}^{(k)}) + \lambda\mathbf{I})\Delta\mathbf{m} = -\mathbf{J}(\mathbf{m}^{(k)})^T\mathbf{F}(\mathbf{m}^{(k)}) \ , \tag{9.30}$$

and the positive parameter λ is adjusted during the course of iterations to ensure convergence. One important reason for this modification is to guarantee that the matrix on the left hand side of (9.30) is nonsingular. Since the matrix in this system of equations is symmetric and positive definite, we can use the Cholesky factorization to efficiently solve the system for the model update steps $\Delta\mathbf{m}$.

For very large values of λ,

$$\mathbf{J}(\mathbf{m}^{(k)})^T\mathbf{J}(\mathbf{m}^{(k)}) + \lambda\mathbf{I} \approx \lambda\mathbf{I} \tag{9.31}$$

and the solution to (9.30) is

$$\Delta \mathbf{m} \approx -\frac{1}{\lambda} \nabla \chi^2(\mathbf{m}) \ . \tag{9.32}$$

This is called a **gradient descent** step, meaning that the algorithm simply moves down-gradient to most rapidly reduce $\chi^2(\mathbf{m})$. The gradient descent approach provides very slow but certain convergence to a local minimum. Conversely, for very small values of λ, the LM method reverts to the GN method (9.29), which gives potentially fast but uncertain convergence.

A challenge associated with implementing the LM method is determining the optimal value of λ. The general strategy is to use small values of λ in situations where the GN method is working well, but to increase λ when the GN method fails to make progress in reducing the residual norm. A simple approach is to start with a small value of λ, and then adjust it in every iteration. If the LM method leads to a residual norm reduction, then update $\mathbf{m}$ and decrease λ by a constant multiplicative factor (a factor of two, for example) before the next iteration. Conversely, if the LM method does not improve the solution, we then increase λ by a constant factor and try again, repeating this process until a model update is found that decreases the residual norm. Robust implementations of the LM method use more sophisticated strategies for adjusting λ, but even this simple strategy works surprisingly well. In practice, a careful LM implementation offers the good performance of the GN method as well as robust convergence properties, and LM is usually the method of choice for small- to medium-sized nonlinear least squares problems.

Note that although the LM stabilization term $\lambda \mathbf{I}$ in (9.30) resembles expressions used elsewhere (e.g., Chapter 4) for Tikhonov regularization purposes, it does not alter the ultimate model achieved at convergence. The $\lambda \mathbf{I}$ term is used to stabilize the solution of the linear system of equations, which determines the search direction to be used. Because the $\lambda \mathbf{I}$ term is only used as a way to improve the convergence of the algorithm, and does *not* enter into the residual norm objective function that is being minimized, it does not regularize the nonlinear least squares problem. We discuss the regularization of nonlinear problems in Chapter 10.

9.4. STATISTICAL ASPECTS OF NONLINEAR LEAST SQUARES

Recall from Appendix B that if a vector $\mathbf{d}$ has a multivariate normal distribution, and $\mathbf{A}$ is an appropriately sized matrix, then $\mathbf{Ad}$ also has a multivariate normal distribution with an associated covariance matrix

$$\text{Cov}(\mathbf{Ad}) = \mathbf{A}\text{Cov}(\mathbf{d})\mathbf{A}^T \ . \tag{9.33}$$

In Chapter 2, we applied this formula to the linear least squares problem for $\mathbf{Gm} = \mathbf{d}$, which we solved by the normal equations. The resulting formula for $\text{Cov}(\mathbf{m})$ was

$$\text{Cov}(\mathbf{m}_{L_2}) = (\mathbf{G}^T\mathbf{G})^{-1}\mathbf{G}^T\text{Cov}(\mathbf{d})\mathbf{G}(\mathbf{G}^T\mathbf{G})^{-1} . \tag{9.34}$$

In the simplest case, where $\text{Cov}(\mathbf{d}) = \sigma^2\mathbf{I}$, (9.34) simplified to

$$\text{Cov}(\mathbf{m}_{L_2}) = \sigma^2(\mathbf{G}^T\mathbf{G})^{-1} . \tag{9.35}$$

For the nonlinear regression problem we no longer have a linear relationship between the data and the estimated model parameters, so we cannot assume that the estimated model parameters have a multivariate normal distribution, and cannot use the above formulas. However, we can obtain useful corresponding formulas by linearization if the data errors are not too large.

If small data perturbations result in small model perturbations in a nonlinear system, we can consider a linearization of the misfit function (9.15) about a solution $\mathbf{m}^*$

$$\mathbf{F}(\mathbf{m}^* + \Delta\mathbf{m}) \approx \mathbf{F}(\mathbf{m}^*) + \mathbf{J}(\mathbf{m}^*)\Delta\mathbf{m} . \tag{9.36}$$

Under this approximation, there is a linear relationship between changes in $\mathbf{F}$ and changes in the parameters $\mathbf{m}$

$$\mathbf{F}(\mathbf{m}^* + \Delta\mathbf{m}) - \mathbf{F}(\mathbf{m}^*) = \Delta\mathbf{F} \approx \mathbf{J}(\mathbf{m}^*)\Delta\mathbf{m} . \tag{9.37}$$

To the extent that the residual misfit terms (the elements of $\mathbf{F}(\mathbf{m}^*)$) are small, the Hessian can be approximated by (9.28). In this case $\mathbf{J}(\mathbf{m}^*)$ in nonlinear regression can take the place of $\mathbf{G}$ in linear problems (e.g., (9.35)) to estimate the covariance of the model parameters. Eq. (9.13) incorporates the respective data element standard deviations σ_i into the formula for the residual norm $\chi^2(\mathbf{m})$, and explicitly weights the nonlinear constraint equations. $\text{Cov}(\mathbf{d})$, in the case of independent data errors, is thus the identity matrix in such a weighted system. In this case we have

$$\text{Cov}(\mathbf{m}^*) \approx (\mathbf{J}(\mathbf{m}^*)^T\mathbf{J}(\mathbf{m}^*))^{-1} . \tag{9.38}$$

As in Chapter 2, we can construct a 95% confidence ellipsoid for the fitted parameters in terms of this covariance matrix,

$$(\mathbf{m} - \mathbf{m}^*)^T\mathbf{J}(\mathbf{m}^*)^T\mathbf{J}(\mathbf{m}^*)(\mathbf{m} - \mathbf{m}^*) \leq \Delta^2 \tag{9.39}$$

where Δ^2 is the 95%-tile of a χ^2 distribution with n degrees of freedom. It is important to reiterate that this confidence ellipsoid only defines an approximate confidence region because we linearized $\mathbf{F}(\mathbf{m})$ in (9.37).

As in Chapter 2, there is an alternative derivation of the 95% confidence region for the fitted parameters in terms of $\chi^2(\mathbf{m}) - \chi^2(\mathbf{m}^*)$. If we let

$$\chi^2(\mathbf{m}) = \|\mathbf{F}(\mathbf{m})\|_2^2 = \mathbf{F}(\mathbf{m})^T\mathbf{F}(\mathbf{m}) \tag{9.40}$$

then the inequality

$$\chi^2(\mathbf{m}) - \chi^2(\mathbf{m}^*) \leq \Delta^2 \tag{9.41}$$

defines an approximate 95% confidence region for $\mathbf{m}^*$. Because of the nonlinearity in the $\chi^2(\mathbf{m})$ function, this 95% confidence region is typically not an ellipsoid. Unfortunately, this confidence region is not exact, because this development is still based on the assumption of multivariate normality, and $\mathbf{m}^*$, because it is not a linear combination of the data vector elements, will not in general have a multivariate normal distribution in nonlinear problems.

Should (9.39) or (9.41) be used in practice? If $\mathbf{G}(\mathbf{m})$ is not too strongly nonlinear, then there should be very little difference between the two confidence regions. However, if $\mathbf{G}(\mathbf{m})$ is more strongly nonlinear, then the difference between the two confidence regions may be large and neither confidence region can really be trusted. In such situations, Monte Carlo methods are a more useful and appropriate way to estimate parameter uncertainty. This is discussed further in Chapter 11.

As with linear regression, it is possible to apply nonlinear regression when the measurement errors are independent and normally distributed and the standard deviations are unknown but assumed to be equal (Section 2.3). We set the σ_i to 1 and minimize the sum of squared residuals, defining the residual vector $\mathbf{r}$ with elements

$$r_i = \mathbf{G}(\mathbf{m}^*)_i - d_i \quad i = 1, 2, \ldots, m . \tag{9.42}$$

Our estimate of the measurement standard deviation is then

$$s = \sqrt{\frac{\|\mathbf{r}\|_2^2}{m - n}} \tag{9.43}$$

and the corresponding approximate covariance matrix for the estimated model parameters is

$$\text{Cov}(\mathbf{m}^*) = s^2(\mathbf{J}(\mathbf{m}^*)^T\mathbf{J}(\mathbf{m}^*))^{-1} . \tag{9.44}$$

Given $\mathbf{m}^*$ and $\text{Cov}(\mathbf{m}^*)$, we can establish confidence intervals for the model parameters using the methods described in Chapter 2. As with any parameter estimation problem, it is also important to examine the residuals for systematic patterns or deviations from normality. If we have not estimated the measurement standard deviation s from the fitted residuals, then it is also important to test the χ^2 value for goodness of fit and perform an associated p-value test (2.22) if we have normally distributed data errors. The appropriateness of this test will depend on how well the nonlinear model is approximated by the Jacobian linearization for points near the optimal parameter values.

Example 9.1

A classic method in hydrology for determining the transmissivity and storage coefficient of an aquifer is called the "slug test" [61].

A known volume Q of water (the slug) is injected into a well, and the resulting effects on the head h at an observation well a distance d away from the injection well are observed at various times t. The head measured at the observation well typically increases rapidly and then decreases more slowly. We wish to determine the dimensionless storage coefficient S and the transmissivity T.

The mathematical model for the slug test is

$$h = \frac{Q}{4\pi\, Tt}e^{-d^2 S/(4\, Tt)} \, . \tag{9.45}$$

We know the parameters $Q = 50$ m^3 and $d = 60$ m, and the times t at which the head h is measured. The data are given in the file **slugdata.mat**. Here the head measurements are roughly accurate to 0.01 m ($\sigma_i = 0.01$ m).

The optimal parameter values are $S = 0.00207$ and $T = 0.585$ m^2/hr. The observed χ^2 value is 2.04, with a corresponding p-value of 0.73. Thus this fit passes the χ^2 test. The data points and fitted curve are shown in Fig. 9.1.

Using the Jacobian evaluated at the optimal parameter values, we computed an approximate covariance matrix for the fitted parameters. The resulting 95% confidence intervals for S and T are

$$S = 0.00207 \pm 0.00012 \tag{9.46}$$

$$T = 0.585 \pm 0.029 \text{ m}^2/\text{hr} \, . \tag{9.47}$$

A contour plot of the χ^2 surface obtained by varying S and T is shown in Fig. 9.2. Note that, unlike our earlier linear regression problems, the contours are not even approximately elliptical because of the nonlinearity of the problem. However, if we zoom in to the immediate vicinity of the optimal parameters (Fig. 9.3), we find that the

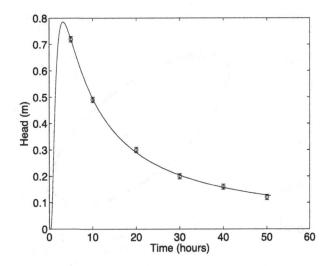

Figure 9.1 Data, with one standard deviation error bars, and fitted model for the slug test.

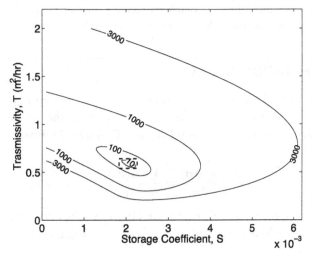

Figure 9.2 χ^2 contour plot for the slug test for a wide range of parameters. The small box shows the close-up axis range displayed in Fig. 9.3.

χ^2 contours are approximately elliptical at this scale. The approximate ellipticity of the contours indicates that the linear approximation of $\mathbf{G(m)}$ around the optimal parameter values is a reasonable approximation for model perturbations in this range.

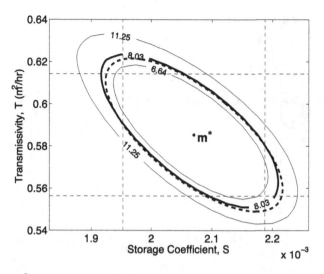

Figure 9.3 Closeup χ^2 contour plot for the immediate vicinity of the optimal parameters $\mathbf{m}^*$ for the slug test example, showing nonlinear 90%, 95% (bold) and 99% confidence contours. The nearly elliptical contours are indicative of a approximately linear behavior at this scale. Dashed ellipse shows the linearized approximation to the 95% confidence contour calculated using (9.38). 95% confidence intervals for the parameters calculated from (9.38) are shown as dashed lines.

9.5. IMPLEMENTATION ISSUES

Iterative methods for solving nonlinear problems require the computation of the functions $f_i(\mathbf{m})$ and their partial derivatives with respect to the model parameters m_j. These partial derivatives in turn depend on the **Fréchet derivatives** of the nonlinear $\mathbf{G}$

$$\frac{\partial f_i(\mathbf{m})}{\partial m_j} = \frac{1}{\sigma_i}\frac{\partial \mathbf{G}(\mathbf{m})_i}{\partial m_j}.$$ (9.48)

In some cases, we have explicit formulas for $\mathbf{G}(\mathbf{m})$ and its derivatives. In other cases, $\mathbf{G}(\mathbf{m})$ exists only as a **black box** subroutine that can be called as required to evaluate the function.

When an explicit formula for $\mathbf{G}(\mathbf{m})$ is available, and the number of parameters is relatively small, we can differentiate analytically. There also exist **automatic differentiation** software packages that can translate the source code of a program that computes $\mathbf{G}(\mathbf{m})$ into a program that computes the derivatives of $\mathbf{G}(\mathbf{m})$.

Another approach is to use **finite differences** to approximate the derivatives of $\mathbf{G}(\mathbf{m})_i$. A simple first-order scheme is to employ the linearization

$$\frac{\partial \mathbf{G}(\mathbf{m})_i}{\partial m_j} \approx \frac{\mathbf{G}(\mathbf{m}+h\mathbf{e}_j)_i - \mathbf{G}(\mathbf{m})_i}{h}$$ (9.49)

where $\mathbf{e}_j$ is the jth standard basis vector, and the scalar h is chosen to be suitably small. However, if h becomes very small, significant roundoff error in the numerator of (9.49) may occur. A good rule of thumb is to set $h = \sqrt{\epsilon}$, where ϵ is the relative accuracy of the evaluations of $\mathbf{G}(\mathbf{m})_i$. For example, if the function evaluations are accurate to 0.0001, then an appropriate choice of h would be about 0.01. Determining the actual accuracy of function evaluations can be difficult, especially when $\mathbf{G}$ is a black box routine. One useful assessment technique is to plot function values as a parameter of interest is varied over a small range. These plots should be linear at the scale of h. When $\mathbf{G}$ is available only as a black box subroutine that can be called with particular values of $\mathbf{m}$, and the source code for the subroutine is not available, then the only available approach is to use finite differences.

In practice, many difficulties in solving nonlinear regression problems can be traced back to incorrect derivative computations. It is thus a good idea to cross-check any available analytical formulas for the derivative with finite-difference approximations. Many software packages for nonlinear regression include options for checking the accuracy of derivative formulas.

A second important issue in the implementation of the GN and LM methods is deciding when to terminate the iterations. We would like to stop when the gradient $\nabla \chi^2(\mathbf{m})$ is approximately $\mathbf{0}$ and $\mathbf{m}$ has stopped changing substantially from one iteration to the next. Because of scaling issues, it is not possible to set an absolute tolerance on $\|\nabla \chi^2(\mathbf{m})\|_2$ that would be appropriate for all problems. Similarly, it is difficult to pick a single absolute tolerance on $\|\mathbf{m}^{(k+1)} - \mathbf{m}^{(k)}\|_2$ or $|\chi^2(\mathbf{m}^{(k+1)}) - \chi^2(\mathbf{m}^{(k)})|$.

The following convergence tests have been normalized so that they will work well on a wide variety of problems. We assume that values of $\mathbf{G}(\mathbf{m})$ can be calculated with a relative accuracy of ϵ. To ensure that the gradient of $\chi^2(\mathbf{m})$ is approximately $\mathbf{0}$, we require that

$$\|\nabla \chi^2(\mathbf{m}^{(k)})\|_2 < \sqrt{\epsilon}(1 + |\chi^2(\mathbf{m}^{(k)})|) . \tag{9.50}$$

To ensure that successive values of $\mathbf{m}$ are close, we require

$$\|\mathbf{m}^{(k)} - \mathbf{m}^{(k-1)}\|_2 < \sqrt{\epsilon}(1 + \|\mathbf{m}^{(k)}\|_2) . \tag{9.51}$$

Finally, to make sure that the values of $\chi^2(\mathbf{m})$ have stopped changing, we require that

$$|\chi^2(\mathbf{m}^{(k)}) - \chi^2(\mathbf{m}^{(k-1)})| < \epsilon(1 + |\chi^2(\mathbf{m}^{(k)})|) . \tag{9.52}$$

There are a number of additional problems that can arise during the solution of a nonlinear regression problem by the GN or LM methods related to the functional behavior of $\chi^2(\mathbf{m})$.

The first issue is that our methods, because they involve calculation of the Jacobian and the Hessian, assume that $\chi^2(\mathbf{m})$ is a smooth function. This means not only that

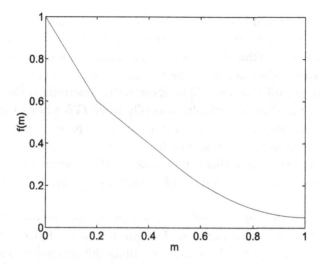

Figure 9.4 An example of a nonsmooth function.

$\chi^2(\mathbf{m})$ must be continuous, but also that its first and second partial derivatives with respect to the parameters must be continuous. Fig. 9.4 shows a function which is itself continuous, but has discontinuities in the first derivative at $m = 0.2$ and the second derivative at $m = 0.5$. When $\mathbf{G}(\mathbf{m})$ is given by an explicit formula, it is usually easy to verify that $\chi^2(\mathbf{m})$ is smooth, but when $\mathbf{G}(\mathbf{m})$ is implemented as a black box routine it may be difficult.

A second issue is that $\chi^2(\mathbf{m})$ may have a "flat bottom." See Fig. 9.5. In such cases, there are many values of $\mathbf{m}$ that come close to fitting the data, and it is difficult to determine the optimal $\mathbf{m}^*$. In practice, this condition is seen to occur when $\mathbf{J}(\mathbf{m}^*)^T\mathbf{J}(\mathbf{m}^*)$ is nearly singular. Because of this ill-conditioning, computing accurate confidence intervals for the model parameters can be effectively impossible. We will address this difficulty in Chapter 10 by applying regularization.

The final problem that we will consider is that $\chi^2(\mathbf{m})$ may be nonconvex and therefore have multiple local minima. See Fig. 9.6. The GN and LM methods are designed to converge to a local minimum, but depending on where we begin the search, there is no way to be certain that such a solution will be a global minimum. Depending on the particular problem, the optimization algorithm might well converge to a locally, rather than globally, optimal solution.

Global optimization methods have been developed to deal with this issue [21,95, 96,179]. Deterministic global optimization procedures can be used on problems with a very small number of variables, whereas stochastic search procedures can be applied to large-scale problems but do not find the global optimum with certainty.

However, even a deterministic global optimization procedure is not a panacea. In the context of nonlinear regression, if the nonlinear least squares problem has multiple

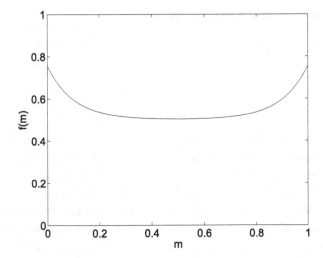

Figure 9.5 An example of a function with a flat bottom.

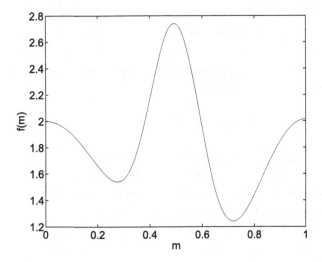

Figure 9.6 An example of a function with multiple local minima.

locally optimal solutions with similar objective function values, then each of these solutions will correspond to a statistically likely solution. We cannot simply report one globally optimal solution as our best estimate and construct confidence intervals using (9.38), because this would mean ignoring other likely solutions. However, if we could show that there is one globally optimal solution and other locally optimal solutions have very small p-values, then it would be appropriate to report the globally optimal solution and corresponding confidence intervals.

Table 9.1 Locally optimal solutions for the sample problem.

Solution number	m_1	m_2	m_3	m_4	χ^2	p-value
1	0.9874	−0.5689	1.0477	−0.7181	17.3871	0.687
2	1.4368	0.1249	−0.5398	−0.0167	40.0649	0.007
3	1.5529	−0.1924	−0.1974	−0.1924	94.7845	2×10^{-11}

Although a thorough discussion of global optimization is beyond the scope of this book, we will discuss one simple global optimization procedure called the **multistart method** in the next example. In the multistart method, we randomly generate a large number of initial solutions, and then apply the LM method starting with each of these initial solutions. We then examine the local minimum solutions found by the procedure, and examine ones with acceptable values of $\chi^2(\mathbf{m})$. The multistart approach has two important practical advantages. First, by potentially identifying multiple locally optimal solutions, we can determine whether there is more than one statistically likely solution. Second, we can make effective use of the fast convergence of the LM method to find the locally optimal solutions.

Example 9.2

Consider the problem of fitting a model of two superimposed exponential decay functions characterized by four parameters

$$y_i = m_1 e^{m_2 x_i} + m_3 x_i e^{m_4 x_i} \tag{9.53}$$

to a set of observations. The true model parameters are $m_1 = 1.0$, $m_2 = -0.5$, $m_3 = 1.0$, and $m_4 = -0.75$, and the x_i values are 25 evenly spaced points between 1 and 7. We compute corresponding y_i values and add independent normally distributed noise with a standard deviation of 0.01 to obtain a synthetic data set.

We next apply the LM method to solve the problem 20 times, using random initial solutions, with each initial parameter uniformly distributed between −1 and 1. This produces a total of three different locally optimal solutions (Table 9.1).

Solution number 1, with a χ^2 value of approximately 17, has an acceptable p-value of about 0.69 for a regression with 21 degrees of freedom. The other two solutions have unreasonably large χ^2 values, and hence much lower p-values. We will thus analyze only the first solution. Fig. 9.7 shows the data points with $1 - \sigma$ error bars and the fitted curve for solution number 1 and Fig. 9.8 shows the corresponding residuals normalized by the data standard deviations. Note that the majority of the residuals are within 0.5 standard deviations, with a few residuals as large as 1.9 standard deviations. There is no obvious residual trend as x ranges from 1 to 7.

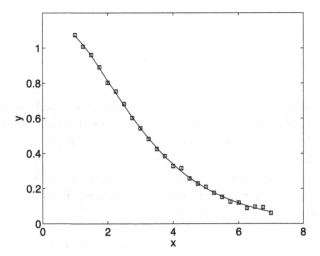

Figure 9.7 Data points and fitted curve.

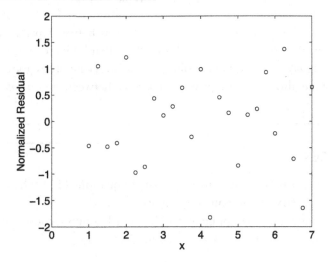

Figure 9.8 Normalized residuals corresponding to Fig. 9.7.

Next, we compute the approximate covariance matrix for the model parameters using (9.38). The square roots of the diagonal elements of the covariance matrix are standard deviations for the individual model parameters. These are then used to compute 95% confidence intervals for model parameters. The solution parameters with 95% confidence intervals are

$$m_1 = 0.98 \pm 0.22$$
$$m_2 = -0.57 \pm 0.77$$
$$m_3 = 1.05 \pm 0.50 \tag{9.54}$$
$$m_4 = -0.72 \pm 0.20 \ .$$

The true parameters (1, -0.5, 1, and -0.75) are all covered by these confidence intervals. However, there is a large degree of uncertainty. This is an example of a poorly conditioned nonlinear regression problem in which the data do not strongly constrain the parameter values.

The correlation matrix provides some insight into the nature of the ill-conditioning in this example by quantifying strong parameter tradeoffs. For our preferred solution, the correlation matrix calculated using (9.38) is

$$\rho = \begin{bmatrix} 1.00 & -0.84 & 0.68 & 0.89 \\ -0.84 & 1.00 & -0.96 & -0.99 \\ 0.68 & -0.96 & 1.00 & 0.93 \\ 0.89 & -0.99 & 0.93 & 1.00 \end{bmatrix} . \tag{9.55}$$

Note the strong positive and negative correlations between parameter pairs. The high negative correlation between m_1 and m_2 tells us that by increasing m_1 and simultaneously decreasing m_2 we can obtain a solution that is very nearly as good as our optimal solution. There are also strong negative correlations between m_2 and m_3 and between m_2 and m_4.

9.6. EXERCISES

1. Show that (9.29) is equivalent to the normal equations (2.3) when $\mathbf{G(m)} = \mathbf{Gm}$, where $\mathbf{G}$ is a matrix of constant coefficients.
2. A recording instrument sampling at 50 Hz records a noisy sinusoidal voltage signal in a 40-s-long record. The data are to be modeled using

$$y(t) = A \sin(2\pi f_0 t + \chi^2) + c + \sigma \eta(t) \tag{9.56}$$

where $\eta(t)$ is believed to be unit standard deviation, independent, and normally distributed noise, and σ is an unknown standard deviation. Using the data in the MATLAB data file **instdata.mat**, solve for the parameters (A, f_0, χ^2, c), using the LM method. Show that it is critical to choose a good initial solution (suitable initial parameters can be found by examining a plot of the time series by eye). Once you are satisfied that you have found a good solution, use it to estimate the noise amplitude σ. Use your solution and estimate of σ to find corresponding covariance and correlation matrices and 95% parameter confidence intervals. Which pair of parameters is most strongly correlated? Are there multiple equally good solutions for this problem?

3. In hydrology, the van Genuchten model is often used to relate the volumetric water content in an unsaturated soil to the head [67]. The model is

$$\theta(h) = \theta_r + \frac{\theta_s - \theta_r}{(1 + (-\alpha h)^n)^{(1-1/n)}} \qquad (9.57)$$

where θ_s is the volumetric water content at saturation, θ_r is the residual volumetric water content at a very large negative head, and α and n are two parameters that can be fit to laboratory measurements.

The file **vgdata.mat** contains measurements for a loam soil at the Bosque del Apache National Wildlife Refuge in New Mexico [87]. Fit the van Genuchten model to the data. The volumetric water content at saturation is $\theta_s = 0.44$, and the residual water content is $\theta_r = 0.09$. You may assume that the measurements of $\theta(h)$ are accurate to about 2% of the measured values.

You will need to determine appropriate values for σ_i, write functions to compute $\theta(h)$ and its derivatives, and then use the LM method to estimate the parameters. In doing so, you should consider whether or not this problem might have local minima. It will be helpful to know that typical values of α range from about 0.001 to 0.02, and typical values of n run from 1 to 10.

4. An alternative version of the LM method stabilizes the GN method by multiplicative damping. Instead of adding $\lambda \mathbf{I}$ to the diagonal of $\mathbf{J}(\mathbf{m}^{(k)})^T \mathbf{J}(\mathbf{m}^{(k)})$, this method multiplies the diagonal of $\mathbf{J}(\mathbf{m}^{(k)})^T \mathbf{J}(\mathbf{m}^{(k)})$ by a factor of $(1 + \lambda)$. Show that this method can fail by producing an example in which the modified $\mathbf{J}(\mathbf{m}^{(k)})^T \mathbf{J}(\mathbf{m}^{(k)})$ matrix is singular, no matter how large λ becomes.

5. A cluster of 10 small earthquakes occurs in a shallow geothermal reservoir. The field is instrumented with nine seismometers, eight of which are at the surface and one of which is 300 m down a borehole. The P-wave velocity of the fractured granite medium is thought to be an approximately uniform 2 km/s. The station locations (in meters relative to a central origin) are given in the MATLAB data file **stmat.mat**.

The arrival times of P-waves from the earthquakes are carefully measured at the stations, with an estimated error of approximately 1 ms. The arrival time estimates for each earthquake e_i at each station (in seconds relative to an arbitrary reference) are given in the data file **eqdata.mat**.

a. Apply the LM method to this data set to estimate least squares locations of the earthquakes.

b. Estimate the uncertainties in x, y, z (in m) and origin time (in s) for each earthquake using the diagonal elements of the appropriate covariance matrix. Do the earthquake locations follow any discernible trend?

Table 9.2 Data for the lightning mapping array problem.

Station	t (s)	x (km)	y (km)	z (km)
1	0.0922360280	−24.3471411	2.14673146	1.18923667
2	0.0921837940	−12.8746056	14.5005985	1.10808551
3	0.0922165500	16.0647214	−4.41975194	1.12675062
4	0.0921199690	0.450543748	30.0267473	1.06693166
6	0.0923199800	−17.3754105	−27.1991732	1.18526730
7	0.0922839580	−44.0424408	−4.95601205	1.13775547
8	0.0922030460	−34.6170855	17.4012873	1.14296361
9	0.0922797660	17.6625731	−24.1712580	1.09097830
10	0.0922497250	0.837203704	−10.7394229	1.18219520
11	0.0921672710	4.88218031	10.5960946	1.12031719
12	0.0921702350	16.9664920	9.64835135	1.09399160
13	0.0922357370	32.6468622	−13.2199767	1.01175261

6. The Lightning Mapping Array (LMA) is a portable system that locates the sources of lightning radio-frequency radiation in three spatial dimensions and time [170]. The system measures the arrival times of impulsive radiation events with an uncertainty specified by a standard deviation of 7×10^{-2} µs. Measurements are made at multiple locations, typically in a region 40 to 60 km in diameter. Each station records the peak radiation event in successive 100 µs time intervals; from such data, several hundred to over a thousand distinct radiation sources may be typically located per lightning discharge. Example data from the LMA are shown in Table 9.2 and are found in the MATLAB data file **lightningdata.mat**.

    a. Use the arrival times at stations 1, 2, 4, 6, 7, 8, 10, and 13 to find the time and location of the associated source. Assume that radio wave signals travel along straight paths at the speed of light in a vacuum (2.997×10^8 m/s).

    b. A challenge in dealing with the large number of detections from the LMA is to disentangle overlapping signals from multiple events. Locate using subsets of the data set to find the largest subset of the data for 9 or more stations that gives a good solution, and compare it to the station subset from part (a).

7. The Brune model is often used to analyze earthquake seismogram spectra to characterize the seismic source [28]. In this model, the displacement spectral amplitude $A(f)$ of the seismogram at frequency f is given by

$$A(f) = \frac{A_0 e^{-\pi t^* f}}{1 + (\frac{f}{f_c})^2} \tag{9.58}$$

where the amplitude A_0, attenuation constant t^*, and corner frequency, f_c, are parameters to be estimated.

Values of $A(f)$ (in units of nm/Hz) for a simulated earthquake are given for 20 frequencies f (in Hz), in the file **brunedata.mat**. Fit the parameters A_0, t^*, and f_c to these data. Estimate the noise level in the data from the residuals. Produce covariance and correlation matrices for the fitted parameters. Compute confidence intervals for the fitted parameters and discuss your results. You may find it useful to apply a logarithmic transformation to simplify the problem.

8. In cosmogenic nuclide dating we attempt to determine the exposure age of a rock sample by measuring the concentration of isotopes produced in the rock by inter-actions with cosmic rays [163]. We consider the problem of estimating the erosion rate and age using concentrations of ^{10}Be in quartz samples taken at several depths below a surface.

   We assume that the quartz has been long buried so that any prior ^{10}Be has decayed. The radioactive decay rate of this isotope of Be is $\lambda = 4.998 \times 10^{-7}$/yr. The rate of production at depth z below the surface is

   $$P(x) = P(0)e^{-\mu z} \tag{9.59}$$

   where $P(0)$ is the production rate at the surface and μ is a constant that gives the rate of attenuation of the cosmic rays with depth. For this problem, we will use $P(0) = 3.2$ atoms ^{10}Be/gram quartz/yr, and $\mu = 0.0166$/cm. It can be shown that the concentration at depth z cm below the final eroded surface after T years of exposure with steady erosion at rate ϵ (cm/yr) is

   $$N(z, T) = \frac{P(0)}{\lambda + \mu\epsilon} e^{-\mu z} (1 - e^{-(\lambda + \mu\epsilon)T}) . \tag{9.60}$$

   The file **be10.mat** contains ^{10}Be concentrations (in atoms ^{10}Be per gram of quartz), measurement standard deviations σ, and measurement depths (in cm).

   a. Construct a grid with erosion rates ranging from 5×10^{-6} cm/yr to 1×10^{-3} cm/yr in steps of 1×10^{-5} cm/yr and ages from 500 years to 199,500 years in steps of 1000 years. Evaluate χ^2 for each point on this grid and find the age and erosion rate that minimize χ^2.

   b. Construct a 95% confidence region for the age and erosion rate using (9.41). Explain why it would be a bad idea to use Eq. (9.39) to construct a confidence region for this particular solution.

9.7. NOTES AND FURTHER READING

Newton's method is central to the field of optimization [51,76,109,110,150], and is, because of its speed, the basis for many methods of nonlinear optimization. A number of modifications to the method are used to ensure convergence to a minimum of $\chi^2(\mathbf{x})$, which may be local or global [76,150]. One important difficulty in Newton's method is that, for very large problems, it may be impractical to store the Hessian matrix. Specialized methods have been developed for the solution of such large-scale optimization problems [76,150].

The GN and LM methods are discussed in more detail in [19,76,150]. Statistical aspects of nonlinear regression are discussed in [7,52,142]. A more detailed discussion of the termination criteria for the LM method described in Section 9.4 can be found in [76]. There are a number of freely available and commercial software packages for nonlinear regression, including GaussFit [102], MINPACK [140], and ODRPACK [22]. Although the Gauss–Newton and Levenberg–Marquardt methods are very commonly used to solve (9.13), other optimization algorithms can sometimes be faster in practice. A comparison of the Gauss–Newton method and a quasi-Newton method in 2-D resistivity inversion is given in [128]. Automatic differentiation has applications in many areas of numerical computing, including optimization and numerical solution of ordinary and partial differential equations. Two books that survey this topic are [46,75]. Global optimization is a large field of research. Some basic references include [21,95, 96]. A survey of global optimization methods in geophysical inversion is [179].

CHAPTER TEN

Nonlinear Inverse Problems

Synopsis

The nonlinear regression approaches of Chapter 9 are generalized to problems requiring regularization. The Tikhonov regularization and Occam's inversion approaches are introduced. Seismic tomography and electrical conductivity inversion examples are used to illustrate the application of these methods. Resolution analysis for nonlinear problems is addressed. We introduce the nonlinear conjugate gradient method for solving large systems of nonlinear equations. The discrete adjoint method is described and illustrated with a regularized heat flow example that utilizes the nonlinear conjugate gradient method in its solution.

10.1. REGULARIZING NONLINEAR LEAST SQUARES PROBLEMS

As with linear problems, the nonlinear least squares approaches can run into difficulty with ill-conditioned problems. This typically happens as the number of model parameters grows. Here, we will discuss regularization of nonlinear inverse problems and algorithms for computing a regularized solution to a nonlinear inverse problem.

The basic ideas of Tikhonov regularization can be extended to nonlinear problems. Suppose that we are given a nonlinear discrete inverse problem where an n-element model $\mathbf{m}$ and an m-element data vector $\mathbf{d}$ are related by a nonlinear system of equations $\mathbf{G}(\mathbf{m}) = \mathbf{d}$. For convenience, we will assume that the m nonlinear constraint equations have been scaled to incorporate the measurement standard deviations σ_i. We seek the solution with the smallest seminorm $\|\mathbf{Lm}\|_2$ that comes sufficiently close to fitting the data vector, where $\mathbf{L}$ is an appropriate roughening matrix (e.g., (4.28)).

We can formulate this problem as

$$\min \|\mathbf{Lm}\|_2$$
$$\|\mathbf{G}(\mathbf{m}) - \mathbf{d}\|_2 \leq \delta . \tag{10.1}$$

Note that the form of the problem is virtually identical to that considered in the linear case (e.g., (4.25)), with the only difference being that we now have a general function $\mathbf{G}(\mathbf{m})$ instead of a matrix–vector multiplication $\mathbf{Gm}$. As in the linear case, we can reformulate this problem in terms of minimizing the misfit subject to a constraint on $\|\mathbf{Lm}\|_2$

$$\min \|\mathbf{G}(\mathbf{m}) - \mathbf{d}\|_2$$
$$\|\mathbf{Lm}\|_2 \leq \epsilon \tag{10.2}$$

Parameter Estimation and Inverse Problems
https://doi.org/10.1016/B978-0-12-804651-7.00015-8

or as a regularized (often referred to as "damped") least squares problem

$$\min \|\mathbf{G}(\mathbf{m}) - \mathbf{d}\|_2^2 + \alpha^2 \|\mathbf{Lm}\|_2^2 . \tag{10.3}$$

All three versions of the regularized least squares problem can be solved by applying standard nonlinear optimization software. In particular, (10.3) is a nonlinear least squares problem, so we can apply the LM or GN methods to it. Of course, any such approach will still have to deal with the possibility of local minima that are not global minima. In some cases, it is possible to show that nonlinear least squares problems are convex, and thus possess only global minima. In other cases we will have to employ multistart or some other global optimization strategy to determine whether there are multiple minima.

To apply the GN method to (10.3), we rewrite it as

$$\min \left\| \begin{array}{c} \mathbf{G}(\mathbf{m}) - \mathbf{d} \\ \alpha \mathbf{Lm} \end{array} \right\|_2^2 . \tag{10.4}$$

The Jacobian of (10.4) for the kth iteration is

$$\mathbf{K}(\mathbf{m}^{(k)}) = \left[\begin{array}{c} \mathbf{J}(\mathbf{m}^{(k)}) \\ \alpha \mathbf{L} \end{array} \right] \tag{10.5}$$

where $\mathbf{J}(\mathbf{m}^{(k)})$ is the Jacobian of $\mathbf{G}(\mathbf{m}^{(k)})$. A GN model step is obtained by applying (9.29) and solving

$$\mathbf{K}(\mathbf{m}^{(k)})^T \mathbf{K}(\mathbf{m}^{(k)}) \Delta \mathbf{m} = -\mathbf{K}(\mathbf{m}^{(k)})^T \left[\begin{array}{c} \mathbf{G}(\mathbf{m}^{(k)}) - \mathbf{d} \\ \alpha \mathbf{Lm}^{(k)} \end{array} \right] \tag{10.6}$$

or, combining (10.5) and (10.6), by solving

$$\left(\mathbf{J}(\mathbf{m}^{(k)})^T \mathbf{J}(\mathbf{m}^{(k)}) + \alpha^2 \mathbf{L}^T \mathbf{L} \right) \Delta \mathbf{m} = -\mathbf{J}(\mathbf{m}^{(k)})^T (\mathbf{G}(\mathbf{m}^{(k)}) - \mathbf{d}) - \alpha^2 \mathbf{L}^T \mathbf{Lm}^{(k)} . \tag{10.7}$$

Eq. (10.7) resembles the LM method (9.30). Note, however, that α in (10.7) now appears in the objective function being minimized (10.4) and thus introduces regularization. To further stabilize the iterations, as in the LM method, a variable $\lambda \mathbf{I}$ term could be added to the matrix term of the left hand side of (10.7) to steer iterative updates towards the direction of gradient descent. This will not be necessary if the explicit regularization of (10.7) sufficiently stabilizes the system of equations.

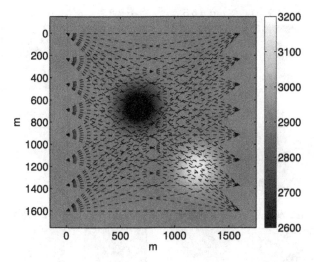

Figure 10.1 True velocity model and the corresponding ray paths for the bent-ray cross-well tomography example.

Example 10.1

Consider a modified version of the cross-well tomography example from Exercise 4.3. We introduce nonlinearity by employing a more realistic forward model that incorporates ray path bending due to seismic velocity changes. The two-dimensional velocity structure is parameterized using a 64-element matrix of uniformly-spaced slowness nodes on an 8 by 8 grid spanning a 1600 m by 1600 m region.

We apply an approximate ray bending technique to estimate refracted ray paths within the slowness model and to estimate travel times and their partial derivatives with respect to the model parameters [212]. Fig. 10.1 shows the true velocity model and a corresponding set of 64 ray paths. The true model consists of a background velocity of 2.9 km/s with large embedded fast (+10%) and slow (−15%) Gaussian-shaped anomalies. The data set consists of the 64 travel times between each pair of opposing sources and receivers with $N(0, (0.001 \text{ s})^2)$ noise added.

Note that refracted ray paths tend to be "repelled" from low-velocity regions (dark shading) and are, conversely, "attracted" to high-velocity regions (light shading) in accordance with Fermat's least-time principle. In practice this effect makes low-velocity regions more difficult to resolve in such studies because they will be less well sampled by ray paths.

A discrete approximation of the two-dimensional Laplacian operator is used to regularize this problem. Iterative GN (10.7) solutions were obtained for a range of 16 values

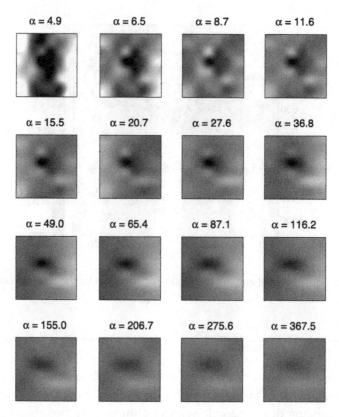

Figure 10.2 Suite of GN (10.7), second-order regularized solutions, ranging from least (upper left) to most (lower right) regularized, and associated α values. The physical dimensions and gray scale are identical to those of Figs. 10.1 and 10.4.

of α ranging logarithmically from approximately 4.9 to 367. Fig. 10.2 shows the suite of solutions after five iterations. An L-curve plot of seminorm versus data misfit is plotted in Fig. 10.3, along with the discrepancy principle value $\delta = 0.001 \cdot \sqrt{64} = 0.008$. Note that the two most lightly regularized solutions are out of their expected monotonically decreasing residual norm positions on the L-curve. This is because the linearization-based GN method is unable to accurately solve the corresponding poorly conditioned least squares systems (10.7) in these cases. Such solutions could be improved by modifying the system of equations with an LM stabilizing term as described above. The solution best satisfying the discrepancy principle corresponds to $\alpha \approx 37$ (Fig. 10.4).

Because we know the true model in this example, it is instructive to examine how well the regularized solutions of Fig. 10.2 compare to it. Fig. 10.5 shows the 2-norm

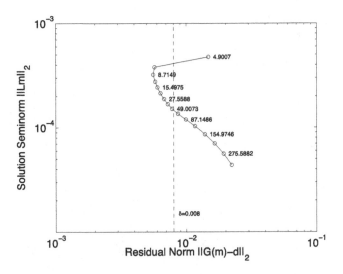

Figure 10.3 L-curve and corresponding α values for the solutions of Fig. 10.2.

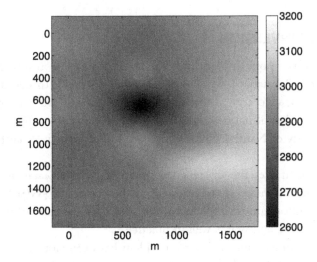

Figure 10.4 Best solution velocity structure (m/s), α selected using the discrepancy principle, $\alpha \approx 37$.

model misfit as a function of α, and demonstrates that the discrepancy principle solution for this problem, and for this particular noise realization, is indeed close to the minimum in $\|\mathbf{m} - \mathbf{m}_{\text{true}}\|_2$. Note that the solution shown in Fig. 10.4 exhibits resolution artifacts that are common in regularized solutions, such as streaking, side lobes, and amplitude underestimation (see Example 10.3).

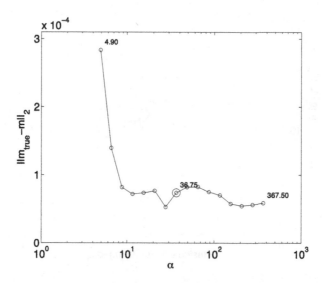

Figure 10.5 Model misfit 2-norm as a function of regularization parameter α, with preferred model highlighted.

10.2. OCCAM'S INVERSION

Occam's inversion is a popular algorithm for nonlinear inversion introduced by Constable, Parker, and Constable [45]. The name refers to the 14th century philosopher William of Ockham, who argued that simpler explanations should always be preferred to more complicated explanations. A similar statement occurs as Rule 1 in Newton's "Rules for the Study of Natural Philosophy" [149]. This principle has become known as "Occam's razor."

Occam's inversion uses the discrepancy principle, and, at each iteration, searches for a solution that minimizes $\|\mathbf{Lm}\|_2$ subject to the constraint $\|\mathbf{G(m)} - \mathbf{d}\|_2 \leq \delta$. The algorithm is straightforward to implement, requires only the nonlinear forward model $\mathbf{G(m)}$ and its Jacobian, and works well in practice.

We assume that our nonlinear inverse problem has been cast in the form of (10.1). The roughening matrix $\mathbf{L}$ can be $\mathbf{I}$ to implement zeroth-order Tikhonov regularization, or it can be a finite difference approximation of a first (4.27) or second (4.28) derivative for higher-order regularization. In practice, Occam's inversion is often used on two- or three-dimensional problems where $\mathbf{L}$ is a discrete approximation of the Laplacian operator.

As usual, we will assume that the measurement errors in $\mathbf{d}$ are independent and normally distributed. For convenience, we will also assume that the system of equations $\mathbf{G(m)} = \mathbf{d}$ has been weighted so that the corresponding standard deviations on each weighted data point σ_i are equal.

The basic idea behind Occam's inversion is an iteratively applied local linearization. Given a trial model $\mathbf{m}^{(k)}$, Taylor's theorem is applied to obtain the local approximation

$$\mathbf{G}(\mathbf{m}^{(k)} + \Delta\mathbf{m}) \approx \mathbf{G}(\mathbf{m}^{(k)}) + \mathbf{J}(\mathbf{m}^{(k)})\Delta\mathbf{m} \qquad (10.8)$$

where $\mathbf{J}(\mathbf{m}^{(k)})$ is the Jacobian

$$\mathbf{J}(\mathbf{m}^{(k)}) = \left[\begin{array}{ccc} \frac{\partial G_1(\mathbf{m})}{\partial m_1} & \cdots & \frac{\partial G_1(\mathbf{m})}{\partial m_n} \\ \vdots & \ddots & \vdots \\ \frac{\partial G_m(\mathbf{m})}{\partial m_1} & \cdots & \frac{\partial G_m(\mathbf{m})}{\partial m_n} \end{array} \right]_{\mathbf{m}=\mathbf{m}^{(k)}} . \qquad (10.9)$$

Using (10.8), the regularized least squares problem (10.3) becomes

$$\min \|\mathbf{G}(\mathbf{m}^{(k)}) + \mathbf{J}(\mathbf{m}^{(k)})\Delta\mathbf{m} - \mathbf{d}\|_2^2 + \alpha^2\|\mathbf{L}(\mathbf{m}^{(k)} + \Delta\mathbf{m})\|_2^2 \qquad (10.10)$$

where the variable is $\Delta\mathbf{m}$ and $\mathbf{m}^{(k)}$ is constant. Reformulating (10.10) as a problem in which the variable is $\mathbf{m}^{(k+1)} = \mathbf{m}^{(k)} + \Delta\mathbf{m}$, we have

$$\min \| \mathbf{J}(\mathbf{m}^{(k)})(\mathbf{m}^{(k)} + \Delta\mathbf{m}) - (\mathbf{d} - \mathbf{G}(\mathbf{m}^{(k)}) + \mathbf{J}(\mathbf{m}^{(k)})\mathbf{m}^{(k)})\|_2^2 + \alpha^2\|\mathbf{L}(\mathbf{m}^{(k)} + \Delta\mathbf{m})\|_2^2 . \qquad (10.11)$$

Finally, letting

$$\hat{\mathbf{d}}(\mathbf{m}^{(k)}) = \mathbf{d} - \mathbf{G}(\mathbf{m}^{(k)}) + \mathbf{J}(\mathbf{m}^{(k)})\mathbf{m}^{(k)} \qquad (10.12)$$

gives

$$\min \| \mathbf{J}(\mathbf{m}^{(k)})\mathbf{m}^{(k+1)} - \hat{\mathbf{d}}(\mathbf{m}^{(k)})\|_2^2 + \alpha^2\|\mathbf{L}(\mathbf{m}^{(k+1)})\|_2^2 . \qquad (10.13)$$

Because $\mathbf{J}(\mathbf{m}^{(k)})$ and $\hat{\mathbf{d}}(\mathbf{m}^{(k)})$ are constant within a given iteration, (10.13) is in the form of a regularized linear least squares problem. If the system is of full rank, the solution is given by

$$\mathbf{m}^{(k+1)} = \mathbf{m}^{(k)} + \Delta\mathbf{m} = \left(\mathbf{J}(\mathbf{m}^{(k)})^T\mathbf{J}(\mathbf{m}^{(k)}) + \alpha^2\mathbf{L}^T\mathbf{L}\right)^{-1}\mathbf{J}(\mathbf{m}^{(k)})^T\hat{\mathbf{d}}(\mathbf{m}^{(k)}) . \qquad (10.14)$$

Note that this method is similar to the GN method applied to the regularized least squares problem (10.3) (see Exercise 10.1). The difference is that in Occam's inversion the parameter α is dynamically adjusted during each iteration, selecting the largest value for which $\chi^2 \leq \delta^2$ or, if this is not possible, selecting a value of α for which χ^2 is minimized. At the end of the procedure, we hope to converge to a solution that minimizes $\|\mathbf{L}\mathbf{m}\|_2$ among all solutions with $\chi^2 \leq \delta^2$. In practice, the constraint will typically be active at optimality and the optimal solution will have $\chi^2 = \delta^2$. However, if the unconstrained minimum of $\|\mathbf{L}\mathbf{m}\|_2$ has $\chi^2 < \delta^2$, the method will converge to that solution.

Furthermore, if no solution exists with $\chi^2 \leq \delta^2$, then the procedure will converge to a solution that minimizes χ^2. We can now state the algorithm.

Algorithm 10.1 Occam's Inversion

Beginning with an initial solution, $\mathbf{m}^{(0)}$, repeat the following steps to compute a sequence of solutions $\mathbf{m}^{(k)}$. Stop when the sequence converges.

1. Calculate the Jacobian $\mathbf{J}(\mathbf{m}^{(k)})$ and the vector $\hat{\mathbf{d}}(\mathbf{m}^{(k)})$.
2. Calculate updated models corresponding to a range of regularization parameter values using (10.14).
3. Find the particular $\mathbf{m}^{(k+1)}$ with the largest value of α such that $\chi^2(\mathbf{m}^{(k+1)}) \leq \delta^2$. If this condition cannot be satisfied, then use a value of α that minimizes $\chi^2(\mathbf{m}^{(k+1)})$.
4. Let $k = k + 1$.

Example 10.2

We consider the problem of estimating subsurface electrical conductivities from above ground electromagnetic induction measurements. The instrument used in this example is the Geonics EM-38 ground conductivity meter. A description of the instrument and the mathematical model of its response can be found in [88]. The mathematical model is complicated, but we will treat it as a black box, and concentrate on the inverse problem.

Measurements are taken at heights of 0, 10, 20, 30, 40, 50, 75, 100, and 150 cm above the surface, with the coils oriented in both the vertical and horizontal orientations. There are a total of 18 observations. We will assume measurement standard deviations of 0.1 mS/m. The measurements are in the data file **EMdata.mat**.

We discretize the subsurface electrical conductivity profile into 10 20 cm-thick layers and a semi-infinite layer below 2 m, giving us 11 parameters to estimate. The forward problem function $\mathbf{G}(\mathbf{m})$ is available to us as a subroutine. Since we do not have simple formulas for $\mathbf{G}(\mathbf{m})$, we cannot write down analytic expressions for the elements of the Jacobian. However, for sufficiently small problems, we can use finite difference approximations (9.49) to estimate the necessary partial derivatives.

We first apply the LM method to estimate unregularized model parameters. After 50 iterations, the LM method produced the model shown in Fig. 10.6. The χ^2 value for this model is 9.62 and there are $18 - 11 = 7$ degrees of freedom, so the model actually fits the data reasonably well. However, the least squares problem is very badly conditioned, with a condition number for $\mathbf{J}^T\mathbf{J}$ of approximately 2×10^{17}. Furthermore, the resulting model is unrealistic because it includes negative electrical conductivities and exhibits the high amplitudes and high frequency oscillations that are characteristic of under-regularized solutions to inverse problems.

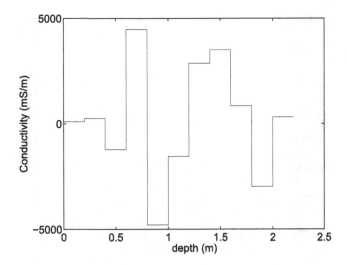

Figure 10.6 LM solution.

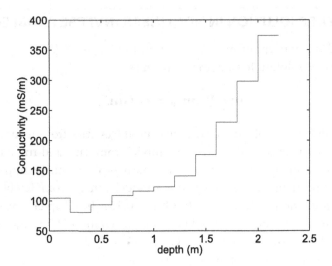

Figure 10.7 Occam's inversion solution.

We next apply Occam's inversion with second-order regularization and a discrepancy principle value of $\delta = 0.1 \cdot \sqrt{18} \approx 0.424$. The resulting model is shown in Fig. 10.7, and Fig. 10.8 shows the true model. The Occam's inversion method provides a fairly good reproduction of the true model.

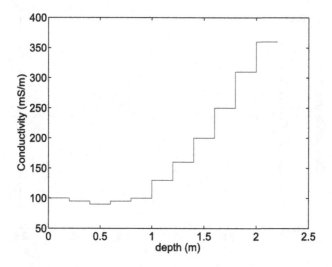

Figure 10.8 True model.

10.3. MODEL RESOLUTION IN NONLINEAR INVERSE PROBLEMS

We introduced the concept of model resolution in Chapter 3 (3.62) by expressing the generalized inverse solution for a linear problem as

$$\mathbf{m} = \mathbf{R}_m \mathbf{m}_{\text{true}} = \mathbf{G}^\dagger \mathbf{G} \mathbf{m}_{\text{true}} \qquad (10.15)$$

where $\mathbf{G}$ is a (forward problem) matrix that produces data from a model, $\mathbf{G}^\dagger$ is the generalized inverse matrix used to recover a model from data, and $\mathbf{m}$ and $\mathbf{m}_{\text{true}}$ are the recovered and true models, respectively. In a linear problem, the action of mapping a model to data and then back to a recovered model can be completely characterized by the model resolution matrix $\mathbf{R}_m = \mathbf{G}^\dagger \mathbf{G}$ in (10.15). In Chapter 4 we saw that this concept could easily be extended to Tikhonov regularization (4.20), by substituting the corresponding Tikhonov inverse matrix, $\mathbf{G}^\sharp$, for $\mathbf{G}^\dagger$ in (10.15).

We can recast (10.15) for nonlinear problems as

$$\mathbf{m} = G^{-1}(G(\mathbf{m}_{\text{true}})) \qquad (10.16)$$

where G^{-1} and G are inverse and forward operators. However, the combined action of the forward and inverse operations is not representable as a matrix, such as $\mathbf{R}_m$ in (10.16), because the forward operator is a nonlinear function or algorithm, and the inverse operator is typically realized with an iterative method based on stepwise linearization, such as GN or Occam's inversion. Furthermore, model resolution for a nonlinear inversion will not only depend on the physics, model parameterization, and data collection specifics, as was the case for linear problems, but may furthermore depend on the choice of starting model used in the solution algorithm, chosen convergence criteria, and possibly on the existence of multiple equally good solutions. Finally, as with linear methods, nonlinear model resolution will depend on the level and nature of the imposed regularization.

Because of these complexities, nonlinear resolution is typically analyzed using resolution tests. In seismic tomography, for example, it is common to evaluate the effects of non–ideal resolution by generating noise-free synthetic data from a spike, checkerboard, or other test model using an identical source and receiver geometry as for the actual data in the problem of interest. A model is then recovered using the identical inverse methodology as was used for the actual data, and is compared to the test model to evaluate inversion artifacts. If there are specific features of the true model that will affect the resolution, such as a known strong velocity gradient with depth in seismic problems that significantly affects the curvature of the ray paths, those features should also be incorporated into the resolution test model. Because the problem is nonlinear, resolution analysis results will also potentially be dependent on test model amplitudes. A second resolution analysis strategy, sometimes referred to as a "squeeze" test, restricts the freedom of the model in regions that are suspected to be unnecessary or marginal for achieving an acceptable fit to the data. One example would be to modify the regularization constraints to strongly penalize model variations in the deeper part of a tomographic model to assess if an acceptable data fit can still be achieved when structure is predominantly restricted to shallower depths.

Example 10.3

Revisiting Example 10.1, we calculate noise-free synthetic data for a checkerboard velocity structure, using an identical starting model and ray path end points. The checkerboard model (Fig. 10.9) consists of a 8 by 8 node 2.9 km/s background model with alternating 10% variations for the 36 interior nodes. Inverting these data using the identical methodology as in Example 10.1 for the same regularization constraints and range of regularization parameters, we obtain the suite of models shown in Fig. 10.10. For lower levels of regularization, the checkerboard is more apparent, whereas for higher levels, the horizontally varying velocity structure is substantially smoothed out. For the level of regularization chosen from the discrepancy principle for Example 10.1, $\alpha \approx 37$, we note substantial smearing that makes it difficult to discern the full checkerboard pattern, indicating that structural variations of this character and spatial scale will be difficult to recover in some parts of the model without additional data. If we did not know the true model in Example 10.1, this test would helpfully show that we would expect significant horizontal smearing for the inversion result shown

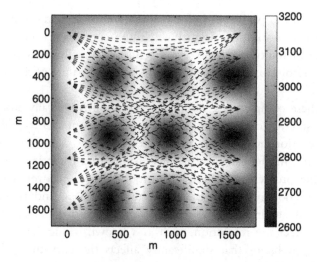

Figure 10.9 Checkerboard test model and ray paths for the cross well tomography problem of Example 10.1.

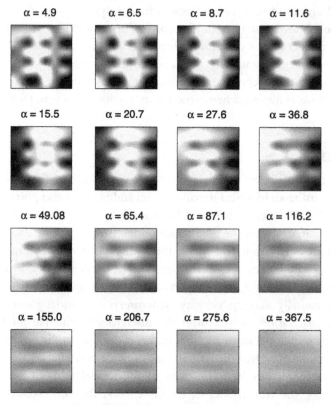

Figure 10.10 Suite of recovered models for a range of regularization parameter values (same as in Fig. 10.2) for the checkerboard test model of Fig. 10.9.

in Fig. 10.4, and that the true model anomalies are thus not necessarily horizontally elongated.

10.4. THE NONLINEAR CONJUGATE GRADIENT METHOD

For problems with a large number of parameters, using the Levenberg–Marquardt or Gauss–Newton methods to solve (10.3) or solving the system of equations in (10.14) may be impractical due to the very large size of the matrix. For example, if our parameter estimation problem utilizes a three-dimensional partial differential equation model that has been discretized on a 500 by 500 by 500 grid, the system of equations would have 125 million equations and 125 million variables. In such a case, it would be impractical to store or factor $\mathbf{J}^T(\mathbf{m}^{(k)})\mathbf{J}(\mathbf{m}^{(k)})$ or to compute and store a model resolution matrix. For such large problems, we require optimization techniques that do not involve the direct solution of a large linear system of equations.

The method of conjugate gradients applied to least squares problems (CGLS) was introduced in Chapter 6 for linear systems of equations. This method has been extended by a number of authors for use in minimizing a general nonlinear function, $f(\mathbf{x})$. In this section we introduce the nonlinear conjugate gradient method of Fletcher and Reeves [63]. The conjugate gradient method for minimizing a quadratic function generates a sequence of vectors $\mathbf{p}^{(k)}$ that are conjugate with respect to a matrix $\mathbf{A}$ (6.34). For a nonlinear optimization problem that is approximately quadratic in a region of interest, we can use the Hessian matrix $\nabla^2 f(\mathbf{x}^{(k)})$ as a substitute for $\mathbf{A}$. Because the Hessian will be changing over iterations, the $\mathbf{p}^{(k)}$ vectors will be only approximately conjugate. In the CGLS method applied to the normal equations, the residual $\mathbf{G}^T(\mathbf{d} - \mathbf{Gm})$ was proportional to the negative of the objective function gradient. In the nonlinear conjugate gradient method we use the gradient of $f(\mathbf{x})$.

The conjugate gradient method for minimizing a quadratic function also made use of a step size α_k that exactly minimized the objective function in the search direction (6.27). In the nonlinear conjugate gradient method, a one-dimensional line search optimization is performed to find the best step size. A variety of formulas have been proposed for computing the conjugate gradient coefficient β_k in the nonlinear case. The Fletcher–Reeves method uses

$$\beta_k = \frac{\|\nabla f(\mathbf{x}^{(k)})\|_2^2}{\|\nabla f(\mathbf{x}^{(k-1)})\|_2^2} .$$ (10.17)

With these modifications to the linear theory we can state the Fletcher–Reeves nonlinear conjugate gradient algorithm.

Algorithm 10.2 Nonlinear Conjugate Gradient Method, Fletcher–Reeves

---

Given a differentiable function $f(\mathbf{x})$, and an initial solution $\mathbf{x}^{(0)}$, let $\beta_0 = 0$, $\mathbf{p}^{(-1)} = \mathbf{0}$, and $k = 0$.

1. Let $\mathbf{g}^{(k)} = \nabla f(\mathbf{x}^{(k)})$.
2. If $k > 0$, let $\beta_k = \dfrac{\mathbf{g}^{(k)^T}\mathbf{g}^{(k)}}{\mathbf{g}^{(k-1)^T}\mathbf{g}^{(k-1)}}$.
3. Let $\mathbf{p}^{(k)} = -\mathbf{g}^{(k)} + \beta_k \mathbf{p}^{(k-1)}$.
4. Find α_k to minimize $f(\mathbf{x}^{(k)} + \alpha_k \mathbf{p}^{(k)})$.
5. Let $\mathbf{x}^{(k+1)} = \mathbf{x}^{(k)} + \alpha_k \mathbf{p}^{(k)}$.
6. Let $k = k + 1$.
7. Repeat the previous steps until convergence.

This algorithm is implemented in the library function **conjg**. The use of the function is illustrated in Example 10.4.

10.5. THE DISCRETE ADJOINT METHOD

In many situations the forward problem involves the numerical solution of a partial differential equation boundary value problem. The corresponding inverse problem is to determine the coefficients of the partial differential equation, or its initial or boundary conditions. In this section we consider an efficient approach to computing derivatives of the solution to a partial differential equation boundary value problem with respect to its parameters. These derivatives can then be used by an optimization algorithm to solve the inverse problem.

Let $\mathbf{A}$ be an m by n matrix, let $\mathbf{x}$ be an n by 1 vector, and let $\mathbf{y}$ be an m by 1 vector. Consider the dot product or inner product of the vector $\mathbf{Ax}$ with $\mathbf{y}$. We can write this using the angle bracket notation for the inner product or using vector–matrix multiplication notation

$$\langle \mathbf{Ax}, \mathbf{y} \rangle = (\mathbf{Ax})^T \mathbf{y} . \tag{10.18}$$

By transposing $\mathbf{Ax}$, we can write this as

$$\langle \mathbf{Ax}, \mathbf{y} \rangle = \mathbf{x}^T \mathbf{A}^T \mathbf{y} = \langle \mathbf{x}, \mathbf{A}^T \mathbf{y} \rangle . \tag{10.19}$$

As long as we know both $\mathbf{A}$ and $\mathbf{A}^T$, we can switch back and forth between these two ways of writing the inner product. In this context, $\mathbf{A}^T$ is called the **adjoint** of $\mathbf{A}$.

Next, assume that a forward problem has been discretized as the solution of a linear system of equations

$$\mathbf{Au} = \mathbf{b} \tag{10.20}$$

where $\mathbf{A}$ is now a square, m by m, matrix. We would like to compute the partial derivatives of some of the solution vector elements u_i with respect to the right hand side elements b_j, $j = 1, 2, \ldots, m$. Consider a small change $\Delta \mathbf{u}$ to $\mathbf{u}$, and the corresponding change $\Delta \mathbf{b}$ to $\mathbf{b}$. We then have

$$\mathbf{A} \Delta \mathbf{u} = \Delta \mathbf{b} . \tag{10.21}$$

To find the change in u_i, we take the inner product of the ith standard basis vector, $\mathbf{e}_i$, and $\Delta \mathbf{u}$, so that

$$\begin{aligned} \Delta u_i &= \langle \mathbf{e}_i, \Delta \mathbf{u} \rangle \\ &= \langle \mathbf{e}_i, \mathbf{A}^{-1} \Delta \mathbf{b} \rangle . \end{aligned} \tag{10.22}$$

Using the adjoint rule for a dot product (10.19), this can be written as

$$\Delta u_i = \langle \mathbf{A}^{-T} \mathbf{e}_i, \Delta \mathbf{b} \rangle . \tag{10.23}$$

Let $\mathbf{v} = \mathbf{A}^{-T} \mathbf{e}_i$ or equivalently, let $\mathbf{v}$ be the solution to

$$\mathbf{A}^T \mathbf{v} = \mathbf{e}_i . \tag{10.24}$$

Then

$$\Delta u_i = \langle \mathbf{v}, \Delta \mathbf{b} \rangle . \tag{10.25}$$

Taking the limit $\Delta \mathbf{b} \to 0$, we have

$$\frac{\partial u_i}{\partial b_j} = v_j \quad j = 1, 2, \ldots, m . \tag{10.26}$$

Another way to see this is that since

$$\Delta \mathbf{u} = \mathbf{A}^{-1} \Delta \mathbf{b} \tag{10.27}$$

the desired partial derivatives are the ith row of $\mathbf{A}^{-1}$. It is not easy to compute the ith row of $\mathbf{A}^{-1}$ directly, but we can find the ith column of $\mathbf{A}^{-T}$ by solving $\mathbf{A}^T \mathbf{v} = \mathbf{e}_i$.

This method saves no effort if we need to compute the partial derivatives of u_i with respect to b_j for all i and j. In that case we really need all of the entries of $\mathbf{A}^{-1}$. However, the adjoint method becomes appealing when we need the partial derivatives for a small number of model parameters with respect to every data point.

In some situations it is desirable to compute a weighted sum of the derivatives of several of the entries in $\mathbf{u}$ with respect to entries in $\mathbf{b}$. If the desired sum is

$$v_j = \sum_{i=1}^{m} c_i \frac{\partial u_i}{\partial b_j} \tag{10.28}$$

then we can obtain $\mathbf{v}$ by solving the adjoint equation

$$\mathbf{A}^T \mathbf{v} = \mathbf{c} . \tag{10.29}$$

The discrete adjoint method can also be applied when $\mathbf{u}$ and $\mathbf{b}$ are related to each other by a nonlinear system of equations. If

$$A(\mathbf{u}) = \mathbf{b} \tag{10.30}$$

we can write

$$A(\mathbf{u} + \Delta\mathbf{u}) = \mathbf{b} + \Delta\mathbf{b} \tag{10.31}$$

and approximate $A(\mathbf{u} + \Delta\mathbf{u})$ by

$$A(\mathbf{u} + \Delta\mathbf{u}) \approx A(\mathbf{u}) + J(\mathbf{u})\Delta\mathbf{u}, \tag{10.32}$$

where $J(\mathbf{u})$ is the Jacobian of $A(\mathbf{u})$. Using this approximation,

$$A(\mathbf{u}) + J(\mathbf{u})\Delta\mathbf{u} \approx \mathbf{b} + \Delta\mathbf{b} \tag{10.33}$$

and

$$J(\mathbf{u})\Delta\mathbf{u} = \Delta\mathbf{b} \tag{10.34}$$

in the limit as $\Delta\mathbf{u}$ goes to 0. This equation has exactly the same form as (10.21). Thus we can solve

$$J(\mathbf{u})^T\mathbf{v} = \mathbf{e}_i \tag{10.35}$$

and obtain the partial derivatives

$$\frac{\partial u_i}{\partial b_j} = v_j \quad j = 1, 2, \ldots, m. \tag{10.36}$$

In many cases a forward problem is implemented as the discretization of a time-dependent partial differential equation in which a linear system of equations is solved at each time step. We begin with an initial solution $\mathbf{u}^{(0)}$ at time $t = 0$, and then propagate forward in time by solving

$$\begin{aligned} \mathbf{A}^{(1)}\mathbf{u}^{(1)} &= \mathbf{u}^{(0)} \\ \mathbf{A}^{(2)}\mathbf{u}^{(2)} &= \mathbf{u}^{(1)} \\ &\vdots \\ \mathbf{A}^{(n)}\mathbf{u}^{(n)} &= \mathbf{u}^{(n-1)}. \end{aligned} \tag{10.37}$$

The individual matrices $\mathbf{A}^{(k)}$ might vary with time, or there might simply be one matrix $\mathbf{A}$ used in all time steps.

In this scheme, we can write $\mathbf{u}^{(n)}$ as

$$\mathbf{u}^{(n)} = \mathbf{A}^{(n)-1}\mathbf{A}^{(n-1)-1}\cdots\mathbf{A}^{(1)-1}\mathbf{u}^{(0)}. \tag{10.38}$$

Although (10.38) is useful in the following analysis, it is typically faster in computational practice to solve the sequence of systems of equations given by (10.37).

To find the derivative of an entry in $\mathbf{u}^{(n)}$ with respect to an entry in $\mathbf{u}^{(0)}$, we can again use the discrete adjoint method. If $\Delta\mathbf{u}^{(0)}$ is a change in the initial solution, then the corresponding change in the final solution, $\Delta\mathbf{u}^{(n)}$ is given by

$$\Delta\mathbf{u}^{(n)} = \mathbf{A}^{(n)-1}\mathbf{A}^{(n-1)-1}\cdots\mathbf{A}^{(1)-1}\Delta\mathbf{u}^{(0)} . \tag{10.39}$$

The ith component of $\Delta\mathbf{u}^{(n)}$ is given by

$$\begin{aligned}
\Delta u_i^{(n)} &= \langle\Delta\mathbf{u}^{(n)}, \mathbf{e}_i\rangle \\
&= \langle\mathbf{A}^{(n)-1}\mathbf{A}^{(n-1)-1}\cdots\mathbf{A}^{(1)-1}\Delta\mathbf{u}^{(0)}, \mathbf{e}_i\rangle .
\end{aligned} \tag{10.40}$$

Using the adjoint, we have

$$\Delta u_i^{(n)} = \langle\Delta\mathbf{u}^{(0)}, \mathbf{A}^{(1)-T}\mathbf{A}^{(2)-T}\cdots\mathbf{A}^{(n)-T}\mathbf{e}_i\rangle . \tag{10.41}$$

Again, it is undesirable to explicitly compute the inverses of all n $\mathbf{A}^{(k)}$ matrices and multiply their transposes together. Rather, we can more efficiently compute the product

$$\mathbf{v} = \mathbf{A}^{(1)-T}\mathbf{A}^{(2)-T}\cdots\mathbf{A}^{(n)-T}\mathbf{e}_i \tag{10.42}$$

by solving a series of systems of equations,

$$\begin{aligned}
\mathbf{A}^{(n)T}\mathbf{v}^{(n)} &= \mathbf{e}_i \\
\mathbf{A}^{(n-1)T}\mathbf{v}^{(n-1)} &= \mathbf{v}^{(n)} \\
&\vdots \\
\mathbf{A}^{(1)T}\mathbf{v} &= \mathbf{v}^{(2)} .
\end{aligned} \tag{10.43}$$

In solving the systems of equations (10.43), we are effectively moving backward rather than forward in time. In other words, we evaluate the adjoint by a process of time reversal.

Once we have solved for $\mathbf{v}$, we have

$$\Delta u_i^{(n)} = \langle\mathbf{v}, \Delta\mathbf{u}^{(0)}\rangle \tag{10.44}$$

or

$$\frac{\partial u_i^{(n)}}{\partial u_j^{(0)}} = v_j . \tag{10.45}$$

Using this approach, we can compute the gradient of $\mathbf{u}_i^{(n)}$ with respect to $\mathbf{u}^{(0)}$ by solving an adjoint problem that is equivalent in difficulty to solving the forward problem.

Example 10.4

Consider the one-dimensional time-dependent heat equation describing the spatial and temporal variation of temperature $u(x, t)$ in a uniform material under heat diffusion

$$\frac{\partial u(x, t)}{\partial t} = D\frac{\partial^2 u(x, t)}{\partial x^2} \tag{10.46}$$

where D is the thermal diffusivity, with boundary conditions

$$u(0, t) = u(1, t) = 0 \tag{10.47}$$

and an initial temperature

$$u(x, 0) = u^{(0)}(x) . \tag{10.48}$$

We perform a laboratory experiment in one spatial dimension. The material has a thermal diffusivity of $D = 1 \times 10^{-6}$ m^2/s (within the range of common geological specimens near room temperature), and the experiment extends over $x = [-1, 1]$. We measure a noisy temperature perturbation signal in degrees Kelvin, $\Delta T = u(t, x)$, at $t = 10{,}000$ s. Measurements are taken at the 39 points $x = -0.95, 0.90, \ldots, 0.95$ m. We seek to recover the temperature perturbation at $t = 0$.

We discretize (10.46) using the Crank–Nicolson finite difference scheme [112]. The spatial grid consists of 1000 points, $x = -0.998, 0.996, \ldots, 0.998, 1.000$ m, and we use time steps of $\Delta t = 50$ s, with 200 time steps thus required to simulate the heat anomaly at $t = 10{,}000$ s.

To iterate time steps, the Crank–Nicolson method solves a system of equations of the form

$$\mathbf{A}\mathbf{u}^{(i+1)} = \mathbf{B}\mathbf{u}^{(i)} \tag{10.49}$$

where $\mathbf{A}$ and $\mathbf{B}$ are tridiagonal matrices. Solving the forward problem thus requires the repeated solution of (10.49), and the adjoint problem can be solved in the same manner by time reversal. To obtain a regularized inverse solution, we combine this Jacobian from the adjoint equation with the Jacobian of the regularization term using (10.5).

Fig. 10.11 shows noisy temperature anomaly data at $t = 10{,}000$ s with normally distributed independent data errors characterized by a standard deviation of 0.1°K. The nonlinear conjugate gradient method with discrete adjoint derivatives was used to solve the inverse problem with zeroth-order Tikhonov regularization.

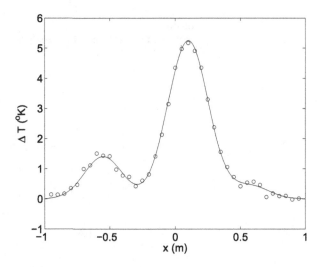

Figure 10.11 Temperature anomaly data at $t = 10{,}000$ s. The noise-free temperature anomaly is shown as a smooth curve.

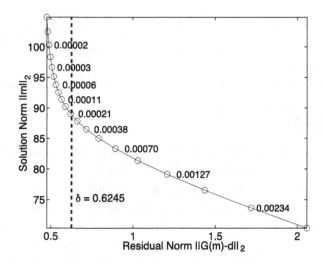

Figure 10.12 L-curve of solutions for the adjoint heat flow example, showing values of the Tikhonov regularization parameter and the discrepancy principle value of the residual norm.

The regularization parameter was chosen to satisfy the discrepancy principle assuming normally distributed independent errors, which gives a residual 2-norm target value of $0.01 \cdot \sqrt{39} \approx 0.6245°K$. The L-curve for this problem is shown in Fig. 10.12 and the corresponding discrepancy principle solution, along with the true $u^{(0)}(x)$ initial condition, is shown in Fig. 10.13. The true model consists of three rectangular heat anomalies,

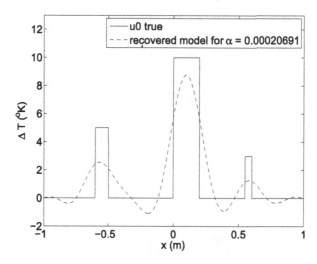

Figure 10.13 True initial condition and inverse solution with regularization parameter determined from the discrepancy principle as shown in Fig. 10.12.

and the zeroth-order Tikhonov regularized solution recovers their general locations, but with limited resolution, resulting in decreased anomaly amplitudes, broader widths, and oscillatory side lobes.

10.6. EXERCISES

1. Show that, for a given value of α, the GN (10.7) and Occam's inversion model (10.14) update steps are mathematically equivalent.

2. Recall Example 1.5, in which we had gravity anomaly observations above a density perturbation of variable depth $m(x)$ and fixed density $\Delta\rho$. Use Occam's inversion to solve an instance of this inverse problem. Consider a gravity anomaly along a 1 km section, with observations taken every 50 m, and density perturbation of $\Delta\rho = 200$ kg/m^3 (0.2 g/cm^3). The perturbation is expected to be at a depth of roughly 200 m.

   The MATLAB data file **gravprob.mat** contains a vector **x** of observation locations. Use the same coordinates for your discretization of the model. The vector **obs** contains the actual observations. Assume that the observations are accurate to about 1.0×10^{-12}.

   a. Derive a formula for the elements of the Jacobian.

   b. Write MATLAB routines to compute the model predictions and the Jacobian for this problem.

c. Use the supplied implementation of Occam's inversion from Example 10.2 to solve the inverse problem with second-order regularization.

d. Discuss your results. What features in the inverse solution appear to be real? What is the resolution of your solution? Were there any difficulties with local minima?

e. What would happen if the true density perturbation was instead at about 1000 m depth?

3. Apply the GN method with explicit regularization to the EM inversion problem by modifying the MATLAB code from Example 10.2. Compare your solution with the solution obtained by Occam's inversion. Which method required more computational effort?

4. Apply Occam's inversion to a cross-well bent-ray tomography problem with identical geometry to Example 10.1. Use the example MATLAB subroutine **getj** used in the example to forward model the travel times, calculate the Jacobian, and implement second-order regularization.

Travel-time data and subroutine control parameters are contained in the MATLAB data file **benddata.mat**. Start with the uniform 2900 m/s velocity 8 by 8 node initial velocity model in **benddata.mat**, and assume independent and normally distributed data errors with $\sigma = 0.001$ ms.

Hint: A search range of α^2 between 10 and 10^5 is appropriate for this problem. MATLAB code for generating a second-order roughening matrix that approximates a two-dimensional Laplacian operator can be found in **makerough.m**.

5. Apply the nonlinear conjugate gradient method with explicit regularization to the EM inversion problem by modifying the MATLAB code from Example 10.2. Compare your solution with the solution obtained in the example by Occam's inversion. Which method required more computational effort?

6. Modify the code from Example 10.4 to count the number of calls to the objective function, gradient, forward solve, and adjoint solve routines. Write a routine to compute the gradient using a finite difference approximation and the forward solve routine. Compare the performance of your finite difference routine for computing the gradient with the performance of the discrete adjoint method. How much slower would the solution of the problem be if you used these finite difference derivatives?

10.7. NOTES AND FURTHER READING

For some problems, it may be impractical to use direct factorization to solve the systems of equations (10.7) or (10.14) involved in computing the Gauss–Newton or Occam step. One approach in this case is to use an iterative method such as conjugate gradients to solve the linear systems of equations [76].

Useful references on adjoint methods include [23,69,93,133,195]. We have introduced the discrete adjoint method in which the differential equation is first discretized and then the adjoint of the discretized equation is used to compute derivatives of the numerical solution with respect to the parameters. In the alternative continuous adjoint approach the adjoint of the differential equation is derived as a differential equation which is then discretized. The relative advantages of discrete and continuous adjoint methods are discussed in [69,182].

In inverse problems with high-dimensional model and data spaces, the most computationally demanding task is often computing derivatives of $\mathbf{G}(\mathbf{m})$ with respect to the model parameters, often generally referred to as Fréchet derivatives [136]. Computation of derivatives via analytic formulas may be unavailable, and straightforward finite-difference derivative estimates (9.49) may be computationally intractable, given that the number of derivative estimates is equal to the product of the data and model space dimensions, and that each Fréchet derivative estimate using (9.49) requires two forward problem calculations. Adjoint methodologies can be applied to some such problems to estimate the desired derivatives very efficiently, and thus have significant utility for iterative solution of very large nonlinear inverse problems. A notable example is full-waveform seismic inversion. In this case the derivatives necessary for each iterative model update can be estimated from just two seismic wavefield calculations in the current model, specifically a forward simulation, and a time-reversed (adjoint) simulation that applies the residual seismograms as pseudo sources at respective seismic station locations [58,198,210].

Bayesian Methods

Synopsis

Following a review of the classical least squares approach to solving inverse problems, we introduce the Bayesian approach, which treats the model as a random variable with a probability distribution that we seek to estimate. A prior distribution for the model parameters is combined with the data to produce a posterior distribution for the model parameters. In special cases, the Bayesian approach produces solutions that are equivalent to the least squares, maximum likelihood, and Tikhonov regularization solutions. Several examples of the Bayesian approach are presented. Markov Chain Monte Carlo methods for sampling from the posterior distribution are presented and demonstrated.

11.1. REVIEW OF THE CLASSICAL APPROACH

In the classical approach to parameter estimation and inverse problems with discrete data and models, we begin with a mathematical model of the form $\mathbf{Gm} = \mathbf{d}$ in the linear case or $G(\mathbf{m}) = \mathbf{d}$ in the nonlinear case. We assume that there exists a true model $\mathbf{m}_{\text{true}}$ and a true data set $\mathbf{d}_{\text{true}}$ such that $\mathbf{Gm}_{\text{true}} = \mathbf{d}_{\text{true}}$. We acquire an actual data set $\mathbf{d}$, which is generally the sum of $\mathbf{d}_{\text{true}}$ and measurement noise. Our goal is to recover $\mathbf{m}_{\text{true}}$ from the noisy data.

For well-conditioned linear problems, under the assumption of independent and normally distributed data errors, the theory is well developed. In Chapter 2 it was shown that the maximum likelihood principle leads to the least squares solution, which is found by minimizing the 2-norm of the residual, $\|\mathbf{Gm} - \mathbf{d}\|_2$. Since there is noise in the data, we should expect some misfit between the data predictions of the forward model and the data, so that observed values of the square of the 2-norm of the inverse standard deviation-weighted residual, χ^2_{obs}, will not typically be zero. We saw that the χ^2 distribution can be used to test the goodness-of-fit of a least squares solution. We showed that the least squares solution, $\mathbf{m}_{L_2}$, is an unbiased estimate of $\mathbf{m}_{\text{true}}$. We were also able to compute a covariance matrix for the estimated parameters

$$\text{Cov}(\mathbf{m}_{L_2}) = (\mathbf{G}^T\mathbf{G})^{-1}\mathbf{G}^T\text{Cov}(\mathbf{d})\mathbf{G}(\mathbf{G}^T\mathbf{G})^{-1} \tag{11.1}$$

to compute confidence intervals for and correlations between the estimated parameters.

This approach works very well for linear regression problems where the least squares problem is well-conditioned. We found, however, that in many cases the least squares problem is not well-conditioned. In such situations, the set of solutions that adequately

Parameter Estimation and Inverse Problems
https://doi.org/10.1016/B978-0-12-804651-7.00016-X

fits the data is large and diverse, and commonly contains physically unreasonable models.

In Chapters 4 through 8, we discussed a number of approaches to regularizing the least squares problem. These approaches pick one "best" solution out of those that adequately fit the data, based on a preference for what sort of model features constitute a good solution. Zeroth-order Tikhonov regularization selects the model that minimizes the model 2-norm $\|\mathbf{m}\|_2$ subject to the residual norm constraint, $\|\mathbf{Gm} - \mathbf{d}\|_2 < \delta$, whereas higher-order Tikhonov regularization selects the model that minimizes a model seminorm $\|\mathbf{Lm}\|_2$ subject to $\|\mathbf{Gm} - \mathbf{d}\|_2 < \delta$. We also introduced L_1 regularization and the related technique of total variation regularization.

For relatively small linear problems, straightforward, insightful, and robust computation of regularized solutions can be performed with the help of the SVD. For large sparse linear problems, iterative methods such as CGLS or LSQR are widely used.

For nonlinear problems, as discussed in Chapters 9 and 10, the Gauss–Newton, Levenberg–Marquardt, or Occam's inversion methods can be used to find a local minimum of the nonlinear least squares problem. We showed how approximate confidence intervals for the fitted parameters can be obtained by linearizing the nonlinear model around the best fit parameters. As in linear inverse problems, the nonlinear least squares problem can be badly conditioned, in which case regularization may be required to obtain a stable solution.

Unfortunately nonlinear problems may have a large number of local minimum solutions, and finding the global minimum can be difficult. Furthermore, if there are several local minimum solutions with acceptable data fits, then it may be difficult to select a single "best" solution.

How can we more generally justify selecting one solution from the set of models that adequately fit the data? One justification is Occam's razor, which is the philosophy that when we have several different hypotheses to consider, we should select the simplest. Solutions selected by regularization are in some sense the simplest models that fit the data. However, this approach is not by itself entirely satisfactory because different choices of the regularization term used in obtaining regularized solutions can result in very different models, and the specific choice of regularization may be subjective.

Recall from Chapter 4 (e.g., Example 4.3) that once we have regularized a least squares problem, we lose the ability to obtain statistically useful confidence intervals for the parameters because regularization introduces bias. In particular, the expected value of the regularized solution is not generally equal to the true model. In practice this regularization bias is often much more significant than the effect of noise in the data. Bounds on the error in Tikhonov-regularized solutions were discussed in Section 4.8. However, these estimates require knowledge of the true model that is typically not available in practice.

11.2. THE BAYESIAN APPROACH

The Bayesian approach is named after Thomas Bayes, an 18th century pioneer in probability theory. The methodology is based on philosophically different ideas than we have considered so far. However, as we will see, it frequently results in similar solutions.

The most fundamental difference between the classical and Bayesian approaches is in the conceptualization of the solution. In the classical approach, there is a specific but unknown model $\mathbf{m}_{\text{true}}$ that we would like to uncover. In the Bayesian approach the model is not deterministic, but is rather a random variable, and the solution takes the form of a probability distribution for the model parameters called the **posterior distribution**. Once we have this probability distribution, we can use it to answer probabilistic questions about the model, for example, "What is the probability that m_5 is less than 1?" In the classical approach such questions do not make sense, since the true model that we seek is not a random variable.

A second very important difference between the classical and Bayesian approaches is that the Bayesian approach naturally incorporates prior information about the solution, ranging from hard additional constraints to experience-based intuition. This information is expressed mathematically as a **prior distribution** for the model. Once data have been collected, they are combined with the prior distribution using Bayes' theorem (B.53) to produce the desired posterior distribution for the model parameters.

If no other information is available, then under the **principle of indifference**, we may alternatively pick a prior distribution where all model parameter values have equal likelihood. Such a prior distribution is said to be **uninformative**.

It should be pointed out that, in the common case where the model parameters are contained in the range $(-\infty, \infty)$, the uninformative prior is not a proper probability distribution. This is because there does not exist a probability density function $f(x)$ satisfying (B.4) so that

$$\int_{-\infty}^{\infty} f(x)\ dx = 1 \tag{11.2}$$

and $f(x)$ is constant. In practice, the use of this improper prior distribution in Bayesian methods can nevertheless be justified because the resulting posterior distribution for the model is a proper distribution.

One of the main objections to the Bayesian approach is that the method is "unscientific" because it allows the analyst to incorporate subjective judgments into the model that are not solely based on the data. Proponents of the approach rejoin that there are also subjective aspects to the classical approach embodied in the choice of regularization biases, and that, furthermore, one is free to choose an uninformative prior distribution.

We denote the prior distribution by $p(\mathbf{m})$ and assume that we can compute the conditional probability distribution, $f(\mathbf{d}|\mathbf{m})$, that, given a particular model, corresponding data, $\mathbf{d}$, will be observed. Given a prior distribution, we then seek the conditional posterior distribution of the model parameters given the data. We will denote this posterior probability distribution for the model parameters by $q(\mathbf{m}|\mathbf{d})$. Bayes' theorem relates the prior and posterior distributions in a way that makes the computation of $q(\mathbf{m}|\mathbf{d})$ possible, and can be stated as follows.

Theorem 11.1.

$$q(\mathbf{m}|\mathbf{d}) = \frac{f(\mathbf{d}|\mathbf{m})p(\mathbf{m})}{c} \tag{11.3}$$

where

$$c = \int_{\text{all models}} f(\mathbf{d}|\mathbf{m})p(\mathbf{m}) \, d\mathbf{m} . \tag{11.4}$$

Note that the constant c simply normalizes the conditional distribution $q(\mathbf{m}|\mathbf{d})$ so that its integral in model space is one.

For some purposes, knowing the normalization constant, c, is not necessary. For example, we can compare two models $\hat{\mathbf{m}}$ and $\bar{\mathbf{m}}$ by computing the likelihood ratio

$$\frac{q(\hat{\mathbf{m}}|\mathbf{d})}{q(\bar{\mathbf{m}}|\mathbf{d})} = \frac{f(\mathbf{d}|\hat{\mathbf{m}})p(\hat{\mathbf{m}})}{f(\mathbf{d}|\bar{\mathbf{m}})p(\bar{\mathbf{m}})} . \tag{11.5}$$

A very small likelihood ratio would indicate that the model $\bar{\mathbf{m}}$ is far more likely than the model $\hat{\mathbf{m}}$. Because c is not always needed, (11.3) is often written as a statement of proportionality

$$q(\mathbf{m}|\mathbf{d}) \propto f(\mathbf{d}|\mathbf{m})p(\mathbf{m}) . \tag{11.6}$$

However, there are many other situations in which knowing c in (11.3) is required. In particular, c is required to compute any posterior probabilities and to compute the expected value and variance of the posterior distribution.

It is important to reemphasize that the probability distribution $q(\mathbf{m}|\mathbf{d})$ does not provide a single model that we can consider to be the "answer." However, in cases where we want to single out a representative model, it may be appropriate to identify the one corresponding to the largest value of $q(\mathbf{m}|\mathbf{d})$. This model is referred to as the **maximum a posteriori (MAP)** model. Another possibility is to select the mean of the posterior distribution. In situations where the posterior distribution is normal, the MAP and posterior mean models will be identical.

In general, the computation of a posterior distribution using (11.3) can be difficult. The chief difficulty lies in evaluating the integral in (11.4). This integral often has very high dimensionality, and numerical integration techniques may thus be computationally daunting.

Fortunately, there are a number of useful special cases in which the computation of the posterior distribution is greatly simplified. One simplification occurs when the prior distribution $p(\mathbf{m})$ is uninformative, in which case (11.6) simplifies to

$$q(\mathbf{m}|\mathbf{d}) \propto f(\mathbf{d}|\mathbf{m}) \tag{11.7}$$

and the posterior distribution is precisely the likelihood function, $L(\mathbf{m}|\mathbf{d})$. Under the maximum likelihood principle, we would select the model $\mathbf{m}_{\text{ML}}$ that maximizes $L(\mathbf{m}|\mathbf{d})$, which is the MAP model.

A further simplification occurs when data noise elements are independent and normally distributed with standard deviation σ. Because the data noise is independent, we can write the likelihood function as

$$L(\mathbf{m}|\mathbf{d}) = f(\mathbf{d}|\mathbf{m}) = f(d_1|\mathbf{m}) \cdot f(d_2|\mathbf{m}) \cdots f(d_m|\mathbf{m}) . \tag{11.8}$$

If the data points d_i are normally distributed with expected values given by the (linear or nonlinear) operation $(G(\mathbf{m}))_i$, and if each has standard deviation σ, we can write,

$$f(d_i|\mathbf{m}) = \frac{1}{\sigma\sqrt{2\pi}} e^{-\frac{((G(\mathbf{m}))_i - d_i)^2}{2\sigma^2}} \tag{11.9}$$

and

$$L(\mathbf{m}|\mathbf{d}) = \left(\frac{1}{\sigma\sqrt{2\pi}}\right)^m e^{-\sum_{i=1}^{m} \frac{((G(\mathbf{m}))_i - d_i)^2}{2\sigma^2}} . \tag{11.10}$$

We can maximize (11.10) by maximizing the exponent or equivalently minimizing the negative of the exponent.

$$\min \sum_{i=1}^{m} \frac{((G(\mathbf{m}))_i - d_i)^2}{2\sigma^2} . \tag{11.11}$$

This is a weighted least squares problem. Thus we have shown that when we have independent and normally distributed data and we use an uninformative prior, the MAP solution is the least squares solution.

Example 11.1

Consider a very simple parameter estimation problem where we perform repeated weightings of a microscopic object to determine its mass in μg. The measurement errors are normally distributed with zero mean and standard deviation $\sigma = 1$ μg. Our goal is to estimate the mass of the object.

For the specified normally distributed and zero mean measurement error, the probability density function for a measurement d given m is

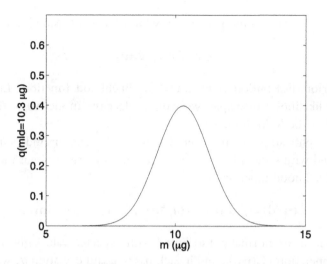

Figure 11.1 Posterior distribution $q(m|d_1 = 10.3\ \mu g)$, uninformative prior.

$$f(d|m) = \frac{1}{\sqrt{2\pi}}e^{-(m-d)^2/2} . \tag{11.12}$$

Suppose we weigh the mass once and obtain a measurement of $d_1 = 10.3\ \mu g$. What do we now know about m? For an uninformative prior, (11.7) gives

$$q(m|d_1 = 10.3) \propto f(10.3|m) = \frac{1}{\sqrt{2\pi}}e^{-(m-10.3)^2/2} . \tag{11.13}$$

Because (11.13) itself a normal probability distribution, the constant of proportionality in (11.3) is one, and the posterior distribution for the mass in μg (Fig. 11.1) is therefore

$$q(m|d_1 = 10.3) = \frac{1}{\sqrt{2\pi}}e^{-(m-10.3)^2/2} . \tag{11.14}$$

Next, suppose that we obtain a second statistically independent measurement of $d_2 = 10.1\ \mu g$. We can then use the distribution (11.14) estimated from the first measurement as an informative prior distribution to compute a revised posterior distribution

$$q(m|d_1 = 10.3,\ d_2 = 10.1) \propto f(d_2 = 10.1|m)q(m|d_1 = 10.3)$$

$$= \frac{1}{\sqrt{2\pi}}e^{-(m-10.1)^2/2}\frac{1}{\sqrt{2\pi}}e^{-(m-10.3)^2/2} . \tag{11.15}$$

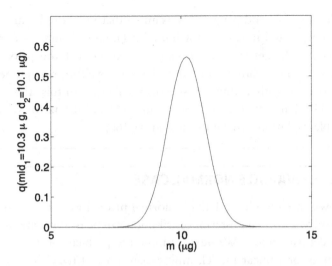

Figure 11.2 Posterior distribution $q(m|d_1 = 10.3 \text{ µg}, d_2 = 10.1 \text{ µg})$, uninformative prior.

Combining the exponents and absorbing the $1/\sqrt{2\pi}$ factors into the constant of proportionality gives

$$q(m|d_1 = 10.3, \ d_2 = 10.1) \propto e^{-((m-10.3)^2 + (m-10.1)^2)/2} . \tag{11.16}$$

Finally, we can simplify the exponent by combining terms and completing the square to obtain

$$(m - 10.3)^2 + (m - 10.1)^2 = 2(m - 10.2)^2 + 0.02 . \tag{11.17}$$

Thus

$$q(m|d_1 = 10.3, \ d_2 = 10.1) \propto e^{-(2(m-10.2)^2 + 0.02)/2} . \tag{11.18}$$

The constant $e^{-0.02/2}$ can be absorbed into the constant of proportionality, giving

$$q(m|d_1 = 10.3, \ d_2 = 10.1) \propto e^{-(10.2-m)^2} . \tag{11.19}$$

Normalizing (11.19) gives a normal posterior distribution

$$q(m|d_1 = 10.3, \ d_2 = 10.1) = \frac{1}{(1/\sqrt{2})\sqrt{2\pi}} e^{-\frac{(10.2-m)^2}{2(1/\sqrt{2})^2}} \tag{11.20}$$

with mean 10.2 µg and $\sigma = 1/\sqrt{2}$ µg (Fig. 11.2). Since we used an uninformative prior and the measurement errors were independent and normally distributed, the MAP solution is precisely the least squares solution for this problem.

It is notable in the second part of this example that we started with a normal prior distribution, incorporated normally distributed data, and obtained a normal posterior distribution (11.20). In general, we should not expect that the prior and posterior distributions will both be familiar distributions with well known properties. A prior distribution associated with a simple posterior distribution in this way is called a **conjugate prior**. There are other families of conjugate distributions for various parameter estimation problems, but in general this is unusual [66].

11.3. THE MULTIVARIATE NORMAL CASE

The result shown in Example 11.1 that a normal prior distribution and normally distributed data lead to a normal posterior distribution can be readily extended to problems with many model parameters. We next examine the problem of determining the posterior distribution for a linear model, multivariate normal (MVN) data and an MVN prior distribution.

Let $\mathbf{d}_{\text{obs}}$ be the observed data, $\mathbf{C}_D$ be the corresponding data covariance matrix, $\mathbf{m}_{\text{prior}}$ be the mean of the prior distribution, and $\mathbf{C}_M$ be the covariance matrix for the prior distribution. The prior distribution is thus, by (B.61),

$$p(\mathbf{m}) \propto e^{-\frac{1}{2}(\mathbf{m}-\mathbf{m}_{\text{prior}})^T \mathbf{C}_M^{-1}(\mathbf{m}-\mathbf{m}_{\text{prior}})} \tag{11.21}$$

and the conditional distribution of the data, given $\mathbf{m}$, is

$$f(\mathbf{d}|\mathbf{m}) \propto e^{-\frac{1}{2}(\mathbf{Gm}-\mathbf{d})^T \mathbf{C}_D^{-1}(\mathbf{Gm}-\mathbf{d})} . \tag{11.22}$$

Thus, (11.6) gives

$$q(\mathbf{m}|\mathbf{d}) \propto e^{-\frac{1}{2}((\mathbf{Gm}-\mathbf{d})^T \mathbf{C}_D^{-1}(\mathbf{Gm}-\mathbf{d})+(\mathbf{m}-\mathbf{m}_{\text{prior}})^T \mathbf{C}_M^{-1}(\mathbf{m}-\mathbf{m}_{\text{prior}}))} . \tag{11.23}$$

Tarantola [199] showed that this can be simplified to

$$q(\mathbf{m}|\mathbf{d}) \propto e^{-\frac{1}{2}(\mathbf{m}-\mathbf{m}_{\text{MAP}})^T \mathbf{C}_{M'}^{-1}(\mathbf{m}-\mathbf{m}_{\text{MAP}})} \tag{11.24}$$

where $\mathbf{m}_{\text{MAP}}$ is the MAP solution, and

$$\mathbf{C}_{M'} = (\mathbf{G}^T \mathbf{C}_D^{-1} \mathbf{G} + \mathbf{C}_M^{-1})^{-1} . \tag{11.25}$$

The MAP solution can be found by maximizing the exponent in (11.23), or equivalently by minimizing its negative

$$\min (\mathbf{Gm} - \mathbf{d})^T \mathbf{C}_D^{-1}(\mathbf{Gm} - \mathbf{d}) + (\mathbf{m} - \mathbf{m}_{\text{prior}})^T \mathbf{C}_M^{-1}(\mathbf{m} - \mathbf{m}_{\text{prior}}) . \tag{11.26}$$

The key to minimizing (11.26) is to rewrite it in terms of the matrix square roots of $\mathbf{C}_M^{-1}$ and $\mathbf{C}_D^{-1}$. Note that every covariance matrix is positive definite and has a unique positive definite matrix square root, which may be calculated in MATLAB using the **sqrtm** routine. This minimization problem can then be reformulated as

$$
\min (\mathbf{C}_D^{-1/2}(\mathbf{Gm} - \mathbf{d}))^T (\mathbf{C}_D^{-1/2}(\mathbf{Gm} - \mathbf{d})) + \\
(\mathbf{C}_M^{-1/2}(\mathbf{m} - \mathbf{m}_{\mathrm{prior}}))^T (\mathbf{C}_M^{-1/2}(\mathbf{m} - \mathbf{m}_{\mathrm{prior}}))
\tag{11.27}
$$

or as the standard least squares problem

$$
\min \left\| \begin{bmatrix} \mathbf{C}_D^{-1/2}\mathbf{G} \\ \mathbf{C}_M^{-1/2} \end{bmatrix} \mathbf{m} - \begin{bmatrix} \mathbf{C}_D^{-1/2}\mathbf{d} \\ \mathbf{C}_M^{-1/2}\mathbf{m}_{\mathrm{prior}} \end{bmatrix} \right\|_2^2 .
\tag{11.28}
$$

Examining the right hand terms in (11.28), note that

$$
\mathrm{Cov}(\mathbf{C}_D^{-1/2}\mathbf{d}) = \mathbf{C}_D^{-1/2}\mathbf{C}_D(\mathbf{C}_D^{-1/2})^T = \mathbf{I} .
\tag{11.29}
$$

The multiplication of $\mathbf{C}_D^{-1/2}$ times $\mathbf{d}$ in (11.28) can thus be conceptualized as a data transformation that both makes the resulting elements independent and normalizes the standard deviations. In the model space, multiplication by $\mathbf{C}_M^{-1/2}$ has the same effect.

An interesting limiting case is where the prior distribution provides essentially no information. Consider an MVN prior distribution with a covariance matrix $\mathbf{C}_M = \alpha^2 \mathbf{I}$, in the limit where α is extremely large. In this case, the diagonal elements of $\mathbf{C}_M^{-1}$ will be extremely small, and the posterior covariance matrix (11.25) will be well-approximated by

$$
\mathbf{C}_{M'} \approx (\mathbf{G}^T \mathrm{Cov}(\mathbf{d})^{-1} \mathbf{G})^{-1} .
\tag{11.30}
$$

If the data covariance matrix is $\mathbf{C}_D = \sigma^2 \mathbf{I}$, then

$$
\mathbf{C}_{M'} \approx \sigma^2 (\mathbf{G}^T\mathbf{G})^{-1}
\tag{11.31}
$$

which is precisely the covariance matrix for the least squares model parameters in (11.1). Furthermore, when we solve (11.28) to obtain the MAP solution, we find that it simplifies to the least squares problem of minimizing $\|\mathbf{Gm} - \mathbf{d}\|_2^2$. Thus, under the common assumption of normally distributed and independent data errors with constant variance, a very broad prior distribution leads to a MAP solution that is the unregularized least squares solution.

It is also worthwhile to consider what happens in the special case where $\mathbf{C}_D = \sigma^2 \mathbf{I}$, and $\mathbf{C}_M = \alpha^2 \mathbf{I}$. In this case the corresponding matrix square roots are also proportional to identity matrices, and (11.28) simplifies to

$$
\min (1/\sigma)^2 \|(\mathbf{Gm} - \mathbf{d})\|_2^2 + (1/\alpha)^2 \|\mathbf{m} - \mathbf{m}_{\mathrm{prior}}\|_2^2
\tag{11.32}
$$

which is a modified optimization problem for zeroth-order Tikhonov regularization (4.4), where the 2-norm regularization term is evaluated relative to $\mathbf{m}_{\mathrm{prior}}$ and the equivalent Tikhonov regularization parameter is σ/α. Thus, the MAP solution obtained by using a prior with independent and normally distributed model parameters is precisely the zeroth-order Tikhonov regularized solution obtained by solving (11.32). However, this does not mean that the Bayesian approach is entirely equivalent to Tikhonov regularization, because the Bayesian solution is a probability distribution, whereas the Tikhonov solution is a single model from that distribution.

Once we have obtained the posterior distribution, it is straightforward to generate corresponding model realizations. We may wish to do this to numerically assess model features, such as estimating the probability that parameters within some region will all occupy a specified range. Following the procedure outlined in Example B.10, we compute the Cholesky factorization of the posterior distribution covariance matrix

$$\mathbf{C}_{M'} = \mathbf{R}^T \mathbf{R} \tag{11.33}$$

and generate a random solution

$$\mathbf{m} = \mathbf{R}^T \mathbf{s} + \mathbf{m}_{\text{MAP}} \tag{11.34}$$

where the vector $\mathbf{s}$ consists of independent and normally distributed random numbers with zero mean and unit standard deviation. We can then statistically analyze the random samples from the posterior distribution produced using (11.34).

Example 11.2

We consider Bayesian solutions to the Shaw problem that was previously considered in Examples 3.3, 4.1, 4.3, and 4.8.

We first use a relatively uninformative MVN prior distribution with mean 0.5, standard deviation 0.5 and zero covariances, so that $\mathbf{C}_M = 0.25\mathbf{I}$. As in the previous examples, the measurement noise has standard deviation 1.0×10^{-6}, so that $\mathbf{C}_D = 1.0 \times 10^{-12}\mathbf{I}$. Solving (11.28) produces the $\mathbf{m}_{\text{MAP}}$ solution shown in Fig. 11.3. Fig. 11.4 shows this same solution with probability bounds. These bounds are not classical 95% confidence intervals (e.g., Fig. 4.9). Rather, they are 95% probability intervals calculated from the MVN posterior distribution, so that there is 95% probability that each model parameter lies within the corresponding symmetric interval around $\mathbf{m}_{\text{MAP}}$.

Fig. 11.5 shows a random solution generated from the posterior distribution, using (11.34). This solution varies considerably from the true model, and demonstrates the large degree of uncertainty in the inverse solution, consistent with the large probability bounds in Fig. 11.3. The roughness of this solution realization is a consequence of the fact that the prior distribution $\mathbf{C}_M$ had zero covariances, so model realizations from this posterior distribution have no preference for smoothness.

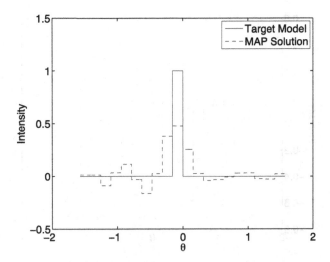

Figure 11.3 The MAP solution and the true model for the Shaw example using an MVN prior distribution with mean 0.5, standard deviation 0.5, and zero covariance.

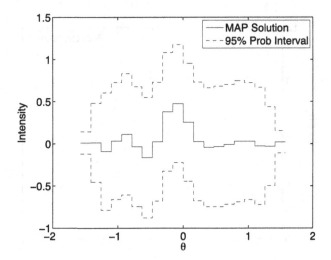

Figure 11.4 The MAP solution of Fig. 11.3, with 95% probability intervals.

Next, consider a more restrictive prior distribution. Suppose we have reason to believe that the largest amplitudes in the solution should be near the center of the model. We thus choose the bell-shaped zero-covariance prior distribution depicted in Fig. 11.6. Figs. 11.7 and 11.8 show the resulting MAP model and its probability intervals. The solution recovery is, not surprisingly, improved in this particular case by our more restrictive prior model given that the true model is highly consistent with the prior distribution (Fig. 11.3).

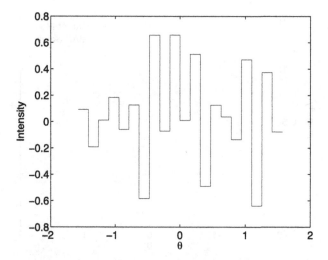

Figure 11.5 A model realization for the Shaw example using an MVN prior distribution with mean 0.5, standard deviation 0.5, and zero covariance.

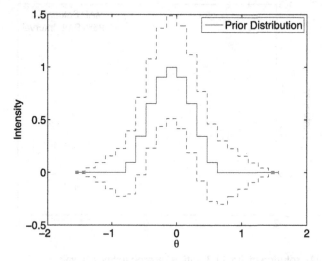

Figure 11.6 A more restrictive zero covariance prior distribution for the Shaw problem with 95% probability intervals.

These results illustrate a principal issue with applying the Bayesian approach to poorly conditioned problems. To obtain a tight posterior distribution in such cases, we will have to make strong prior assumptions. Conversely, if such assumptions are not made, then we cannot recover the true model features well. This situation is analo-

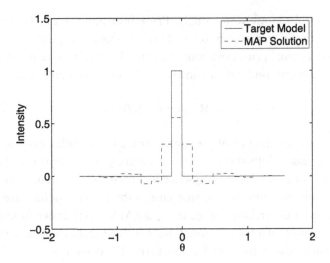

Figure 11.7 The MAP solution for the Shaw example using the prior distribution shown in Fig. 11.6.

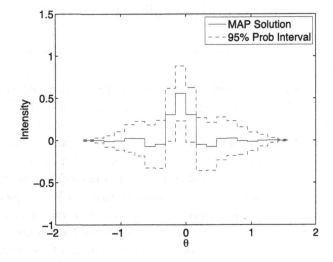

Figure 11.8 The MAP solution of Fig. 11.7 with 95% probability intervals.

gous to that of Tikhonov regularization, which must also impose strong and consistent model bias to produce "good" solutions and must also make strong model assumptions to estimate solution error bounds (Section 4.8).

In the previous example, we applied a prior that preferentially concentrated model structure in a particular region (the center) by imposing a zero prior with small standard deviations near the model edges. Because this prior distribution had zero covariances, resulting model realizations were rough. Prior distributions can be readily designed to enforce smoothness constraints on realizations from the posterior distribution by specifying a nondiagonal prior covariance matrix. A straightforward way to accomplish this for a 1-dimensional model is to construct a correlation matrix with columns

$$\mathbf{R}_{i,.} = \text{shift}(a_j, i) \tag{11.35}$$

where a_j is the desired sequence of parameter correlations, with a zero lag correlation of one. The shift operator shifts the sequence (truncating as necessary) so that the zero–lag (unit) maximum of the correlation sequence is centered on element i, and hence on the diagonal of $\mathbf{R}$. Suitable sequences a_i that ensure the positive definiteness of $\mathbf{R}$ can be constructed using autocorrelation (e.g., using the MATLAB **xcorr** function). Here, we use the autocorrelation of a triangle function, which produces a cubic approximation to a Gaussian function. Given the correlation matrix $\mathbf{R}$, a corresponding prior distribution covariance matrix with uniform parameter variances σ_M can then be constructed as

$$\mathbf{C}_M = \sigma_M^2 \mathbf{R} \ . \tag{11.36}$$

Example 11.3

Consider the Vertical Seismic Profile (VSP) problem first introduced in Example 1.3, which was solved and analyzed using Tikhonov regularization in Examples 4.4, 4.5, 4.6, and 4.7. We revisit this problem as a Bayesian problem, implementing an MVN prior with a covariance matrix of the form of (11.36). The model consists of 50 equally spaced slowness intervals along a 1000 m vertical borehole with an interval length of 20 m. Seismic travel time data are collected at 50 equally spaced depths with independent zero-mean normal errors with standard deviation of $\sigma_D = 2 \times 10^{-4}$ s, producing a data covariance matrix $\mathbf{C}_D = \sigma_D^2 \mathbf{I}$. We apply a prior distribution that is consistent with a seismic slowness decrease (velocity increase) with depth and has a constant gradient between known seismic slownesses from the top and the bottom of the borehole. We first impose a prior distribution standard deviation of $\sigma_M = 2 \times 10^{-5}$ s/m on all parameters to represent an estimated variability in model slownesses, and utilize a prior correlation function that falls off with a scale length of 5 model intervals (i.e., a correlation of $1/e$ at a model parameter lag of approximately 5 model parameters or 100 m). The prior distribution and its parameter standard deviations are shown in Fig. 11.9, and the corresponding parameter correlation function is shown in Fig. 11.10. The resulting posterior distribution and its standard deviations are shown in Fig. 11.11. We next apply a prior with twice the correlation length (Fig. 11.12). The resulting posterior distribution is shown in Fig. 11.13.

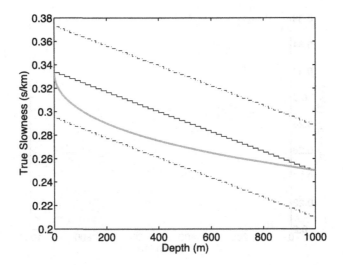

Figure 11.9 A constant-slowness gradient prior distribution and its 95% distribution intervals for the VSP problem. The true model is shown as the gray smooth curve.

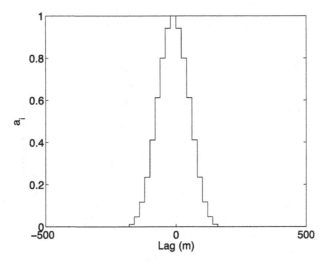

Figure 11.10 A correlation function for the prior distribution of Fig. 11.9 with a $1/e$ correlation length of approximately 5 parameters (100 m).

The approach described in this section can be extended to nonlinear problems. To find the MAP solution, solve the nonlinear least squares problem

$$\min \ (\mathbf{G}(\mathbf{m}) - \mathbf{d})^T \mathbf{C}_D^{-1} (\mathbf{G}(\mathbf{m}) - \mathbf{d}) + (\mathbf{m} - \mathbf{m}_{\text{prior}})^T \mathbf{C}_M^{-1} (\mathbf{m} - \mathbf{m}_{\text{prior}}) \ . \qquad (11.37)$$

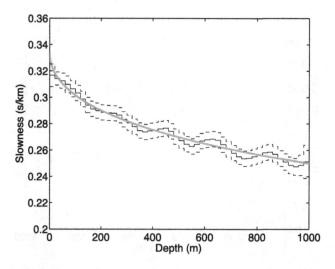

Figure 11.11 MAP model obtained from the posterior distribution, and its 95% distribution intervals, using the prior distribution described in Figs. 11.9 and 11.10. The true model is shown as the gray smooth curve.

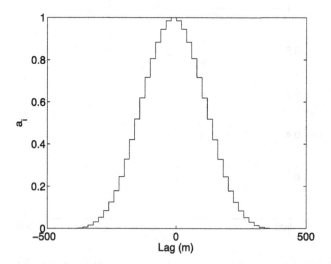

Figure 11.12 A correlation function for the prior distribution of Fig. 11.9 with a $1/e$ correlation length of approximately 10 parameters (200 m).

One can then linearize around the MAP solution to obtain the approximate posterior covariance

$$\mathbf{C}_{M'} = (\mathbf{J}(\mathbf{m}_{\mathrm{MAP}})^T \mathbf{C}_D^{-1} \mathbf{J}(\mathbf{m}_{\mathrm{MAP}}) + \mathbf{C}_M^{-1})^{-1} \qquad (11.38)$$

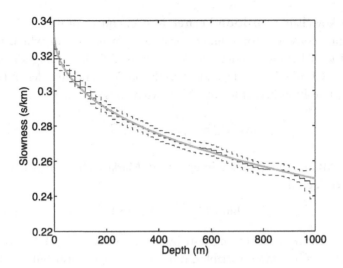

Figure 11.13 MAP model obtained from the posterior distribution, and its 95% distribution intervals, using the prior distribution depicted in Figs. 11.9 and 11.12. The true model is shown as the gray smooth curve.

where $\mathbf{J}(\mathbf{m})$ is the Jacobian. As with other nonlinear optimization problems, we must consider the possibility of multiple local optima. If (11.37) has multiple solutions with comparable likelihoods, then a single MAP solution and associated $\mathbf{C}_{M'}$ from (11.38) will not accurately characterize the posterior distribution.

11.4. THE MARKOV CHAIN MONTE CARLO (MCMC) METHOD

We next introduce the use of Markov Chain Monte Carlo (MCMC) methods to sample from a posterior distribution. Given a sufficient population of such samples, we can use them to characterize the solution of a Bayesian inverse problem. Because MCMC methods depend only on the forward model and associated likelihood calculations, they are easily applied to both linear and nonlinear problems.

A Markov chain is a sequence of random variables

$$X^{(0)}, X^{(1)}, \ldots \tag{11.39}$$

where the probability distribution of $X^{(n+1)}$ depends solely on the previous value, $X^{(n)}$, and not on earlier values of random variables in the sequence. That is,

$$P(X^{(n+1)}|X^{(0)}, X^{(1)}, \ldots, X^{(n)}) = P(X^{(n+1)}|X^{(n)}) \,. \tag{11.40}$$

The particular Markov chains considered here will be time–invariant, so that

$$P(X^{(n)}, X^{(n+1)}) = P(X^{(n+1)}|X^{(n)}) \,, \tag{11.41}$$

called the Markov chain **transition kernel**, is independent of n.

For example, consider a one-dimensional random process in which we start with $X^{(0)} = 0$, and then iteratively compute $X^{(n+1)}$ from $X^{(n)}$ by adding an $N(0, \sigma^2)$ realization to $X^{(n)}$. Clearly $X^{(n+1)}$ depends directly on $X^{(n)}$ and this dependence is time-invariant. The transition kernel for this Markov chain is

$$k(x_1, \ x_2) = \frac{1}{\sqrt{2\pi}\sigma} e^{-\frac{1}{2}(x_1 - x_2)^2 \sigma^2} \ . \tag{11.42}$$

In the MCMC method, we are interested in Markov chains that have limiting distributions, $q(\mathbf{x})$, such that

$$\lim_{n \to \infty} P(X^{(n)} | X^{(0)}) = q(\mathbf{x}) \ . \tag{11.43}$$

Not all Markov chains have limiting distributions. For example, if $X^{(n)}$ is the sum of n independent $N(0, \sigma^2)$ random variables, it has an $N(0, n\sigma^2)$ distribution. This does not approach a limiting probability distribution as $n \to \infty$.

For a general multivariate model, it can be shown that if $q(\mathbf{x})$ and $k(\mathbf{x}, \mathbf{y})$ satisfy the **local balance** equation

$$q(\mathbf{x})k(\mathbf{x}, \mathbf{y}) = q(\mathbf{y})k(\mathbf{y}, \mathbf{x}) \tag{11.44}$$

for all models $\mathbf{x}$ and $\mathbf{y}$, then $q(\mathbf{x})$ is the limiting distribution of the Markov chain. Eq. (11.44) states that the rate of model transitions from $\mathbf{x}$ to $\mathbf{y}$ equals the rate of transitions from $\mathbf{y}$ to $\mathbf{x}$.

The **Metropolis–Hastings Sampler** is an algorithm that simulates a Markov Chain with a specified limiting distribution. We will apply it to produce samples from a posterior distribution $q(\mathbf{m}|\mathbf{d})$ that will tend to densely sample its higher likelihood regions. With enough such samples, we can usefully characterize the posterior distribution of a Bayesian parameter estimation or inverse problem. Since the data vector $\mathbf{d}$ is given, we will simply write $q(\mathbf{m}|\mathbf{d})$ as $q(\mathbf{m})$ throughout the following development.

We begin a Metropolis–Hastings Sampler implementation by picking a **proposal distribution** $r(\mathbf{x}, \mathbf{y})$ that facilitates random steps in the posterior model space. These randomly perturbed samples will subsequently be subjected to a likelihood-based test in the Metropolis–Hastings Sampler. A common choice for the proposal distribution is a multivariate normal distribution with zero covariances and variances σ_i^2, so that

$$r(x_i, y_i) \propto e^{-\frac{1}{2}|x_i - y_i|_2^2/\sigma_i^2} \ . \tag{11.45}$$

Although $r(\mathbf{x}, \mathbf{y})$ cannot be implemented as a transition kernel directly, because it does not satisfy (11.44), this can be remedied by introducing a factor called the **acceptance ratio**

$$\alpha(\mathbf{x}, \mathbf{y}) = \min(1, s) \ , \tag{11.46}$$

where

$$s = \frac{q(\mathbf{y})r(\mathbf{y}, \mathbf{x})}{q(\mathbf{x})r(\mathbf{x}, \mathbf{y})} .$$ (11.47)

Note that $0 \le \alpha(\mathbf{x}, \mathbf{y}) \le 1$. Also note that

$$\alpha(\mathbf{y}, \mathbf{x}) = \min\left(1, s^{-1}\right) .$$ (11.48)

At least one of $\alpha(\mathbf{x}, \mathbf{y})$ and $\alpha(\mathbf{y}, \mathbf{x})$ will thus be equal to one, depending on whether s is greater than or less than one.

Now, let

$$\tilde{k}(\mathbf{x}, \mathbf{y}) = \alpha(\mathbf{x}, \mathbf{y})r(\mathbf{x}, \mathbf{y}) .$$ (11.49)

$\alpha(\mathbf{x}, \mathbf{y}) \le 1$, so

$$\beta(x) = \int_{\text{all models}} \tilde{k}(\mathbf{x}, \mathbf{y}) \, d\mathbf{y} = \int_{\text{all models}} \alpha(\mathbf{x}, \mathbf{y})r(\mathbf{x}, \mathbf{y}) d\mathbf{y} \le 1 .$$ (11.50)

Thus $\tilde{k}(\mathbf{x}, \mathbf{y})$ is not a properly normalized transition kernel. We can, however, obtain a normalized kernel by defining

$$k(\mathbf{x}, \mathbf{y}) = \tilde{k}(\mathbf{x}, \mathbf{y}) + (1 - \beta)\delta(\mathbf{x} - \mathbf{y})$$ (11.51)

so that

$$\int_{\text{all models}} k(\mathbf{x}, \mathbf{y}) d\mathbf{y} = \int_{\text{all models}} \alpha(\mathbf{x}, \mathbf{y})r(\mathbf{x}, \mathbf{y}) \, d\mathbf{y} + \int_{\text{all models}} (1 - \beta)\delta(\mathbf{x} - \mathbf{y}) \, d\mathbf{y}$$
$$= \beta + (1 - \beta)$$
$$= 1 .$$ (11.52)

A simple algorithm can now be used to generate a random value $\mathbf{y}$ from $\mathbf{x}$ that satisfies (11.44).

Algorithm 11.1 Transition Kernel Evaluation
___
1. Generate a candidate $\mathbf{y}$ from $\mathbf{x}$ according to a proposal distribution $r(\mathbf{x}, \mathbf{y})$.
2. Compute $\alpha(\mathbf{x}, \mathbf{y})$.
3. With probability α, return the candidate $\mathbf{y}$.
4. With probability $1 - \alpha$, return the previous value $\mathbf{x}$.

Now, we need to show that $q(\mathbf{x})$ and $k(\mathbf{x}, \mathbf{y})$ as defined above satisfy the local balance equation (11.44). If $\mathbf{y} = \mathbf{x}$, this is obviously true. For $\mathbf{y} \ne \mathbf{x}$, we need to consider the two cases of $\alpha(\mathbf{x}, \mathbf{y}) = 1$ or $\alpha(\mathbf{y}, \mathbf{x}) = 1$. If $\alpha(\mathbf{x}, \mathbf{y}) = 1$, then

$$\alpha(\mathbf{y}, \mathbf{x}) = s^{-1} = \frac{q(\mathbf{x})r(\mathbf{x}, \mathbf{y})}{q(\mathbf{y})r(\mathbf{y}, \mathbf{x})} \tag{11.53}$$

and

$$q(\mathbf{x})k(\mathbf{x}, \mathbf{y}) = q(\mathbf{x})\alpha(\mathbf{x}, \mathbf{y})r(\mathbf{x}, \mathbf{y}) = q(\mathbf{x})r(\mathbf{x}, \mathbf{y}) . \tag{11.54}$$

Also,

$$q(\mathbf{y})k(\mathbf{y}, \mathbf{x}) = q(\mathbf{y})\alpha(\mathbf{y}, \mathbf{x})r(\mathbf{y}, \mathbf{x}) = q(\mathbf{y})\frac{q(\mathbf{x})r(\mathbf{x}, \mathbf{y})}{q(\mathbf{y})r(\mathbf{y}, \mathbf{x})}r(\mathbf{y}, \mathbf{x}) = q(\mathbf{x})r(\mathbf{x}, \mathbf{y}) . \tag{11.55}$$

However, $q(\mathbf{x})r(\mathbf{x}, \mathbf{y}) = q(\mathbf{x})k(\mathbf{x}, \mathbf{y})$, so

$$q(\mathbf{y})k(\mathbf{y}, \mathbf{x}) = q(\mathbf{x})k(\mathbf{x}, \mathbf{y}) , \tag{11.56}$$

thus satisfying (11.44) A similar argument shows that (11.44) is satisfied for the case where $\alpha(\mathbf{y}, \mathbf{x}) = 1$.

There are several important tactics that help to further simplify the method. Because the product of q and r appears in both the numerator and denominator of s, we need only know these factors to constants of proportionality, and thus do not need to normalize q and r in individual calculations. Also note that the posterior distribution $q(\mathbf{m})$ is proportional to the product of the prior $p(\mathbf{m})$ and the likelihood $f(\mathbf{d}|\mathbf{m})$ (11.3). We can thus write (11.46) as

$$\alpha(\mathbf{x}, \mathbf{y}) = \min\left(1, \frac{p(\mathbf{y})f(\mathbf{d}|\mathbf{y})r(\mathbf{y}, \mathbf{x})}{p(\mathbf{x})f(\mathbf{d}|\mathbf{x})r(\mathbf{x}, \mathbf{y})}\right) . \tag{11.57}$$

If $r(\mathbf{x}, \mathbf{y})$ is a symmetric distribution, such as (11.45), then $r(\mathbf{x}, \mathbf{y}) = r(\mathbf{y}, \mathbf{x})$, and we can simplify (11.57) to

$$\alpha(\mathbf{x}, \mathbf{y}) = \min\left(1, \frac{q(\mathbf{y})}{q(\mathbf{x})}\right) . \tag{11.58}$$

In computational practice, numbers in the numerator of (11.58) may be extremely small, and can thus generate floating point underflow problems. This is easily avoided by evaluating $\log \alpha(\mathbf{x}, \mathbf{y})$ instead of $\alpha(\mathbf{x}, \mathbf{y})$. We now have all the components to describe the Metropolis–Hastings Sampler.

Algorithm 11.2 The Metropolis–Hastings Sampler

Given a starting model, $\mathbf{m}^{(0)}$, repeat the following steps for $k = 1, 2, \ldots$ until the posterior distribution is sufficiently sampled by the set of models $\mathbf{m}^{(k)}$.

1. Generate a candidate model $\mathbf{c}$ from the previous model, $\mathbf{m}^{(k)}$, using the proposal distribution $r(\mathbf{m}^{(k)}, \mathbf{c})$.
2. Compute $\log \alpha(\mathbf{m}^{(k)}, \mathbf{c})$.
3. Let t be a realization of a uniformly distributed random variable on $[0, 1]$.
4. If $\log t < \log \alpha(\mathbf{m}^{(k)}, \mathbf{c})$, then accept the candidate model and let $\mathbf{m}^{(k+1)} = \mathbf{c}$; otherwise reject the candidate model and let $\mathbf{m}^{(k+1)} = \mathbf{m}^{(k)}$.

If $\log t$ is sufficiently small, we will occasionally accept a new model that has a small acceptance ratio and thus move towards a model with reduced likelihood. This property of the algorithm helps overcome the problem of becoming trapped near a localized likelihood maximum. The ability to fruitfully and efficiently sample the posterior distribution will also depend on the size of the steps taken in generating candidate models. In the case of the normal formulation for the proposal distribution (11.45), this will be controlled by the size of the σ_i. Smaller steps will result in higher acceptance rates, but the algorithm may be unacceptably slow at usefully exploring the posterior distribution. Conversely, larger steps will result in lower acceptance ratios and will thus be less frequently accepted. Either situation may cause the algorithm to become stuck in a particular region of the posterior distribution. Often, the step size parameters are explored adaptively in multiple runs of the algorithm (which may be run as independent parallel processes on a multi-CPU computer system). Some studies [66] suggest that the algorithm is optimally tuned when the new model acceptance rate is between approximately 20% and 50%.

11.5. ANALYZING MCMC OUTPUT

Although the limiting distribution of the Markov chain sampled by the Metropolis–Hastings algorithm is the desired posterior distribution, there are significant practical challenges in analyzing the output of an MCMC simulation.

First, successive models $\mathbf{m}^{(k)}$, $\mathbf{m}^{(k+1)}$, ..., produced by the simulation are typically strongly correlated with each other, but most statistical techniques require independent samples. For example, Eq. (B.74) for a 95% confidence interval assumes independence and can fail spectacularly when the samples are positively correlated.

In practice this complication can be avoided by analyzing a subset of samples that are far apart in the sample sequence. For example, if we examine $\mathbf{m}^{(k)}$, $\mathbf{m}^{(k+10,000)}$, $\mathbf{m}^{(k+20,000)}$, ..., it is likely that corresponding parameter samples taken 10,000 steps apart will not be highly correlated. We can verify this by plotting the successive model autocorrelations for the history of sampling over some moving window length. In practice we may have to try various lags until the samples are effectively uncorrelated.

Second, early Metropolis–Hastings algorithm samples will be biased by the initial model $\mathbf{m}^{(0)}$, which may not lie in a high likelihood region of the posterior distribution. This issue is commonly addressed by skipping over early samples in the chain to give time for the algorithm to "warm up" and/or by running the process with a variety of starting models, either sequentially or in parallel on multiple CPUs. For example, if it is determined that samples spaced 10,000 steps apart are effectively uncorrelated, then it might be reasonable to let the Metropolis–Hastings sampler establish itself for 10,000 steps before beginning to collect samples.

Once we are confident that the procedure has produced a large enough collection of effectively independent samples, we can use the results to characterize the posterior distribution. For a suitably large sample, the MAP solution can be estimated as the retrieved posterior distribution sample that has the greatest likelihood. The posterior distribution may be approximately multivariate normal, which can be established by examining model histograms and Q–Q plots. In this case we can readily construct probability intervals for describing the posterior distribution from the sample mean and covariance using normal assumptions. However, if the distribution is distinctly non-normal, it will be more difficult to produce a simple summary of its character, particularly for very high dimensional models. A common approach is to produce and evaluate scatter plots and/or histograms that display key features of the posterior distribution and to use counting statistics to establish probability intervals.

Example 11.4

Reconsidering the ill-posed nonlinear parameter estimation problem of Example 9.2, we apply the Metropolis–Hastings algorithm to the problem of fitting four parameters, m_i, to the nonlinear function

$$d_i = (G(\mathbf{m}))_i = m_1 e^{m_2 x_i} + m_3 x_i e^{m_4 x_i} \tag{11.59}$$

given a set of observations, d_i with specified independent normally distributed data noise, specified by corresponding standard deviations σ_i. As in Example 9.2, the true model parameters are $m_1 = 1.0$, $m_2 = -0.5$, $m_3 = 1.0$, and $m_4 = -0.75$, data are produced at 25 equally spaced points, x_i, on the interval $[1, 7]$, and $N(0, 0.01^2)$ independent noise is added to each data element.

The likelihood function is specified by

$$f(\mathbf{d}|\mathbf{m}) \propto \prod_{i=1}^{m} e^{-\frac{1}{2}(d_i - G(\mathbf{m})_i)^2/\sigma_i^2} \tag{11.60}$$

which, after taking the natural logarithm, is

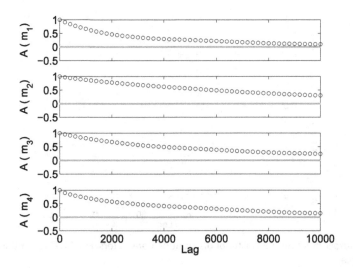

Figure 11.14 Autocorrelations for posterior distribution parameters prior to thinning for autocorrelation reduction.

$$\log(f(\mathbf{d}|\mathbf{m})) = -\frac{1}{2} \sum_{i=1}^{m} (d_i - G(\mathbf{m})_i)^2 / \sigma_i^2 + C \qquad (11.61)$$

where C is the logarithm of the constant of proportionality in (11.60).

The procedure was initiated using a random model selected from a 4-dimensional uniform distribution bounded by $[1, -1]$ in each parameter direction and applying a uniform prior for the region $m_1 = [0, 2]$, $m_2 = [-1, 0]$, $m_3 = [0, 2]$, and $m_4 = [-1, 0]$.

In each of the MCMC steps, we applied independent normally distributed random perturbations with standard deviations of 0.005 in each of the four model parameter directions, accepting or rejecting the corresponding candidate models according to the Metropolis–Hastings algorithm. The procedure produced an acceptance rate of approximately 39.5%, which is in the nominally acceptable range of not being too large or too small [66].

Fig. 11.14 shows that there are positive autocorrelations out to a lag of about 10,000. We decided to warm up the simulation for 100,000 steps and take one sample every 10,000 steps after this to reduce parameter autocorrelation. We ultimately sampled the posterior distribution using a total of 4,100,000 steps to produce a low-autocorrelation set of 400 posterior distribution samples. Fig. 11.15 shows that these samples are effectively decorrelated.

Fig. 11.16 shows scatter plots and histograms of the resulting sampled posterior distribution, along with the corresponding MAP solution and 95% probability intervals for each parameter direction, which well enclose the true solution. Fig. 11.17 shows the

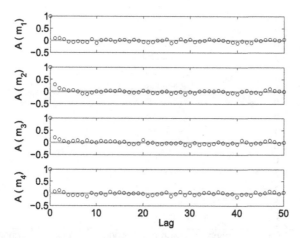

Figure 11.15 Autocorrelations for posterior distribution parameters (Fig. 11.17) after thinning to every 10,000th sample.

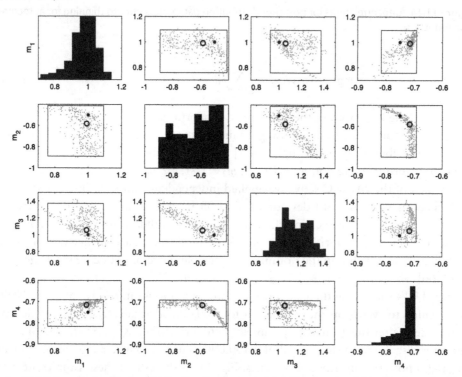

Figure 11.16 Sampled posterior distribution for Example 11.4. The true model is shown as the large black dot and the MAP model estimated from the maximum likelihood posterior distribution sample is indicated by the open circle. 400 retrieved samples of the posterior distribution (every 10,000th calculated sample from the MCMC output) are shown as gray dots. 95% probability intervals estimated from the MCMC posterior distribution samples are shown by boxes.

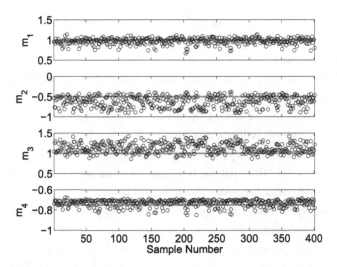

Figure 11.17 MCMC history of the thinned posterior samples plotted in Fig. 11.16. True parameters are shown as gray lines.

time history of the corresponding 400 samples. The prominent quasi-linear scattering trends in the output (for example, between m_2 and m_3) in Fig. 11.16 are indicative of a high degree of anti-correlation between some parameter pairs. This anti-correlation is also apparent in the sequence of parameter samples in Fig. 11.17, for example in the anti-correlated trends of m_2 and m_3.

11.6. EXERCISES

1. Reanalyze the data in Example 11.1 using a prior distribution that is uniform on the interval [9, 11]. Compute the posterior distribution after the first measurement of 10.3 µg and after the second measurement of 10.1 µg. What is the posterior mean?

2. Consider the estimation problem $\mathbf{d} = \mathbf{m}$ (i.e., where $\mathbf{G} = \mathbf{I}$) in two dimensions. The data, $\mathbf{d} = [5\ 15]^T$, have identical and independent normal errors with standard deviations of $\sqrt{2}$. Apply a zero-mean MVN prior characterized by a covariance matrix where m_1 and m_2 have a correlation coefficient of 0.9, and equal standard deviations of $\sqrt{5}$. Calculate the MAP model, and compute and plot the 50%, 90%, and 95% contours of the MVN distributions $\mathbf{d}$, $\mathbf{m}_{\text{prior}}$ and the posterior model.

3. In writing (11.28) we made use of the matrix square root.
   a. Suppose that $\mathbf{A}$ is a symmetric and positive definite matrix. Using the SVD, find an explicit formula for the matrix square root. Your square root should itself be a symmetric and positive definite matrix.

    b. Show that instead of using the matrix square roots of $\mathbf{C}_D^{-1}$ and $\mathbf{C}_M^{-1}$, we could have used the Cholesky factorizations of $\mathbf{C}_D^{-1}$ and $\mathbf{C}_M^{-1}$ in formulating the least squares problem.

4. Consider the following coin tossing experiment. We repeatedly toss a coin, and each time record whether it comes up heads (0), or tails (1). The bias b of the coin is the probability that it comes up heads. We have reason to believe that this is not a fair coin, so we will not assume that $b = 1/2$. Instead, we will begin with a uniform prior distribution $p(b) = 1$, for $0 \le b \le 1$.

    a. What is $f(d|b)$? Note that the only possible data are 0 and 1, so this distribution will involve delta functions at $d = 0$, and $d = 1$.

    b. Suppose that on our first flip, the coin comes up heads. Compute the posterior distribution $q(b|d_1 = 0)$.

    c. The second, third, fourth, and fifth flips are 1, 1, 1, and 1. Find the posterior distribution $q(b|d_1 = 0, \ d_2 = 1, \ d_3 = 1, \ d_4 = 1, \ d_5 = 1)$. Plot the posterior distribution.

    d. What is your MAP estimate of the bias?

    e. Now, suppose that you initially felt that the coin was at least close to fair, with

$$p(b) \propto e^{-10(b-0.5)^2} \quad 0 \le b \le 1 . \tag{11.62}$$

Repeat the analysis of the five coin flips described above.

5. Apply the Bayesian method to Exercise 2 in Chapter 4. Select what you consider to be a reasonable prior. How sensitive is your solution to the prior mean and covariance?

6. Apply the Bayesian method to Exercise 9.3. Assume an MVN prior distribution for α and n, with α and n independent. The prior for α should have a mean of 0.01 and a standard deviation of 0.005, and the prior for n should have a mean of 5 and a standard deviation of 3. Compare your solution with the solution that you obtained to Exercise 9.3. How sensitive is your solution to the choice of the prior?

7. Repeat Exercise 11.6, using MCMC to estimate the posterior distribution. Compare your solution to the solution that you obtained for Exercise 11.6.

8. Apply the Metropolis–Hastings sampler to produce a sampled posterior distribution for the nonlinear parameter estimation problem of Example 9.1. Use a prior distribution that is uniform on $S = [0, 0.01]$ and $T = [0, 2]$, a zero covariance multivariate normal proposal distribution, and a starting model of $(S, T) = (5 \times 10^{-3}, 1.0)$. Generate 200,000 samples using a 10,000 sample warm up and explore independent step sizes for the two parameters in your proposal distribution to obtain a Metropolis–Hastings sampler acceptance rate between 10% and 50%.

Extract every 1000th sample for analysis and establish that these 191 samples are not highly dependent by examining sample autocorrelation functions. Examine the sampled distribution to obtain the MAP model and empirical 95% probability intervals on S and T. Apply a Q–Q plot and assess the multivariate normality of the sampled posterior distribution and compare normal assumption and empirical estimates of the 95% probability intervals.

9. Apply the Metropolis–Hastings sampler to produce a sampled posterior distribution for the nonlinear inverse problem for gravity observations above a buried density perturbation with an unknown variable depth $m(x)$ and a fixed density perturbation $\Delta\rho$, as described in Exercise 10.2. Your prior should be selected to favor smooth models (specified by nonzero parameter correlations).

10. Recall the cosmogenic nuclide dating Exercise 9.8. In this exercise you will perform a Bayesian analysis of the same data.

    a. Convert the grid of χ^2 values into likelihood values. Using a flat prior, $p(\epsilon, T) = 1$ compute the posterior probability density on your grid. You will need to normalize the posterior distribution properly. Make a surface plot of the result.

    b. Integrate your posterior with respect to erosion rate to get a marginal posterior distribution for exposure age. Integrate the posterior with respect to age to get a marginal posterior distribution for the erosion rate. Make sure that these are properly normalized, and plot both posterior distributions.

    c. Repeat parts (a) and (b) using the prior

    $$p(\epsilon, T) \propto e^{-(\epsilon - 0.0005)^2/(2 \times 0.0002^2)} \ .$$

    d. Discuss your results.

11. Repeat Exercise 11.10 using MCMC.

11.7. NOTES AND FURTHER READING

The arguments for and against the use of Bayesian methods in statistics and inverse problems have raged for decades. Some classical references that provide context for these arguments include [47,54,100,101,175]. Sivia's book [183] is a good general introduction to Bayesian ideas for scientists and engineers. The book by Calvetti and Somersalo introduces Bayesian methods for inverse problems including Markov Chain Monte Carlo sampling and includes MATLAB examples [30]. The book by Kaipio and Somersalo provides a more detailed theoretical treatment of Bayesian methods for inverse problems and includes some interesting case studies [105]. An early paper by Tarantola and Vallete on the application of the Bayesian approach was quite influential [200], and [199] is a longstanding standard reference work on Bayesian methods for inverse problems. The book by Rodgers [171] focuses on application of the Bayesian approach to problems in atmospheric sounding. The paper of Gouveia and Scales [74] discusses the relative advantages and disadvantages of Bayesian and classical methods for inverse problems. The draft textbook by Scales and Smith [176] takes a Bayesian approach to inverse problems.

In many cases the solution to an inverse problem will be used in making a decision, with measurable consequences for making the "wrong" decision. Statistical decision theory can be helpful in determining the optimal decision. The paper by Evans and Stark provides a good introduction to the application of statistical decision theory to inverse problems [60].

CHAPTER TWELVE

Epilogue

The theme of this book is obtaining and analyzing solutions to discretized parameter estimation problems using classical and Bayesian approaches. We have discussed computational procedures for both linear and nonlinear problems. Classical procedures produce estimates of the parameters and their associated uncertainties. In Bayesian methods, the model is a random variable, and the solution is its probability distribution.

However, there are critical issues that arise in solving these problems. When we discretize a continuous problem, the choice of the discretization scheme, basis functions, and grid spacing can have large effects on the behavior of the discretized problem and its solutions, and these effects will not be reflected in the statistical analysis of the solution of the discretized problem. The discretization errors in the solution could potentially be far larger than any explicitly computed statistical uncertainty. Thus it is important to ensure that the discretization provides an adequate approximation to a continuous problem. If no formal analysis is performed, it is at least desirable to see whether varying the discretization has a significant effect on the solutions obtained.

For well-conditioned problems with normally distributed measurement errors, we can use the classical least squares approach. This results in unbiased parameter estimates and associated confidence intervals. For ill-conditioned problems, and for problems where we have good reason to prefer a specific bias in the character of the solution, Tikhonov and other regularization methods can be applied to obtain a solution. However, regularization introduces solution bias, and it is impossible to even bound this bias without making additional assumptions about the true model.

Although Bayesian approaches are also applicable to well-conditioned situations, they are particularly interesting in the context of ill-conditioned problems. By selecting a prior distribution we make explicit assumptions about the model. The resulting posterior distribution is not affected by regularization bias, but is statistically influenced by the prior distribution. In the multivariate normal case for linear problems the Bayesian approach is no more difficult computationally than the least squares approach.

Various efforts have been made to avoid the use of subjective priors in the Bayesian approach. Principles such as maximum entropy can be used to derive prior distributions that have been claimed to be, in some sense, "objective." However, we do not find these arguments completely convincing, and in many cases the choice of prior ultimately has a similar, although perhaps more readily implemented, effect to the choice of regularization in Tikhonov and other classical approaches. Markov Chain Monte Carlo methods present a useful, but computationally intensive, methodology to sample and characterize Bayesian posterior distributions using forward model-based likelihood calculations.

Both the classical and Bayesian approaches can be extended to nonlinear inverse problems. Computations become substantially more complex and we may encounter multiple locally optimal solutions. In both standard approaches, the statistical analysis is typically performed approximately by analyzing a linearization of the nonlinear model around the estimated parameters. However, the validity of this approach will depend on the data uncertainties and the nonlinearity of the problem. The Bayesian approach can theoretically be applied when measurement errors are not normally distributed. However, the associated analytical computations can be difficult in practice. Markov Chain Monte Carlo methods provide a very general approach for characterizing the posterior distribution that avoids the difficulties of analytic formulations.

APPENDIX A

Review of Linear Algebra

Synopsis

A summary of essential concepts, definitions, and theorems in linear algebra used throughout this book.

A.1. SYSTEMS OF LINEAR EQUATIONS

Recall that a system of linear equations can be solved by the process of **Gaussian elimination**.

Example A.1

Consider the system of equations

$$
\begin{aligned}
x_1 + 2x_2 + 3x_3 &= 14 \\
x_1 + 2x_2 + 2x_3 &= 11 \\
x_1 + 3x_2 + 4x_3 &= 19 .
\end{aligned}
\tag{A.1}
$$

We eliminate x_1 from the second and third equations by subtracting the first equation from the second and third equations to obtain

$$
\begin{aligned}
x_1 + 2x_2 + 3x_3 &= 14 \\
-x_3 &= -3 \\
x_2 + x_3 &= 5 .
\end{aligned}
\tag{A.2}
$$

We would like x_2 to appear in the second equation, so we interchange the second and third equations

$$
\begin{aligned}
x_1 + 2x_2 + 3x_3 &= 14 \\
x_2 + x_3 &= 5 \\
-x_3 &= -3 .
\end{aligned}
\tag{A.3}
$$

Next, we eliminate x_2 from the first equation by subtracting two times the second equation from the first equation

$$
\begin{aligned}
x_1 + x_3 &= 4 \\
x_2 + x_3 &= 5 \\
-x_3 &= -3 .
\end{aligned}
\tag{A.4}
$$

We then multiply the third equation by -1 to get an equation for x_3

$$x_1 + x_3 = 4$$
$$x_2 + x_3 = 5 \tag{A.5}$$
$$x_3 = 3 \ .$$

Finally, we eliminate x_3 from the first two equations

$$x_1 = 1$$
$$x_2 = 2 \tag{A.6}$$
$$x_3 = 3 \ .$$

The solution to the original system of equations is thus $x_1 = 1$, $x_2 = 2$, $x_3 = 3$. Geometrically the constraints specified by the three equations of (A.1) describe three planes that, in this case, intersect in a single point.

In solving (A.1), we used three **elementary row operations**: adding a multiple of one equation to another equation, multiplying an equation by a nonzero constant, and swapping two equations. This process can be extended to solve systems of equations with an arbitrary number of variables.

In performing the elimination process, the actual names of the variables are insignificant. We could have renamed the variables in the above example to a, b, and c without changing the solution in any significant way. Because the actual names of the variables are insignificant, we can save space by writing down the significant coefficients from the system of equations in **matrix** form as an **augmented matrix**. The augmented matrix form is also useful in solving a system of equations in computer algorithms, where the elements of the augmented matrix are stored in an array.

In augmented matrix form (A.1) becomes

$$\left[\begin{array}{ccc|c} 1 & 2 & 3 & 14 \\ 1 & 2 & 2 & 11 \\ 1 & 3 & 4 & 19 \end{array} \right] \ . \tag{A.7}$$

In augmented notation, the elementary row operations become adding a multiple of one row to another row, multiplying a row by a nonzero constant, and interchanging two rows. The Gaussian elimination process is essentially identical to the process used in Example A.1, with the final version of the augmented matrix given by

$$\left[\begin{array}{ccc|c} 1 & 0 & 0 & 1 \\ 0 & 1 & 0 & 2 \\ 0 & 0 & 1 & 3 \end{array} \right] \ . \tag{A.8}$$

Definition A.1. A matrix is said to be in **reduced row echelon form (RREF)** if it has the following properties:

1. The first nonzero element in each row is a one. The first nonzero row elements of the matrix are called **pivot elements**. A column in which a pivot element appears is called a **pivot column**.
2. Except for the pivot element, all elements in pivot columns are zero.
3. Each pivot element is to the right of the pivot elements in previous rows.
4. Any rows consisting entirely of zeros are at the bottom of the matrix.

In solving a system of equations in augmented matrix form, we apply elementary row operations to reduce the augmented matrix to RREF and then convert back to conventional notation to read off the solutions. The process of transforming a matrix into RREF can easily be automated. In MATLAB, this is done by the **rref** command.

It can be shown that any linear system of equations has either no solutions, exactly one solution, or infinitely many solutions [121]. In a system of two dimensions, for example, lines represented by the equations can fail to intersect (no solution), intersect at a point (one solution) or intersect in a line (many solutions). The following example shows how to determine the number of solutions from the RREF of the augmented matrix.

Example A.2

Consider a system of two equations in three variables that has many solutions

$$\begin{aligned} x_1 + x_2 + x_3 &= 0 \\ x_1 + 2x_2 + 2x_3 &= 0 \, . \end{aligned} \tag{A.9}$$

We put this system of equations into augmented matrix form and then find the RREF, which is

$$\left[\begin{array}{ccc|c} 1 & 0 & 0 & 0 \\ 0 & 1 & 1 & 0 \end{array} \right] . \tag{A.10}$$

We can translate this back into equation form as

$$\begin{aligned} x_1 &= 0 \\ x_2 + x_3 &= 0 \, . \end{aligned} \tag{A.11}$$

Clearly, x_1 must be 0 in any solution to the system of equations. However, x_2 and x_3 are not fixed. We could treat x_3 as a **free variable** and allow it to take on any value. Whatever value x_3 takes on, x_2 must be equal to $-x_3$. Geometrically, this system of equations describes the intersection of two planes, where the intersection consists of points on the line $x_2 = -x_3$ in the $x_1 = 0$ plane.

A linear system of equations may have more equation constraints than variables, in which case the system of equations is **over-determined**. Although over-determined systems often have no solutions, it is possible for an over-determined system of equations to have either many solutions or exactly one solution.

Conversely, a system of equations with fewer equations than variables is **under-determined**. Although in many cases under-determined systems of equations have infinitely many solutions, it is also possible for such systems to have no solutions.

A system of equations with all zeros on the right hand side is **homogeneous**. Every homogeneous system of equations has at least one solution, the trivial solution in which all variables are zero. A system of equations with a nonzero right hand side is **nonhomogeneous**.

A.2. MATRIX AND VECTOR ALGEBRA

As we have seen in the previous section, a matrix is a table of numbers laid out in rows and columns. A **vector** is simply a matrix consisting of a single column of numbers. In general, matrices and vectors may contain complex numbers as well as real numbers. With the exception of Chapter 8, all vectors and matrices in this book are real.

There are several important notational conventions used here for matrices and vectors. Bold face capital letters such as **A**, **B**, ... are used to denote matrices. Bold face lower case letters such as **x**, **y**, ... are used to denote vectors. Lower case letters or Greek letters such as m, n, α, β, ... will be used to denote scalars.

At times we will need to refer to specific parts of a matrix. The notation $A_{i,j}$ denotes the element of the matrix **A** in row i and column j. We denote the jth element of the vector **x** by x_j. The notation $\mathbf{A}_{.j}$ is used to refer to column j of the matrix **A**, whereas $\mathbf{A}_{i,.}$ refers to row i of **A**.

We can also construct larger matrices from smaller matrices. The notation $\mathbf{A} = [\mathbf{B}\ \mathbf{C}]$ means that the matrix **A** is composed of the matrices **B** and **C**, with matrix **C** to the right of **B**.

If **A** and **B** are two matrices of the *same size*, we can add them by simply adding corresponding elements. Similarly, we can subtract **B** from **A** by subtracting the corresponding elements of **B** from those of **A**. We can multiply a scalar times a matrix by multiplying the scalar times each matrix element. Because vectors are just n by 1 matrices, we can perform the same arithmetic operations on vectors. A **zero matrix**, **0**, is a matrix composed of all zero elements. A zero matrix plays the same role in matrix algebra as the scalar 0, with

$$\mathbf{A} + \mathbf{0} = \mathbf{A} \tag{A.12}$$
$$= \mathbf{0} + \mathbf{A} . \tag{A.13}$$

Using vector notation, we can write a linear system of equations in **vector form**.

Example A.3

Recall the system of equations (A.9)

$$\begin{aligned} x_1 + x_2 + x_3 &= 0 \\ x_1 + 2x_2 + 2x_3 &= 0 \end{aligned} \tag{A.14}$$

from Example A.2. We can write this in vector form as

$$x_1 \begin{bmatrix} 1 \\ 1 \end{bmatrix} + x_2 \begin{bmatrix} 1 \\ 2 \end{bmatrix} + x_3 \begin{bmatrix} 1 \\ 2 \end{bmatrix} = \begin{bmatrix} 0 \\ 0 \end{bmatrix} . \tag{A.15}$$

The expression on the left hand side of (A.15) where vectors are multiplied by scalars and the results are summed together is called a **linear combination**.

If $\mathbf{A}$ is an m by n matrix, and $\mathbf{x}$ is an n element vector, we can multiply $\mathbf{A}$ times $\mathbf{x}$, where the product is defined by

$$\mathbf{A}\mathbf{x} = x_1 \mathbf{A}_{.,1} + x_2 \mathbf{A}_{.,2} + \cdots + x_n \mathbf{A}_{.,n} . \tag{A.16}$$

Example A.4

Given

$$\mathbf{A} = \begin{bmatrix} 1 & 2 & 3 \\ 4 & 5 & 6 \end{bmatrix} \tag{A.17}$$

and

$$\mathbf{x} = \begin{bmatrix} 1 \\ 0 \\ 2 \end{bmatrix} \tag{A.18}$$

then

$$\mathbf{A}\mathbf{x} = 1 \begin{bmatrix} 1 \\ 4 \end{bmatrix} + 0 \begin{bmatrix} 2 \\ 5 \end{bmatrix} + 2 \begin{bmatrix} 3 \\ 6 \end{bmatrix} = \begin{bmatrix} 7 \\ 16 \end{bmatrix} . \tag{A.19}$$

The formula (A.16) for $\mathbf{Ax}$ is a linear combination much like the one that occurred in the vector form of a system of equations. It is possible to write any linear system of equations in the form $\mathbf{Ax} = \mathbf{b}$, where $\mathbf{A}$ is a matrix containing the coefficients of the variables in the equations, $\mathbf{b}$ is a vector containing the coefficients on the right hand sides of the equations, and $\mathbf{x}$ is a vector containing the variables.

Definition A.2. If $\mathbf{A}$ is a matrix of size m by n, and $\mathbf{B}$ is a matrix of size n by r, then the product $\mathbf{C} = \mathbf{AB}$ is obtained by multiplying $\mathbf{A}$ times each of the columns of $\mathbf{B}$ and assembling the matrix vector products in $\mathbf{C}$

$$\mathbf{C} = \begin{bmatrix} \mathbf{AB}_{.,1} & \mathbf{AB}_{.,2} & \cdots & \mathbf{AB}_{.,r} \end{bmatrix} . \tag{A.20}$$

This approach given in (A.20) for calculating a matrix–matrix product will be referred to as the **matrix–vector method**.

Note that the product (A.20) is only possible if the two matrices are of compatible sizes. If $\mathbf{A}$ has m rows and n columns, and $\mathbf{B}$ has n rows and r columns, then the product $\mathbf{AB}$ exists and is of size m by r. In some cases, it is thus possible to multiply $\mathbf{AB}$ but not $\mathbf{BA}$. It is important to note that when both $\mathbf{AB}$ and $\mathbf{BA}$ exist, $\mathbf{AB}$ is not generally equal to $\mathbf{BA}$!

An alternate way to compute the product of two matrices is the **row–column expansion method**, where the product element $C_{i,j}$ is calculated as the matrix product of row i of $\mathbf{A}$ and column j of $\mathbf{B}$.

Example A.5

Let

$$\mathbf{A} = \begin{bmatrix} 1 & 2 \\ 3 & 4 \\ 5 & 6 \end{bmatrix} \tag{A.21}$$

and

$$\mathbf{B} = \begin{bmatrix} 5 & 2 \\ 3 & 7 \end{bmatrix} . \tag{A.22}$$

The product matrix $\mathbf{C} = \mathbf{AB}$ will be of size 3 by 2. We compute the product using both methods. First, using the matrix–vector approach (A.20), we have

$$\mathbf{C} = \begin{bmatrix} \mathbf{AB}_{.,1} & \mathbf{AB}_{.,2} \end{bmatrix} \tag{A.23}$$

$$= \begin{bmatrix} 5 \begin{bmatrix} 1 \\ 3 \\ 5 \end{bmatrix} + 3 \begin{bmatrix} 2 \\ 4 \\ 6 \end{bmatrix} & 2 \begin{bmatrix} 1 \\ 3 \\ 5 \end{bmatrix} + 7 \begin{bmatrix} 2 \\ 4 \\ 6 \end{bmatrix} \end{bmatrix} \tag{A.24}$$

$$= \begin{bmatrix} 11 & 16 \\ 27 & 34 \\ 43 & 52 \end{bmatrix} . \tag{A.25}$$

Next, we use the row–column approach

$$\mathbf{C} = \begin{bmatrix} 1 \cdot 5 + 2 \cdot 3 & 1 \cdot 2 + 2 \cdot 7 \\ 3 \cdot 5 + 4 \cdot 3 & 3 \cdot 2 + 4 \cdot 7 \\ 5 \cdot 5 + 6 \cdot 3 & 5 \cdot 2 + 6 \cdot 7 \end{bmatrix} \tag{A.26}$$

$$= \begin{bmatrix} 11 & 16 \\ 27 & 34 \\ 43 & 52 \end{bmatrix} . \tag{A.27}$$

Definition A.3. The n by n **identity matrix** $\mathbf{I}_n$ is composed of ones in the diagonal and zeros in the off-diagonal elements.

For example, the 3 by 3 identity matrix is

$$\mathbf{I}_3 = \begin{bmatrix} 1 & 0 & 0 \\ 0 & 1 & 0 \\ 0 & 0 & 1 \end{bmatrix} . \tag{A.28}$$

We often write $\mathbf{I}$ without specifying the size of the matrix in situations where the size of matrix is obvious from the context. It is easily shown that if $\mathbf{A}$ is an m by n matrix, then

$$\mathbf{AI}_n = \mathbf{A} \tag{A.29}$$

$$= \mathbf{I}_m \mathbf{A} . \tag{A.30}$$

Thus, multiplying by $\mathbf{I}$ in matrix algebra is similar to multiplying by 1 in conventional scalar algebra.

We have not defined matrix division, but it is possible at this point to define the matrix algebra equivalent of the reciprocal.

Definition A.4. If $\mathbf{A}$ is an n by n matrix, and there is a matrix $\mathbf{B}$ such that

$$\mathbf{AB} = \mathbf{BA} = \mathbf{I} \tag{A.31}$$

then $\mathbf{B}$ is the **inverse** of $\mathbf{A}$. We write $\mathbf{B} = \mathbf{A}^{-1}$.

How do we compute the inverse of a matrix? If $\mathbf{AB} = \mathbf{I}$, then

$$\begin{bmatrix} \mathbf{AB}_{\cdot,1} & \mathbf{AB}_{\cdot,2} & \cdots & \mathbf{AB}_{\cdot,n} \end{bmatrix} = \mathbf{I} . \tag{A.32}$$

Since the columns of the identity matrix and $\mathbf{A}$ are known, we can solve

$$\mathbf{AB}_{.,1} = \begin{bmatrix} 1 \\ 0 \\ \cdot \\ \cdot \\ \cdot \\ 0 \end{bmatrix} \qquad (\text{A.33})$$

to obtain $\mathbf{B}_1$. We can find the remaining columns of the inverse in the same way. If any of these systems of equations are inconsistent, then $\mathbf{A}^{-1}$ does not exist.

The inverse matrix can be used to solve a system of linear equations with n equations and n variables. Given the system of equations $\mathbf{Ax} = \mathbf{b}$, and $\mathbf{A}^{-1}$, we can multiply $\mathbf{Ax} = \mathbf{b}$ on both sides by the inverse to obtain

$$\mathbf{A}^{-1}\mathbf{Ax} = \mathbf{A}^{-1}\mathbf{b} . \qquad (\text{A.34})$$

Because

$$\mathbf{A}^{-1}\mathbf{Ax} = \mathbf{Ix} \qquad (\text{A.35})$$
$$= \mathbf{x} \qquad (\text{A.36})$$

this gives the solution

$$\mathbf{x} = \mathbf{A}^{-1}\mathbf{b} . \qquad (\text{A.37})$$

This argument shows that if $\mathbf{A}^{-1}$ exists, then for any right hand side $\mathbf{b}$, a system of equations has a *unique* solution. If $\mathbf{A}^{-1}$ does not exist, then the system of equations may have either many solutions or no solution.

Definition A.5. When $\mathbf{A}$ is an n by n matrix, $\mathbf{A}^k$ is the product of k copies of $\mathbf{A}$. By convention, we define $\mathbf{A}^0 = \mathbf{I}$.

Definition A.6. The **transpose** of a matrix $\mathbf{A}$, denoted $\mathbf{A}^T$, is obtained by taking the columns of $\mathbf{A}$ and writing them as the rows of the transpose. We will also use the notation $\mathbf{A}^{-T}$ for $(\mathbf{A}^{-1})^T$.

Example A.6

Let

$$\mathbf{A} = \begin{bmatrix} 2 & 1 \\ 5 & 2 \end{bmatrix} . \qquad (\text{A.38})$$

Then

$$\mathbf{A}^T = \begin{bmatrix} 2 & 5 \\ 1 & 2 \end{bmatrix}. \tag{A.39}$$

Definition A.7. A matrix is **symmetric** if $\mathbf{A} = \mathbf{A}^T$.

Although many elementary textbooks on linear algebra consider only square diagonal matrices, we will have occasion to refer to m by n matrices that have nonzero elements only on the diagonal.

Definition A.8. An m by n matrix $\mathbf{A}$ is **diagonal** if $A_{i,j} = 0$ whenever $i \neq j$.

Definition A.9. An m by n matrix $\mathbf{R}$ is **upper-triangular** if $R_{i,j} = 0$ whenever $i > j$. A matrix $\mathbf{L}$ is **lower-triangular** if $\mathbf{L}^T$ is upper-triangular.

Example A.7

The matrix

$$\mathbf{S} = \begin{bmatrix} 1 & 0 & 0 & 0 & 0 \\ 0 & 2 & 0 & 0 & 0 \\ 0 & 0 & 3 & 0 & 0 \end{bmatrix} \tag{A.40}$$

is diagonal, and the matrix

$$\mathbf{R} = \begin{bmatrix} 1 & 2 & 3 \\ 0 & 2 & 4 \\ 0 & 0 & 5 \\ 0 & 0 & 0 \end{bmatrix} \tag{A.41}$$

is upper-triangular.

Theorem A.1. The following statements are true for any scalars s and t and matrices $\mathbf{A}$, $\mathbf{B}$, and $\mathbf{C}$. It is assumed that the matrices are of the appropriate size for the operations involved and that whenever an inverse occurs, the matrix is invertible.
1. $\mathbf{A} + 0 = 0 + \mathbf{A} = \mathbf{A}$.
2. $\mathbf{A} + \mathbf{B} = \mathbf{B} + \mathbf{A}$.
3. $(\mathbf{A} + \mathbf{B}) + \mathbf{C} = \mathbf{A} + (\mathbf{B} + \mathbf{C})$.
4. $\mathbf{A}(\mathbf{BC}) = (\mathbf{AB})\mathbf{C}$.
5. $\mathbf{A}(\mathbf{B} + \mathbf{C}) = \mathbf{AB} + \mathbf{AC}$.
6. $(\mathbf{A} + \mathbf{B})\mathbf{C} = \mathbf{AC} + \mathbf{BC}$.
7. $(st)\mathbf{A} = s(t\mathbf{A})$.
8. $s(\mathbf{AB}) = (s\mathbf{A})\mathbf{B} = \mathbf{A}(s\mathbf{B})$.

9. $(s+t)\mathbf{A} = s\mathbf{A} + t\mathbf{A}$.
10. $s(\mathbf{A} + \mathbf{B}) = s\mathbf{A} + s\mathbf{B}$.
11. $(\mathbf{A}^T)^T = \mathbf{A}$.
12. $(s\mathbf{A})^T = s(\mathbf{A}^T)$.
13. $(\mathbf{A} + \mathbf{B})^T = \mathbf{A}^T + \mathbf{B}^T$.
14. $(\mathbf{AB})^T = \mathbf{B}^T\mathbf{A}^T$.
15. $(\mathbf{AB})^{-1} = \mathbf{B}^{-1}\mathbf{A}^{-1}$.
16. $(\mathbf{A}^{-1})^{-1} = \mathbf{A}$.
17. $(\mathbf{A}^T)^{-1} = (\mathbf{A}^{-1})^T$.
18. If $\mathbf{A}$ and $\mathbf{B}$ are n by n matrices, and $\mathbf{AB} = \mathbf{I}$, then $\mathbf{A}^{-1} = \mathbf{B}$ and $\mathbf{B}^{-1} = \mathbf{A}$.

The first ten rules in this list are identical to rules of conventional algebra, and you should have little trouble in applying them. The rules involving transposes and inverses are new, but they can be mastered without too much trouble.

Some students have difficulty with the following statements, which would appear to be true on the surface, but that are in fact **false** for at least some matrices.

1. $\mathbf{AB} = \mathbf{BA}$.
2. If $\mathbf{AB} = 0$, then $\mathbf{A} = 0$ or $\mathbf{B} = 0$.
3. If $\mathbf{AB} = \mathbf{AC}$ and $\mathbf{A} \neq 0$, then $\mathbf{B} = \mathbf{C}$.

It is a worthwhile exercise to construct examples of 2 by 2 matrices for which each of these statements is false.

A.3. LINEAR INDEPENDENCE

Definition A.10. The vectors $\mathbf{v}_1$, $\mathbf{v}_2$, ..., $\mathbf{v}_n$ are **linearly independent** if the system of equations

$$c_1\mathbf{v}_1 + c_2\mathbf{v}_2 + \cdots + c_n\mathbf{v}_n = 0 \tag{A.42}$$

has only the trivial solution $\mathbf{c} = 0$. If there are multiple solutions, then the vectors are **linearly dependent**.

Determining whether or not a set of vectors is linearly independent is simple. Just solve the above system of equations (A.42).

Example A.8

Let

$$\mathbf{A} = \begin{bmatrix} 1 & 2 & 3 \\ 4 & 5 & 6 \\ 7 & 8 & 9 \end{bmatrix}. \tag{A.43}$$

Are the columns of $\mathbf{A}$ linearly independent vectors? To determine this we set up the system of equations $\mathbf{Ax} = 0$ in an augmented matrix, and then find the RREF

$$
\left[
\begin{array}{ccc|c}
1 & 0 & -1 & 0 \\
0 & 1 & 2 & 0 \\
0 & 0 & 0 & 0
\end{array}
\right].
\tag{A.44}
$$

The solutions are

$$
\mathbf{x} = x_3 \begin{bmatrix} 1 \\ -2 \\ 1 \end{bmatrix}.
\tag{A.45}
$$

We can set $x_3 = 1$ and obtain the nonzero solution

$$
\mathbf{x} = \begin{bmatrix} 1 \\ -2 \\ 1 \end{bmatrix}.
\tag{A.46}
$$

Thus, the columns of $\mathbf{A}$ are linearly dependent.

There are a number of important theoretical consequences of linear independence. For example, it can be shown that if the columns of an n by n matrix $\mathbf{A}$ are linearly independent, then $\mathbf{A}^{-1}$ exists, and the system of equations $\mathbf{Ax} = \mathbf{b}$ has a unique solution for every right hand side $\mathbf{b}$ [121].

A.4. SUBSPACES OF R^n

So far, we have worked with vectors of real numbers in the n-dimensional space R^n. There are a number of properties of R^n that make it convenient to work with vectors. First, the operation of vector addition always works. We can take any two vectors in R^n and add them together and get another vector in R^n. Second, we can multiply any vector in R^n by a scalar and obtain another vector in R^n. Finally, we have the 0 vector, with the property that for any vector $\mathbf{x}$, $\mathbf{x} + 0 = 0 + \mathbf{x} = \mathbf{x}$.

Definition A.11. A **subspace** W of R^n is a subset of R^n that satisfies the three properties:
1. If $\mathbf{x}$ and $\mathbf{y}$ are vectors in W, then $\mathbf{x} + \mathbf{y}$ is also a vector in W.
2. If $\mathbf{x}$ is a vector in W and s is any real scalar, then $s\mathbf{x}$ is also a vector in W.
3. The 0 vector is in W. A subspace of R^n is **nontrivial** if it contains vectors other than the zero vector.

Example A.9

In R^3, the plane P defined by the equation

$$x_1 + x_2 + x_3 = 0 \tag{A.47}$$

is a subspace of R^n. To see this, note that if we take any two vectors in the plane and add them together, we get another vector in the plane. If we take a vector in this plane and multiply it by any scalar, we get another vector in the plane. Finally, $\mathbf{0}$ is a vector in the plane.

Subspaces are important because they provide an environment within which all rules of matrix–vector algebra apply. An especially important subspace of R^n that we will work with is the **null space** of an m by n matrix.

Definition A.12. Let $\mathbf{A}$ be an m by n matrix. The null space of $\mathbf{A}$, written $N(\mathbf{A})$, is the set of all vectors $\mathbf{x}$ such that $\mathbf{Ax} = \mathbf{0}$.

To show that $N(\mathbf{A})$ is actually a subspace of R^n, we need to show that:

1. If $\mathbf{x}$ and $\mathbf{y}$ are in $N(\mathbf{A})$, then $\mathbf{Ax} = \mathbf{0}$ and $\mathbf{Ay} = \mathbf{0}$. By adding these equations, we find that $\mathbf{A(x+y)} = \mathbf{0}$. Thus $\mathbf{x} + \mathbf{y}$ is in $N(\mathbf{A})$.
2. If $\mathbf{x}$ is in $N(\mathbf{A})$ and s is any scalar, then $\mathbf{Ax} = \mathbf{0}$. We can multiply this equation by s to get $s\mathbf{Ax} = \mathbf{0}$. Thus $\mathbf{A}(s\mathbf{x}) = \mathbf{0}$, and $s\mathbf{x}$ is in $N(\mathbf{A})$.
3. $\mathbf{A0} = \mathbf{0}$, so $\mathbf{0}$ is in $N(\mathbf{A})$.

Computationally, the null space of a matrix can be determined by solving the system of equations $\mathbf{Ax} = \mathbf{0}$.

Example A.10

Let

$$\mathbf{A} = \begin{bmatrix} 3 & 1 & 9 & 4 \\ 2 & 1 & 7 & 3 \\ 5 & 2 & 16 & 7 \end{bmatrix}. \tag{A.48}$$

To find the null space of $\mathbf{A}$, we solve the system of equations $\mathbf{Ax} = \mathbf{0}$. To solve the equations, we put the system of equations into an augmented matrix

$$\left[\begin{array}{cccc|c} 3 & 1 & 9 & 4 & 0 \\ 2 & 1 & 7 & 3 & 0 \\ 5 & 2 & 16 & 7 & 0 \end{array}\right] \tag{A.49}$$

and find the RREF

$$\left[\begin{array}{cccc|c} 1 & 0 & 2 & 1 & 0 \\ 0 & 1 & 3 & 1 & 0 \\ 0 & 0 & 0 & 0 & 0 \end{array}\right] . \tag{A.50}$$

From the augmented matrix, we find that

$$\mathbf{x} = x_3 \begin{bmatrix} -2 \\ -3 \\ 1 \\ 0 \end{bmatrix} + x_4 \begin{bmatrix} -1 \\ -1 \\ 0 \\ 1 \end{bmatrix} . \tag{A.51}$$

Any vector in the null space can be written as a linear combination of the above vectors, so the null space is a two-dimensional plane within R^4.

Now, consider the problem of solving $\mathbf{Ax} = \mathbf{b}$, where

$$\mathbf{b} = \begin{bmatrix} 22 \\ 17 \\ 39 \end{bmatrix} \tag{A.52}$$

and one particular solution is

$$\mathbf{p} = \begin{bmatrix} 1 \\ 2 \\ 1 \\ 2 \end{bmatrix} . \tag{A.53}$$

We can take any vector in the null space of $\mathbf{A}$ and add it to this solution to obtain another solution. Suppose that $\mathbf{x}$ is in $N(\mathbf{A})$. Then

$$\mathbf{A}(\mathbf{x} + \mathbf{p}) = \mathbf{Ax} + \mathbf{Ap}$$
$$\mathbf{A}(\mathbf{x} + \mathbf{p}) = \mathbf{0} + \mathbf{b}$$
$$\mathbf{A}(\mathbf{x} + \mathbf{p}) = \mathbf{b} .$$

For example,

$$\mathbf{x} = \begin{bmatrix} 1 \\ 2 \\ 1 \\ 2 \end{bmatrix} + 2 \begin{bmatrix} -2 \\ -3 \\ 1 \\ 0 \end{bmatrix} + 3 \begin{bmatrix} -1 \\ -1 \\ 0 \\ 1 \end{bmatrix} \tag{A.54}$$

is also a solution to $\mathbf{Ax} = \mathbf{b}$.

In the context of inverse problems, the null space is critical because the presence of a nontrivial null space leads to nonuniqueness in the solution to a linear system of equations.

Definition A.13. A **basis** for a subspace W is a set of vectors $\mathbf{v}_1, \ldots, \mathbf{v}_p$ such that
1. Any vector in W can be written as a linear combination of the basis vectors.
2. The basis vectors are linearly independent.

A particularly simple and useful basis is the **standard basis**.

Definition A.14. The **standard basis** for R^n is the set of vectors $\mathbf{e}_1, \ldots, \mathbf{e}_n$ such that the elements of $\mathbf{e}_i$ are all zero except for the ith element, which is one.

Any nontrivial subspace W of R^n will have many different bases. For example, we can take any basis and multiply one of the basis vectors by 2 to obtain a new basis. It is possible to show that all bases for a subspace W have the same number of basis vectors [121].

Theorem A.2. Let W be a subspace of R^n with basis $\mathbf{v}_1, \mathbf{v}_2, \ldots, \mathbf{v}_p$. Then all bases for W have p basis vectors, and p is the **dimension** of W.

It can be shown that the procedure used in the above example always produces a basis for $N(\mathbf{A})$ [121]. A basis for the null space of a matrix can be found in MATLAB using the **null** command.

Definition A.15. Let $\mathbf{A}$ be an m by n matrix. The **column space** or **range** of $\mathbf{A}$ (written $R(\mathbf{A})$) is the set of all vectors $\mathbf{b}$ such that $\mathbf{Ax} = \mathbf{b}$ has at least one solution. In other words, the column space is the set of all vectors $\mathbf{b}$ that can be written as a linear combination of the columns of $\mathbf{A}$.

The range is important in the context of discrete linear inverse problems, because $R(\mathbf{G})$ consists of all vectors $\mathbf{d}$ for which there is a model $\mathbf{m}$ such that $\mathbf{Gm} = \mathbf{d}$.

To find the column space of a matrix, we consider what happens when we compute the RREF of $[\mathbf{A} \,|\, \mathbf{b}]$. In the part of the augmented matrix corresponding to the left hand side of the equations we always get the same result, namely the RREF of $\mathbf{A}$. The solution to the system of equations may involve some free variables, but we can always set these free variables to 0. Thus when we are able to solve $\mathbf{Ax} = \mathbf{b}$, we can solve the system of equations by using only variables corresponding to the pivot columns in the RREF of $\mathbf{A}$. In other words, if we can solve $\mathbf{Ax} = \mathbf{b}$, then we can write $\mathbf{b}$ as a linear combination of the pivot columns of $\mathbf{A}$. Note that these are columns from the original matrix $\mathbf{A}$, not columns from the RREF of $\mathbf{A}$. An orthonormal (see below) basis for the range of a matrix can be found in MATLAB using the **orth** command.

Example A.11

As in the previous example, let

$$\mathbf{A} = \begin{bmatrix} 3 & 1 & 9 & 4 \\ 2 & 1 & 7 & 3 \\ 5 & 2 & 16 & 7 \end{bmatrix}. \tag{A.55}$$

To find the column space of $\mathbf{A}$, note that the RREF of $\mathbf{A}$ is

$$\begin{bmatrix} 1 & 0 & 2 & 1 \\ 0 & 1 & 3 & 1 \\ 0 & 0 & 0 & 0 \end{bmatrix}. \tag{A.56}$$

Thus whenever we can solve $\mathbf{Ax} = \mathbf{b}$, we can find a solution in which x_3 and x_4 are 0. In other words, whenever there is a solution to $\mathbf{Ax} = \mathbf{b}$, we can write $\mathbf{b}$ as a linear combination of the first two columns of $\mathbf{A}$

$$\mathbf{b} = x_1 \begin{bmatrix} 3 \\ 2 \\ 5 \end{bmatrix} + x_2 \begin{bmatrix} 1 \\ 1 \\ 2 \end{bmatrix}. \tag{A.57}$$

Since these two vectors are linearly independent and span $R(\mathbf{A})$, they form a basis for $R(\mathbf{A})$. The dimension of $R(\mathbf{A})$ is two.

In finding the null space and range of a matrix $\mathbf{A}$ we found that the basis vectors for $N(\mathbf{A})$ corresponded to non-pivot columns of $\mathbf{A}$, while the basis vectors for $R(\mathbf{A})$ corresponded to pivot columns of $\mathbf{A}$. Since the matrix $\mathbf{A}$ had n columns, we obtain the theorem.

Theorem A.3.

$$\dim N(\mathbf{A}) + \dim R(\mathbf{A}) = n. \tag{A.58}$$

In addition to the null space and range of a matrix $\mathbf{A}$, we will often work with the null space and range of the transpose of $\mathbf{A}$. Since the columns of $\mathbf{A}^T$ are rows of $\mathbf{A}$, the column space of $\mathbf{A}^T$ is also called the **row space** of $\mathbf{A}$. Since each row of $\mathbf{A}$ can be written as a linear combination of the nonzero rows of the RREF of $\mathbf{A}$, the nonzero rows of the RREF form a basis for the row space of $\mathbf{A}$. There are exactly as many nonzero rows in the RREF of $\mathbf{A}$ as there are pivot columns. Thus we have the following theorem.

Theorem A.4.

$$\dim(R(\mathbf{A}^T)) = \dim R(\mathbf{A}) . \tag{A.59}$$

Definition A.16. The **rank** of an m by n matrix $\mathbf{A}$ is the dimension of $R(\mathbf{A})$. If $\text{rank}(\mathbf{A}) = \min(m, n)$, then $\mathbf{A}$ has **full rank**. If $\text{rank}(\mathbf{A}) = m$, then $\mathbf{A}$ has **full row rank**. If $\text{rank}(\mathbf{A}) = n$, then $\mathbf{A}$ has **full column rank**. If $\text{rank}(\mathbf{A}) < \min(m, n)$, then $\mathbf{A}$ is **rank deficient**.

The rank of a matrix is readily found in MATLAB by using the **rank** command.

A.5. ORTHOGONALITY AND THE DOT PRODUCT

Definition A.17. Let $\mathbf{x}$ and $\mathbf{y}$ be two vectors in R^n. The **dot product** of $\mathbf{x}$ and $\mathbf{y}$ is

$$\mathbf{x} \cdot \mathbf{y} = \mathbf{x}^T \mathbf{y} = x_1 y_1 + x_2 y_2 + \cdots + x_n y_n . \tag{A.60}$$

Definition A.18. Let $\mathbf{x}$ be a vector in R^n. The **2-norm** or **Euclidean length** of $\mathbf{x}$ is

$$\|\mathbf{x}\|_2 = \sqrt{\mathbf{x}^T \mathbf{x}} = \sqrt{x_1^2 + x_2^2 + \cdots + x_n^2} . \tag{A.61}$$

Later we will introduce two other ways of measuring the "length" of a vector. The subscript 2 is used to distinguish this 2-norm from the other norms.

You may be familiar with an alternative definition of the dot product in which $\mathbf{x} \cdot \mathbf{y} = \|\mathbf{x}\| \|\mathbf{y}\| \cos(\theta)$ where θ is the angle between the two vectors. The two definitions are equivalent. To see this, consider a triangle with sides $\mathbf{x}$, $\mathbf{y}$, and $\mathbf{x} - \mathbf{y}$. See Fig. A.1. The angle between sides $\mathbf{x}$ and $\mathbf{y}$ is θ. By the law of cosines,

$$\|\mathbf{x} - \mathbf{y}\|_2^2 = \|\mathbf{x}\|_2^2 + \|\mathbf{y}\|_2^2 - 2\|\mathbf{x}\|_2 \|\mathbf{y}\|_2 \cos(\theta)$$
$$(\mathbf{x} - \mathbf{y})^T (\mathbf{x} - \mathbf{y}) = \mathbf{x}^T \mathbf{x} + \mathbf{y}^T \mathbf{y} - 2\|\mathbf{x}\|_2 \|\mathbf{y}\|_2 \cos(\theta)$$
$$\mathbf{x}^T \mathbf{x} - 2\mathbf{x}^T \mathbf{y} + \mathbf{y}^T \mathbf{y} = \mathbf{x}^T \mathbf{x} + \mathbf{y}^T \mathbf{y} - 2\|\mathbf{x}\|_2 \|\mathbf{y}\|_2 \cos(\theta)$$
$$-2\mathbf{x}^T \mathbf{y} = -2\|\mathbf{x}\|_2 \|\mathbf{y}\|_2 \cos(\theta)$$
$$\mathbf{x}^T \mathbf{y} = \|\mathbf{x}\|_2 \|\mathbf{y}\|_2 \cos(\theta) .$$

We can also use this formula to compute the angle between two vectors.

$$\theta = \cos^{-1}\left(\frac{\mathbf{x}^T \mathbf{y}}{\|\mathbf{x}\|_2 \|\mathbf{y}\|_2} \right) . \tag{A.62}$$

Definition A.19. Two vectors $\mathbf{x}$ and $\mathbf{y}$ in R^n are **orthogonal**, or equivalently, **perpendicular** (written $\mathbf{x} \perp \mathbf{y}$), if $\mathbf{x}^T \mathbf{y} = 0$.

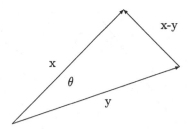

Figure A.1 Relationship between the dot product and the angle between two vectors.

Definition A.20. A set of vectors $\mathbf{v}_1, \mathbf{v}_2, \ldots, \mathbf{v}_p$ is **orthogonal** if each pair of vectors in the set is orthogonal.

Definition A.21. Two subspaces V and W of R^n are **orthogonal** if every vector in V is perpendicular to every vector in W.

If $\mathbf{x}$ is in $N(\mathbf{A})$, then $\mathbf{A}\mathbf{x} = \mathbf{0}$. Since each element of the product $\mathbf{A}\mathbf{x}$ can be obtained by taking the dot product of a row of $\mathbf{A}$ and $\mathbf{x}$, $\mathbf{x}$ is perpendicular to each row of $\mathbf{A}$. Since $\mathbf{x}$ is perpendicular to all columns of $\mathbf{A}^T$, it is perpendicular to $R(\mathbf{A}^T)$. We have the following theorem.

Theorem A.5. Let $\mathbf{A}$ be an m by n matrix. Then

$$N(\mathbf{A}) \perp R(\mathbf{A}^T) \,. \tag{A.63}$$

Furthermore,

$$N(\mathbf{A}) + R(\mathbf{A}^T) = R^n \,. \tag{A.64}$$

That is, any vector $\mathbf{x}$ in R^n can be written uniquely as $\mathbf{x} = \mathbf{p} + \mathbf{q}$ where $\mathbf{p}$ is in $N(\mathbf{A})$ and $\mathbf{q}$ is in $R(\mathbf{A}^T)$.

Definition A.22. A basis in which the basis vectors are orthogonal is an **orthogonal basis**. A basis in which the basis vectors are orthogonal and have length one is an **orthonormal basis**.

Definition A.23. An n by n matrix $\mathbf{Q}$ is **orthogonal** if the columns of $\mathbf{Q}$ are orthogonal and each column of $\mathbf{Q}$ has length one.

With the requirement that the columns of an orthogonal matrix have length one, using the term "orthonormal" would make logical sense. However, the definition of "orthogonal" given here is standard and we will not try to change standard usage.

Orthogonal matrices have a number of useful properties.

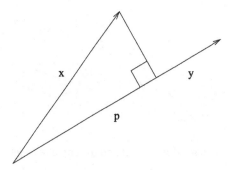

Figure A.2 The orthogonal projection of **x** onto **y**.

Theorem A.6. If $\mathbf{Q}$ is an orthogonal matrix, then:
1. $\mathbf{Q}^T\mathbf{Q} = \mathbf{Q}\mathbf{Q}^T = \mathbf{I}$. In other words, $\mathbf{Q}^{-1} = \mathbf{Q}^T$.
2. For any vector $\mathbf{x}$ in R^n, $\|\mathbf{Q}\mathbf{x}\|_2 = \|\mathbf{x}\|_2$.
3. For any two vectors $\mathbf{x}$ and $\mathbf{y}$ in R^n, $\mathbf{x}^T\mathbf{y} = (\mathbf{Q}\mathbf{x})^T(\mathbf{Q}\mathbf{y})$.

A problem that we will often encounter in practice is projecting a vector $\mathbf{x}$ onto another vector $\mathbf{y}$ or onto a subspace W to obtain a projected vector $\mathbf{p}$. See Fig. A.2. We know that

$$\mathbf{x}^T\mathbf{y} = \|\mathbf{x}\|_2\|\mathbf{y}\|_2\cos(\theta) \tag{A.65}$$

where θ is the angle between $\mathbf{x}$ and $\mathbf{y}$. Also,

$$\cos(\theta) = \frac{\|\mathbf{p}\|_2}{\|\mathbf{x}\|_2} . \tag{A.66}$$

Thus

$$\|\mathbf{p}\|_2 = \frac{\mathbf{x}^T\mathbf{y}}{\|\mathbf{y}\|_2} . \tag{A.67}$$

Since $\mathbf{p}$ points in the same direction as $\mathbf{y}$,

$$\mathbf{p} = \text{proj}_{\mathbf{y}}\mathbf{x} = \frac{\mathbf{x}^T\mathbf{y}}{\mathbf{y}^T\mathbf{y}}\mathbf{y} . \tag{A.68}$$

The vector $\mathbf{p}$ is called the **orthogonal projection** or simply the **projection** of $\mathbf{x}$ onto $\mathbf{y}$.

Similarly, if W is a subspace of R^n with an orthogonal basis $\mathbf{w}_1, \mathbf{w}_2, \ldots, \mathbf{w}_p$, then the **orthogonal projection of x onto** W is

$$\mathbf{p} = \text{proj}_W\mathbf{x} = \frac{\mathbf{x}^T\mathbf{w}_1}{\mathbf{w}_1^T\mathbf{w}_1}\mathbf{w}_1 + \frac{\mathbf{x}^T\mathbf{w}_2}{\mathbf{w}_2^T\mathbf{w}_2}\mathbf{w}_2 + \cdots + \frac{\mathbf{x}^T\mathbf{w}_p}{\mathbf{w}_p^T\mathbf{w}_p}\mathbf{w}_p . \tag{A.69}$$

Note that this equation can be simplified considerably if the orthogonal basis vectors are also orthonormal. In that case, $\mathbf{w}_1^T\mathbf{w}_1$, $\mathbf{w}_2^T\mathbf{w}_2$, ..., $\mathbf{w}_p^T\mathbf{w}_p$ are all 1.

It is inconvenient that the projection formula requires an orthogonal basis. The **Gram–Schmidt orthogonalization process** can be used to turn any basis for a subspace of R^n into an orthogonal basis. We begin with a basis $\mathbf{v}_1$, $\mathbf{v}_2$, ..., $\mathbf{v}_p$. The process recursively constructs an orthogonal basis by taking each vector in the original basis and then subtracting off its projection on the space spanned by the previous vectors. The formulas are

$$\mathbf{w}_1 = \mathbf{v}_1$$

$$\mathbf{w}_2 = \mathbf{v}_2 - \frac{\mathbf{v}_1^T\mathbf{v}_2}{\mathbf{v}_1^T\mathbf{v}_1}\mathbf{v}_1 = \mathbf{v}_2 - \frac{\mathbf{w}_1^T\mathbf{v}_2}{\mathbf{w}_1^T\mathbf{w}_1}\mathbf{w}_1$$

$$\vdots$$

$$\mathbf{w}_p = \mathbf{v}_p - \frac{\mathbf{w}_1^T\mathbf{v}_p}{\mathbf{w}_1^T\mathbf{w}_1}\mathbf{w}_1 - \cdots - \frac{\mathbf{w}_{p-1}^T\mathbf{v}_p}{\mathbf{w}_{p-1}^T\mathbf{w}_{p-1}}\mathbf{w}_{p-1} \ . \tag{A.70}$$

Unfortunately, the Gram–Schmidt process is numerically unstable when applied to large bases. In MATLAB the command **orth** provides a numerically stable way to produce an orthogonal basis from a nonorthogonal basis. An important property of orthogonal projection is that the projection of $\mathbf{x}$ onto W is the point in W that is closest to $\mathbf{x}$. In the special case that $\mathbf{x}$ is in W, the projection of $\mathbf{x}$ onto W is $\mathbf{x}$.

Given an inconsistent system of equations $\mathbf{A}\mathbf{x} = \mathbf{b}$, it is often desirable to find an approximate solution. A natural measure of the quality of an approximate solution is the norm of the difference between $\mathbf{A}\mathbf{x}$ and $\mathbf{b}$, $\|\mathbf{A}\mathbf{x} - \mathbf{b}\|$. A solution that minimizes the 2-norm, $\|\mathbf{A}\mathbf{x} - \mathbf{b}\|_2$, is called a **least squares solution**, because it minimizes the sum of the squares of the residuals.

The least squares solution can be obtained by projecting $\mathbf{b}$ onto $R(\mathbf{A})$. This calculation requires us to first find an orthogonal basis for $R(\mathbf{A})$. There is an alternative approach that does not require finding an orthogonal basis. Let

$$\mathbf{A}\mathbf{x}_{ls} = \text{proj}_{R(\mathbf{A})}\mathbf{b} \ . \tag{A.71}$$

Then, the difference between the projection (A.71) and $\mathbf{b}$, $\mathbf{A}\mathbf{x}_{ls} - \mathbf{b}$, will be perpendicular to $R(\mathbf{A})$ (Fig. A.3). This orthogonality means that each of the columns of $\mathbf{A}$ will be orthogonal to $\mathbf{A}\mathbf{x}_{ls} - \mathbf{b}$. Thus

$$\mathbf{A}^T(\mathbf{A}\mathbf{x}_{ls} - \mathbf{b}) = 0 \tag{A.72}$$

or

$$\mathbf{A}^T\mathbf{A}\mathbf{x}_{ls} = \mathbf{A}^T\mathbf{b} \ . \tag{A.73}$$

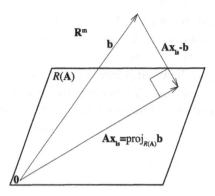

Figure A.3 Geometric conceptualization of the least squares solution to $\mathbf{Ax} = \mathbf{b}$. $\mathbf{b}$ generally lies in R^m, but $R(\mathbf{A})$ is generally a subspace of R^m. The least squares solution $\mathbf{x}_{ls}$ minimizes $\|\mathbf{Ax} - \mathbf{b}\|_2$.

This last system of equations is referred to as the **normal equations** for the least squares problem. It can be shown that if the columns of $\mathbf{A}$ are linearly independent, then the normal equations have exactly one solution for $\mathbf{x}_{ls}$ and this solution minimizes the sum of squared residuals [121].

A.6. EIGENVALUES AND EIGENVECTORS

Definition A.24. An n by n matrix $\mathbf{A}$ has an eigenvalue λ with an associated eigenvector $\mathbf{x}$ if $\mathbf{x}$ is not $\mathbf{0}$, and

$$\mathbf{Ax} = \lambda\mathbf{x} . \tag{A.74}$$

To find eigenvalues and eigenvectors, we rewrite the eigenvector equation (A.74) as

$$(\mathbf{A} - \lambda\mathbf{I})\mathbf{x} = \mathbf{0} . \tag{A.75}$$

To find nonzero eigenvectors, the matrix $\mathbf{A} - \lambda\mathbf{I}$ must be singular. This leads to the **characteristic equation**

$$\det(\mathbf{A} - \lambda\mathbf{I}) = 0 , \tag{A.76}$$

where det denotes the determinant. For small matrices (2 by 2 or 3 by 3), it is relatively simple to solve (A.76) to find the eigenvalues. The eigenvalues can then be substituted into (A.75) and the resulting system can then be solved to find corresponding eigenvectors. Note that the eigenvalues can, in general, be complex. For larger matrices, solving the characteristic equation becomes impractical and more sophisticated numerical methods are used. The MATLAB command **eig** can be used to find eigenvalues and eigenvectors of a matrix.

Suppose that we can find a set of n linearly independent eigenvectors, $\mathbf{v}_i$, of an n by n matrix $\mathbf{A}$ with associated eigenvalues λ_i. These eigenvectors form a basis for R^n. We can use the eigenvectors to **diagonalize** the matrix as

$$\mathbf{A} = \mathbf{P}\mathbf{\Lambda}\mathbf{P}^{-1} \tag{A.77}$$

where

$$\mathbf{P} = \begin{bmatrix} \mathbf{v}_1 & \mathbf{v}_2 & \cdots & \mathbf{v}_n \end{bmatrix} \tag{A.78}$$

and $\mathbf{\Lambda}$ is a diagonal matrix of eigenvalues

$$\Lambda_{ii} = \lambda_i \ . \tag{A.79}$$

To see that this works, simply compute $\mathbf{AP}$

$$\begin{aligned} \mathbf{AP} &= \mathbf{A} \begin{bmatrix} \mathbf{v}_1 & \mathbf{v}_2 & \cdots & \mathbf{v}_n \end{bmatrix} \\ &= \begin{bmatrix} \lambda_1\mathbf{v}_1 & \lambda_2\mathbf{v}_2 & \cdots & \lambda_n\mathbf{v}_n \end{bmatrix} \\ &= \mathbf{P}\mathbf{\Lambda} \ . \end{aligned}$$

Thus $\mathbf{A} = \mathbf{P}\mathbf{\Lambda}\mathbf{P}^{-1}$. Not all matrices are diagonalizable, because not all matrices have n linearly independent eigenvectors. However, there is an important special case in which matrices can always be diagonalized.

Theorem A.7. If $\mathbf{A}$ is a real symmetric matrix, then $\mathbf{A}$ can be written as

$$\mathbf{A} = \mathbf{Q}\mathbf{\Lambda}\mathbf{Q}^{-1} = \mathbf{Q}\mathbf{\Lambda}\mathbf{Q}^{T} \tag{A.80}$$

where $\mathbf{Q}$ is a real orthogonal matrix of eigenvectors of $\mathbf{A}$ and $\mathbf{\Lambda}$ is a real diagonal matrix of the eigenvalues of $\mathbf{A}$.

This **orthogonal diagonalization** of a real symmetric matrix $\mathbf{A}$ will be useful later on when we consider orthogonal factorizations of general matrices.

The eigenvalues of symmetric matrices are particularly important in the analysis of quadratic forms.

Definition A.25. A **quadratic form** is a function of the form

$$f(\mathbf{x}) = \mathbf{x}^T\mathbf{A}\mathbf{x} \tag{A.81}$$

where $\mathbf{A}$ is a symmetric n by n matrix. The quadratic form $f(\mathbf{x})$ is **positive definite (PD)** if $f(x) \geq 0$ for all $\mathbf{x}$ and $f(\mathbf{x}) = 0$ only when $\mathbf{x} = \mathbf{0}$. The quadratic form is **positive semidefinite** if $f(\mathbf{x}) \geq 0$ for all $\mathbf{x}$. Similarly, a symmetric matrix $\mathbf{A}$ is positive definite if the associated quadratic form $f(\mathbf{x}) = \mathbf{x}^T \mathbf{A} \mathbf{x}$ is positive definite. The quadratic form is **negative semidefinite** if $-f(\mathbf{x})$ is positive semidefinite. If $f(\mathbf{x})$ is neither positive semidefinite nor negative semidefinite, then $f(\mathbf{x})$ is **indefinite**.

Positive definite quadratic forms have an important application in analytic geometry. Let $\mathbf{A}$ be a symmetric and positive definite matrix. Then the region defined by the inequality

$$(\mathbf{x} - \mathbf{c})^T \mathbf{A}(\mathbf{x} - \mathbf{c}) \le \delta \tag{A.82}$$

is an ellipsoidal volume, with its center at $\mathbf{c}$. We can diagonalize $\mathbf{A}$ as

$$\mathbf{A} = \mathbf{P}\mathbf{\Lambda}\mathbf{P}^{-1} \tag{A.83}$$

where the columns of $\mathbf{P}$ are normalized eigenvectors of $\mathbf{A}$, and $\mathbf{\Lambda}$ is a diagonal matrix whose elements are the eigenvalues of $\mathbf{A}$. It can be shown that the ith eigenvector of $\mathbf{A}$ points in the direction of the ith semimajor axis of the ellipsoid, and the length of the ith semimajor axis is given by $\sqrt{\delta/\lambda_i}$ [121].

An important connection between positive semidefinite matrices and eigenvalues is the following theorem.

Theorem A.8. A symmetric matrix $\mathbf{A}$ is positive semidefinite if and only if its eigenvalues are greater than or equal to 0. $\mathbf{A}$ is positive definite if and only if its eigenvalues are greater than 0.

This provides a convenient way to check whether or not a symmetric matrix is positive semidefinite or positive definite.

The **Cholesky factorization** provides another way to determine whether or not a symmetric matrix is positive definite.

Theorem A.9. Let $\mathbf{A}$ be an n by n positive definite and symmetric matrix. Then $\mathbf{A}$ can be written uniquely as

$$\mathbf{A} = \mathbf{R}^T \mathbf{R} = \mathbf{L}\mathbf{L}^T \tag{A.84}$$

where $\mathbf{R}$ is a nonsingular upper-triangular matrix and $\mathbf{L} = \mathbf{R}^T$ is a nonsingular lower-triangular matrix. Note that $\mathbf{A}$ can be factored in this way if and only if it is positive definite.

The MATLAB command **chol** can be used to compute the Cholesky factorization.

A.7. VECTOR AND MATRIX NORMS

Although the conventional Euclidean length (A.61) is most commonly used, there are alternative ways to measure the length of a vector.

Definition A.26. Any measure of vector length satisfying the following four conditions is called a **norm**.

1. For any vector $\mathbf{x}$, $\|\mathbf{x}\| \geq 0$.
2. For any vector $\mathbf{x}$ and any scalar s, $\|s\mathbf{x}\| = |s| \|\mathbf{x}\|$.
3. For any vectors $\mathbf{x}$ and $\mathbf{y}$, $\|\mathbf{x} + \mathbf{y}\| \leq \|\mathbf{x}\| + \|\mathbf{y}\|$.
4. $\|\mathbf{x}\| = 0$ if and only if $\mathbf{x} = \mathbf{0}$.

If $\|\ \|$ satisfies conditions 1, 2, and 3, but does not satisfy condition 4, then $\|\ \|$ is called a **seminorm**.

Definition A.27. The p-**norm** of a vector in R^n is defined for $p \geq 1$ by

$$\|x\|_p = (|x_1|^p + |x_2|^p + \cdots + |x_n|^p)^{1/p} . \tag{A.85}$$

It can be shown that for any $p \geq 1$, (A.85) satisfies the conditions of Definition A.26 [72]. The conventional Euclidean length is just the 2-norm, but two other p-norms are also commonly used. The **1-norm** is the sum of the absolute values of the elements in $\mathbf{x}$. The ∞-**norm** is obtained by taking the limit as p goes to infinity. The ∞-norm is the maximum of the absolute values of the elements in $\mathbf{x}$. The MATLAB command **norm** can be used to compute the norm of a vector, and has options for the 1, 2, and infinity norms.

The 2-norm is particularly important because of its natural connection with dot products and projections. The projection of a vector onto a subspace is the point in the subspace that is closest to the vector as measured by the 2-norm. We have also seen in (A.73) that the problem of minimizing $\|\mathbf{A}\mathbf{x} - \mathbf{b}\|_2$ can be solved by computing projections or by using the normal equations. In fact, the 2-norm can be tied directly to the dot product by the formula

$$\|\mathbf{x}\|_2 = \sqrt{\mathbf{x}^T \mathbf{x}} . \tag{A.86}$$

The 1- and ∞-norms can also be useful in finding approximate solutions to over-determined linear systems of equations. To minimize the maximum of the residuals, we minimize $\|\mathbf{A}\mathbf{x} - \mathbf{b}\|_\infty$. To minimize the sum of the absolute values of the residuals, we minimize $\|\mathbf{A}\mathbf{x} - \mathbf{b}\|_1$. Unfortunately, these minimization problems are generally more difficult to solve than least squares problems.

Definition A.28. Any measure of the size or length of an m by n matrix that satisfies the following five properties can be used as a **matrix norm**.

1. For any matrix $\mathbf{A}$, $\|\mathbf{A}\| \geq 0$.
2. For any matrix $\mathbf{A}$ and any scalar s, $\|s\mathbf{A}\| = |s| \|\mathbf{A}\|$.
3. For any matrices $\mathbf{A}$ and $\mathbf{B}$, $\|\mathbf{A} + \mathbf{B}\| \leq \|\mathbf{A}\| + \|\mathbf{B}\|$.
4. $\|\mathbf{A}\| = 0$ if and only if $\mathbf{A} = \mathbf{0}$.
5. For any two matrices $\mathbf{A}$ and $\mathbf{B}$ of compatible sizes, $\|\mathbf{A}\mathbf{B}\| \leq \|\mathbf{A}\| \|\mathbf{B}\|$.

Definition A.29. The *p*-**norm** of a matrix **A** is

$$\|\mathbf{A}\|_p = \max_{\|\mathbf{x}\|_p = 1} \|\mathbf{A}\mathbf{x}\|_p \tag{A.87}$$

where $\|\mathbf{x}\|_p$ and $\|\mathbf{A}\mathbf{x}\|_p$ are vector *p*-norms, and $\|\mathbf{A}\|_p$ is the matrix *p*-norm of **A**.

Solving the maximization problem of (A.87) to determine a matrix *p*-norm could be extremely difficult. Fortunately, there are simpler formulas for the most commonly used matrix *p*-norms, for example

$$\|\mathbf{A}\|_1 = \max_j \sum_{i=1}^m |A_{i,j}| \tag{A.88}$$

$$\|\mathbf{A}\|_2 = \sqrt{\lambda_{\max}(\mathbf{A}^T \mathbf{A})} \tag{A.89}$$

$$\|\mathbf{A}\|_\infty = \max_i \sum_{j=1}^n |A_{i,j}| \tag{A.90}$$

where $\lambda_{\max}(\mathbf{A}^T \mathbf{A})$ denotes the largest eigenvalue of $\mathbf{A}^T \mathbf{A}$. See Exercises A.11, A.12, and C.4.

Definition A.30. The **Frobenius norm** of an *m* by *n* matrix is given by

$$\|\mathbf{A}\|_F = \sqrt{\sum_{i=1}^m \sum_{j=1}^n A_{i,j}^2} \,. \tag{A.91}$$

Definition A.31. A matrix norm $\|\ \|_M$ and a vector norm $\|\ \|_V$ are **compatible** if

$$\|\mathbf{A}\mathbf{x}\|_V \le \|\mathbf{A}\|_M \|\mathbf{x}\|_V \,. \tag{A.92}$$

The matrix *p*-norm is by its definition compatible with the vector *p*-norm from which it was derived. It can also be shown that the Frobenius norm of a matrix is compatible with the vector 2-norm [138]. Thus the Frobenius norm is often used with the vector 2-norm.

In practice, the Frobenius norm, 1-norm, and ∞-norm of a matrix are easy to compute, whereas the 2-norm of a matrix can be difficult to compute for large matrices. The MATLAB command **norm** has options for computing the 1, 2, infinity, and Frobenius norms of a matrix.

A.8. THE CONDITION NUMBER OF A LINEAR SYSTEM

Suppose that we want to solve a system of *n* equations in *n* variables

$$\mathbf{A}\mathbf{x} = \mathbf{b} \,. \tag{A.93}$$

Suppose further that because of measurement errors in $\mathbf{b}$, we actually solve

$$\mathbf{A}\hat{\mathbf{x}} = \hat{\mathbf{b}} \ . \tag{A.94}$$

Can we get a bound on $\|\mathbf{x} - \hat{\mathbf{x}}\|$ in terms of $\|\mathbf{b} - \hat{\mathbf{b}}\|$? Starting with (A.93) and (A.94) we have

$$\mathbf{A}(\mathbf{x} - \hat{\mathbf{x}}) = \mathbf{b} - \hat{\mathbf{b}} \tag{A.95}$$

$$(\mathbf{x} - \hat{\mathbf{x}}) = \mathbf{A}^{-1}(\mathbf{b} - \hat{\mathbf{b}}) \tag{A.96}$$

$$\|\mathbf{x} - \hat{\mathbf{x}}\| = \|\mathbf{A}^{-1}(\mathbf{b} - \hat{\mathbf{b}})\| \tag{A.97}$$

$$\|\mathbf{x} - \hat{\mathbf{x}}\| \leq \|\mathbf{A}^{-1}\|\|\mathbf{b} - \hat{\mathbf{b}}\| \ . \tag{A.98}$$

This formula provides an absolute bound on the error in the solution. It is also worthwhile to compute a relative error bound.

$$\frac{\|\mathbf{x} - \hat{\mathbf{x}}\|}{\|\mathbf{b}\|} \leq \frac{\|\mathbf{A}^{-1}\|\|\mathbf{b} - \hat{\mathbf{b}}\|}{\|\mathbf{b}\|} \tag{A.99}$$

$$\frac{\|\mathbf{x} - \hat{\mathbf{x}}\|}{\|\mathbf{A}\mathbf{x}\|} \leq \frac{\|\mathbf{A}^{-1}\|\|\mathbf{b} - \hat{\mathbf{b}}\|}{\|\mathbf{b}\|} \tag{A.100}$$

$$\|\mathbf{x} - \hat{\mathbf{x}}\| \leq \|\mathbf{A}\mathbf{x}\|\|\mathbf{A}^{-1}\|\frac{\|\mathbf{b} - \hat{\mathbf{b}}\|}{\|\mathbf{b}\|} \tag{A.101}$$

$$\|\mathbf{x} - \hat{\mathbf{x}}\| \leq \|\mathbf{A}\|\|\mathbf{x}\|\|\mathbf{A}^{-1}\|\frac{\|\mathbf{b} - \hat{\mathbf{b}}\|}{\|\mathbf{b}\|} \tag{A.102}$$

$$\frac{\|\mathbf{x} - \hat{\mathbf{x}}\|}{\|\mathbf{x}\|} \leq \|\mathbf{A}\|\|\mathbf{A}^{-1}\|\frac{\|\mathbf{b} - \hat{\mathbf{b}}\|}{\|\mathbf{b}\|} \ . \tag{A.103}$$

The relative error in $\mathbf{b}$ is measured by

$$\frac{\|\mathbf{b} - \hat{\mathbf{b}}\|}{\|\mathbf{b}\|} \ . \tag{A.104}$$

The relative error in $\mathbf{x}$ is measured by

$$\frac{\|\mathbf{x} - \hat{\mathbf{x}}\|}{\|\mathbf{x}\|} \ . \tag{A.105}$$

The constant

$$\text{cond}(\mathbf{A}) = \kappa(\mathbf{A}) = \|\mathbf{A}\|\|\mathbf{A}^{-1}\| \tag{A.106}$$

is called the **condition number** of $\mathbf{A}$.

Note that nothing that we did in the calculation of the condition number depends on which norm we used. The condition number can be computed using the 1-norm, 2-norm, ∞-norm, or Frobenius norm. The MATLAB command **cond** can be used to find the condition number of a matrix. It has options for the 1, 2, infinity, and Frobenius norms.

The condition number provides an upper bound on how inaccurate the solution to a system of equations might be because of errors in the right hand side. In some cases, the condition number greatly overestimates the error in the solution. As a practical matter, it is wise to assume that the error is of roughly the size predicted by the condition number. In practice, double precision floating point arithmetic only allows us to store numbers to about 16 digits of precision. If the condition number is greater than 10^{16}, then by the above inequality, there may be no accurate digits in the computer solution to the system of equations. Systems of equations with very large condition numbers are called **ill-conditioned**.

It is important to understand that ill-conditioning is a property of the system of equations and not of the algorithm used to solve the system of equations. Ill-conditioning cannot be fixed simply by using a better algorithm. Instead, we must either increase the precision of our floating point representation or find a different, better conditioned system of equations to solve.

A.9. THE QR FACTORIZATION

Although the theory of linear algebra can be developed using the reduced row echelon form, there is an alternative computational approach that works better in practice. The basic idea is to compute factorizations of matrices that involve orthogonal, diagonal, and upper-triangular matrices. This alternative approach leads to algorithms that can quickly compute accurate solutions to linear systems of equations and least squares problems. In this section we introduce the QR factorization which is one of the most widely used orthogonal matrix factorizations. Another factorization, the Singular Value Decomposition (SVD), is introduced in Chapter 3.

Theorem A.10. Let **A** be an m by n matrix. **A** can be written as

$$\mathbf{A} = \mathbf{QR} \qquad (A.107)$$

where **Q** is an m by m orthogonal matrix, and **R** is an m by n upper-triangular matrix. This is called the **QR factorization of A**.

The MATLAB command **qr** can be used to compute the QR factorization of a matrix. In a common situation, **A** will be an m by n matrix with $m > n$ and the rank of

A will be n. In this case, we can write

$$\mathbf{R} = \left[\begin{array}{c} \mathbf{R}_1 \\ \mathbf{0} \end{array} \right] \tag{A.108}$$

where $\mathbf{R}_1$ is n by n, and

$$\mathbf{Q} = [\mathbf{Q}_1 \ \mathbf{Q}_2] \tag{A.109}$$

where $\mathbf{Q}_1$ is m by n and $\mathbf{Q}_2$ is m by $m - n$. In this case the QR factorization has some important properties.

Theorem A.11. Let $\mathbf{Q}$ and $\mathbf{R}$ be the QR factorization of an m by n matrix $\mathbf{A}$ with $m > n$ and rank$(\mathbf{A}) = n$. Then
1. The columns of $\mathbf{Q}_1$ are an orthonormal basis for $R(\mathbf{A})$.
2. The columns of $\mathbf{Q}_2$ are an orthonormal basis for $N(\mathbf{A}^T)$.
3. The matrix $\mathbf{R}_1$ is nonsingular.

Now, suppose that we want to solve the least squares problem

$$\min \|\mathbf{Ax} - \mathbf{b}\|_2 . \tag{A.110}$$

Since multiplying a vector by an orthogonal matrix does not change its length, this is equivalent to

$$\min \|\mathbf{Q}^T(\mathbf{Ax} - \mathbf{b})\|_2 . \tag{A.111}$$

But

$$\mathbf{Q}^T\mathbf{A} = \mathbf{Q}^T\mathbf{QR} = \mathbf{R} . \tag{A.112}$$

So, we have

$$\min \|\mathbf{Rx} - \mathbf{Q}^T\mathbf{b}\|_2 \tag{A.113}$$

or

$$\min \left\| \begin{array}{c} \mathbf{R}_1\mathbf{x} - \mathbf{Q}_1^T\mathbf{b} \\ \mathbf{0x} - \mathbf{Q}_2^T\mathbf{b} \end{array} \right\|_2 . \tag{A.114}$$

Whatever value of $\mathbf{x}$ we pick, we will probably end up with nonzero residuals because of the $\mathbf{0x} - \mathbf{Q}_2^T\mathbf{b}$ part of the least squares problem. We cannot minimize the norm of this part of the vector. However, we can find an $\mathbf{x}$ that exactly solves $\mathbf{R}_1\mathbf{x} = \mathbf{Q}_1^T\mathbf{b}$. Thus we can minimize the least squares problem by solving the square system of equations

$$\mathbf{R}_1\mathbf{x} = \mathbf{Q}_1^T\mathbf{b} . \tag{A.115}$$

The advantage of solving this system of equations instead of the normal equations (A.73) is that the normal equations are typically much more badly conditioned than (A.115).

A.10. COMPLEX MATRICES AND VECTORS

Although nearly all mathematical operations in this textbook are done with real numbers, complex numbers do appear in Chapter 8 when we consider the Fourier transform. We assume that the reader is familiar with arithmetic involving complex numbers including addition, subtraction, multiplication, division, and complex exponentials. Most theorems of linear algebra extend trivially from real to complex vectors and matrices. In this section we briefly discuss our notation and some important differences between the real and complex cases.

Given a complex number $z = a + bi$, the **complex conjugate** of z is $z^* = a - bi$. Note that the absolute value of z is

$$|z| = \sqrt{a^2 + b^2} = \sqrt{z^*z} .$$
(A.116)

The main difference between linear algebra on real vectors and complex vectors is in the definition of the dot product of two vectors. We define the dot product of two complex vectors $\mathbf{x}$ and $\mathbf{y}$ to be

$$\mathbf{x} \cdot \mathbf{y} = \mathbf{x}^{*^T}\mathbf{y} .$$
(A.117)

The advantage of this definition is that

$$\mathbf{x}^{*^T}\mathbf{x} = \sum_{k=1}^{n} x_k^* x_k = \sum_{k=1}^{n} |x_k|^2 ,$$
(A.118)

and we can then define the 2-norm of a complex vector by

$$\|x\|_2 = \sqrt{\mathbf{x}^{*^T}\mathbf{x}} .$$
(A.119)

The operation of taking the complex conjugate and transpose, called the **Hermitian transpose**, occurs so frequently that we denote this by

$$\mathbf{x}^H = \mathbf{x}^{*^T} .$$
(A.120)

Note that for a real vector, $\mathbf{x}$, the conjugate is simply $\mathbf{x}^* = \mathbf{x}$, so $\mathbf{x}^H = \mathbf{x}^T$. In MATLAB, the apostrophe denotes the Hermitian transpose.

In general, you will almost never go wrong by using the Hermitian transpose in any linear algebra computation involving complex numbers that would normally involve a transpose when working with real vectors and matrices. For example, if we want to minimize $\|\mathbf{Gm} - \mathbf{d}\|_2$, where $\mathbf{G}$, $\mathbf{m}$, and $\mathbf{d}$ are complex, we can solve the normal equations

$$\mathbf{G}^H\mathbf{Gm} = \mathbf{G}^H\mathbf{d} .$$
(A.121)

A.11. LINEAR ALGEBRA IN SPACES OF FUNCTIONS

So far, we have considered only vectors in R^n. The concepts of linear algebra can be extended to other contexts. In general, as long as the objects that we want to consider can be multiplied by scalars and added together, and as long as they obey the laws of vector algebra, then we have a **vector space** in which we can practice linear algebra. If we can also define a vector product similar to the dot product, then we have what is called an **inner product space**, and we can define orthogonality, projections, and the 2-norm.

There are many different vector spaces used in various areas of science and mathematics. For our work in inverse problems, a very commonly used vector space is the space of functions defined on an interval $[a, b]$.

Multiplying a scalar times a function or adding two functions together clearly produces another function. In this space, the function $z(x) = 0$ takes the place of the $\mathbf{0}$ vector, since $f(x) + z(x) = f(x)$. Two functions $f(x)$ and $g(x)$ are linearly independent if the only solution to

$$c_1 f(x) + c_2 g(x) = z(x) \tag{A.122}$$

is $c_1 = c_2 = 0$.

We can define the dot product of two functions f and g to be

$$f \cdot g = \int_a^b f(x) g(x) dx . \tag{A.123}$$

Another commonly used notation for this dot product or **inner product** of f and g is

$$f \cdot g = \langle f, g \rangle . \tag{A.124}$$

It is easy to show that this inner product has all of the algebraic properties of the dot product of two vectors in R^n. A more important motivation for defining the dot product in this way is that it leads to a useful definition of the 2-norm of a function. Following our earlier formula that $\|x\|_2 = \sqrt{x^T x}$, we have

$$\| f \|_2 = \sqrt{\int_a^b f(x)^2 \, dx} . \tag{A.125}$$

Using this definition, the distance between two functions f and g is

$$\| f - g \|_2 = \sqrt{\int_a^b (f(x) - g(x))^2 \, dx} . \tag{A.126}$$

This measure is obviously zero when $f(x) = g(x)$ everywhere, but can also be 0 when $f(x)$ and $g(x)$ differ at a finite or countably infinite set of points. The measure is only nonzero if $f(x)$ and $g(x)$ differ on an interval.

Using this inner product and norm, we can reconstruct the theory of linear algebra from R^n in our space of functions. This includes the concepts of orthogonality, projections, norms, and least squares solutions.

Definition A.32. Given a collection of functions $f_1(x)$, $f_2(x)$, ..., $f_m(x)$ in an inner product space, the **Gram matrix** of the functions is the m by m matrix Γ, whose elements are given by

$$\Gamma_{i,j} = f_i \cdot f_j . \tag{A.127}$$

The Gram matrix has several important properties. It is symmetric and positive semidefinite. If the functions are linearly independent, then the Gram matrix is also positive definite. Furthermore, the rank of Γ is equal to size of the largest linearly independent subset of the functions $f_1(x), \ldots, f_m(x)$.

A.12. EXERCISES

1. Let $\mathbf{A}$ be an m by n matrix with n pivot columns in its RREF. Can the system of equations $\mathbf{Ax} = \mathbf{b}$ have infinitely many solutions?

2. If $\mathbf{C} = \mathbf{AB}$ is a 5 by 4 matrix, then how many rows does $\mathbf{A}$ have? How many columns does $\mathbf{B}$ have? Can you say anything about the number of columns in $\mathbf{A}$?

3. Suppose that $\mathbf{v}_1$, $\mathbf{v}_2$ and $\mathbf{v}_3$ are three vectors in R^3 and that $\mathbf{v}_3 = -2\mathbf{v}_1 + 3\mathbf{v}_2$. Are the vectors linearly dependent or linearly independent?

4. Let

$$\mathbf{A} = \begin{bmatrix} 1 & 2 & 3 & 4 \\ 2 & 2 & 1 & 3 \\ 4 & 6 & 7 & 11 \end{bmatrix}. \tag{A.128}$$

   Find bases for $N(\mathbf{A})$, $R(\mathbf{A})$, $N(\mathbf{A}^T)$ and $R(\mathbf{A}^T)$. What are the dimensions of the four subspaces?

5. Let $\mathbf{A}$ be an n by n matrix such that $\mathbf{A}^{-1}$ exists. What are $N(\mathbf{A})$, $R(\mathbf{A})$, $N(\mathbf{A}^T)$, and $R(\mathbf{A}^T)$?

6. Let $\mathbf{A}$ be any 9 by 6 matrix. If the dimension of the null space of $\mathbf{A}$ is 5, then what is the dimension of $R(\mathbf{A})$? What is the dimension of $R(\mathbf{A}^T)$? What is the rank of $\mathbf{A}$?

7. Suppose that a nonhomogeneous system of equations with four equations and six unknowns has a solution with two free variables. Is it possible to change the right hand side of the system of equations so that the modified system of equations has no solutions?

8. Let W be the set of vectors $\mathbf{x}$ in R^4 such that the element product $x_1 x_2 = 0$. Is W a subspace of R^4?

9. Let $\mathbf{v}_1$, $\mathbf{v}_2$, $\mathbf{v}_3$ be a set of three nonzero orthogonal vectors. Show that the vectors are also linearly independent.

10. Show that if $\mathbf{x} \perp \mathbf{y}$, then

$$\|\mathbf{x} + \mathbf{y}\|_2^2 = \|\mathbf{x}\|_2^2 + \|\mathbf{y}\|_2^2 . \tag{A.129}$$

11. In this exercise, we will derive the formula (A.88) for the 1-norm of a matrix. Begin with the optimization problem

$$\|\mathbf{A}\|_1 = \max_{\|\mathbf{x}\|_1 = 1} \|\mathbf{A}\mathbf{x}\|_1 . \tag{A.130}$$

   a. Show that if $\|\mathbf{x}\|_1 = 1$, then

$$\|\mathbf{A}\mathbf{x}\|_1 \leq \max_j \sum_{i=1}^{m} |A_{i,j}| . \tag{A.131}$$

   b. Find a vector $\mathbf{x}$ such that $\|\mathbf{x}\|_1 = 1$, and

$$\|\mathbf{A}\mathbf{x}\|_1 = \max_j \sum_{i=1}^{m} |A_{i,j}| . \tag{A.132}$$

   c. Conclude that

$$\|\mathbf{A}\|_1 = \max_{\|\mathbf{x}\|_1 = 1} \|\mathbf{A}\mathbf{x}\|_1 = \max_j \sum_{i=1}^{m} |A_{i,j}| . \tag{A.133}$$

12. Derive the formula (A.90) for the infinity norm of a matrix.

13. Let $\mathbf{A}$ be an m by n matrix.
   a. Show that $\mathbf{A}^T \mathbf{A}$ is symmetric.
   b. Show that $\mathbf{A}^T \mathbf{A}$ is positive semidefinite Hint: Use the definition of positive semidefinite rather than trying to compute eigenvalues.
   c. Show that if $\text{rank}(\mathbf{A}) = n$, then the only solution to $\mathbf{A}\mathbf{x} = 0$ is $\mathbf{x} = 0$.
   d. Use part c to show that if $\text{rank}(\mathbf{A}) = n$, then $\mathbf{A}^T \mathbf{A}$ is positive definite.
   e. Use part d to show that if $\text{rank}(\mathbf{A}) = n$, then $\mathbf{A}^T \mathbf{A}$ is nonsingular.
   f. Show that $N(\mathbf{A}^T \mathbf{A}) = N(\mathbf{A})$.

14. Show that

$$\text{cond}(\mathbf{A}\mathbf{B}) \leq \text{cond}(\mathbf{A})\text{cond}(\mathbf{B}) . \tag{A.134}$$

15. Let $\mathbf{A}$ be a symmetric and positive definite matrix with Cholesky factorization

$$\mathbf{A} = \mathbf{R}^T \mathbf{R} . \tag{A.135}$$

Show how the Cholesky factorization can be used to solve $\mathbf{Ax} = \mathbf{b}$ by solving two systems of equations, each of which has $\mathbf{R}$ or $\mathbf{R}^T$ as its matrix.

16. Let $P_3[0, 1]$ be the space of polynomials of degree less than or equal to 3 on the interval $[0, 1]$. The polynomials $p_1(x) = 1$, $p_2(x) = x$, $p_3(x) = x^2$, and $p_4(x) = x^3$ form a basis for $P_3[0, 1]$, but they are not orthogonal with respect to the inner product

$$f \cdot g = \int_0^1 f(x)g(x) \, dx \, . \tag{A.136}$$

Use the Gram–Schmidt orthogonalization process to construct an orthogonal basis for $P_3[0, 1]$. Once you have your basis, use it to find the third degree polynomial that best approximates $f(x) = e^{-x}$ on the interval $[0, 1]$ in the least squares sense.

A.13. NOTES AND FURTHER READING

Much of this material is typically covered in sophomore level linear algebra courses, and there are an enormous number of textbooks at this level. One good introductory linear algebra textbook is [121]. At a slightly more advanced level, [194] and [138] are both excellent. The book by Strang and Borre [196] reviews linear algebra in the context of geodetic problems.

Fast and accurate algorithms for linear algebra computations are a somewhat more advanced and specialized topic. A classic reference is [72]. Other good books on this topic include [209] and [50].

The extension of linear algebra to spaces of functions is a topic in the subject of functional analysis. Many textbooks on functional analysis assume that the reader has considerable mathematical background. One book that is reasonably accessible to readers with limited mathematical backgrounds is [130].

Review of Probability and Statistics

Synopsis

A brief review is given of the topics in classical probability and statistics that are used throughout the book. Connections between probability theory and its application to the analysis of data with random measurement errors are highlighted. Note that some very different philosophical interpretations of probability theory are discussed in Chapter 11.

B.1. PROBABILITY AND RANDOM VARIABLES

The mathematical theory of probability begins with an **experiment**, which has a set S of possible outcomes. We will be interested in **events** that are subsets A of S.

Definition B.1. The **probability function** P is a function defined on subsets of S with the following properties:
1. $P(S) = 1$.
2. For every event $A \subseteq S$, $P(A) \geq 0$.
3. If events $A_1, A_2, \ldots, A_n$, are pairwise mutually exclusive (i.e., impossible to both occur), so that $A_i \cap A_j$ is empty for all pairs i, j, then

$$P(\cup_{i=1}^{n} A_i) = \sum_{i=1}^{n} P(A_i) .$$
(B.1)

The probability properties given above are fundamental to developing the mathematics of probability theory. However, applying this definition of probability to real world situations frequently requires ingenuity.

Example B.1

Consider the experiment of throwing a dart at a dart board. We will assume that our dart thrower is an expert who always hits the dart board. The sample space S consists of the points on the dart board. We can define an event A that consists of the points in the bullseye, so that $P(A)$ is the probability that the thrower hits the bullseye.

In practice, the outcome of an experiment is often a number rather than an event. Random variables are a useful generalization of the basic concept of probability.

Definition B.2. A **random variable** X is a function $X(s)$ that assigns a value to each outcome s in the sample space S.

Each time we perform an experiment, we obtain a particular value of the random variable. These values are called **realizations** of the random variable.

Example B.2

To continue our previous example, let X be the function that takes a point on the dart board and returns the associated score. Suppose that throwing the dart in the bullseye scores 50 points. Then for each point s in the bullseye, $X(s) = 50$.

In this book we deal frequently with experimental measurements that can include some random measurement error.

Example B.3

Suppose we measure the mass of an object five times to obtain the realizations $m_1 = 10.1$ kg, $m_2 = 10.0$ kg, $m_3 = 10.0$ kg, $m_4 = 9.9$ kg, and $m_5 = 10.1$ kg. We will assume that there is one true mass m, and that the measurements we obtained varied because of random measurement errors e_i, so that

$$m_1 = m + e_1, \quad m_2 = m + e_2, \quad m_3 = m + e_3, \quad m_4 = m + e_4, \quad m_5 = m + e_5 . \quad \text{(B.2)}$$

We can treat the measurement errors as realizations of a random variable E. Equivalently, since the true mass m is just a constant, we could treat the measurements $m_1, m_2, \ldots, m_5$ as realizations of a random variable M. In practice it makes little difference whether we treat the measurements or the measurement errors as random variables.

Note that in a Bayesian approach the mass m of the object would itself be a random variable. This is a viewpoint that we consider in Chapter 11.

The relative probability of realization values for a random variable can be character-ized by a nonnegative **probability density function (PDF)**, $f_X(x)$, with

$$P(X \leq a) = \int_{-\infty}^{a} f_X(x) \, dx \quad \text{(B.3)}$$

and

$$\int_{-\infty}^{\infty} f_X(x) \, dx = 1 . \quad \text{(B.4)}$$

The following definitions give some useful random variables that frequently arise in inverse problems.

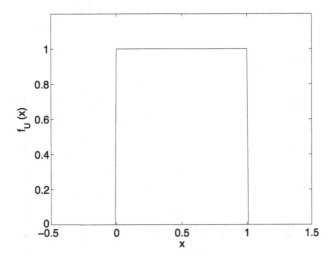

Figure B.1 The PDF for the uniform random variable on [0, 1].

Definition B.3. The **uniform** random variable on the interval $[a, b]$ (Fig. B.1) has the probability density function

$$f_U(x) = \begin{cases} \frac{1}{b-a} & a \leq x \leq b \\ 0 & x < a \\ 0 & x > b . \end{cases} \tag{B.5}$$

Definition B.4. The **normal** or **Gaussian** random variable (Fig. B.2) has the probability density function

$$f_N(x) = \frac{1}{\sigma\sqrt{2\pi}} e^{-\frac{1}{2}(x-\mu)^2/\sigma^2} . \tag{B.6}$$

The notation $N(\mu, \sigma^2)$ is used to denote a normal distribution with parameters μ and σ. The **standard normal** random variable, $N(0, 1)$, has $\mu = 0$ and $\sigma^2 = 1$.

Definition B.5. The **Student's t distribution** with ν degrees of freedom (Fig. B.3) has the probability density function

$$f_t(x) = \frac{\Gamma((\nu+1)/2)}{\Gamma(\nu/2)} \frac{1}{\sqrt{\nu\pi}} \left(1 + \frac{x^2}{\nu}\right)^{-(\nu+1)/2} \tag{B.7}$$

where the **gamma function** is

$$\Gamma(x) = \int_0^\infty \xi^{x-1} e^{-\xi} \, d\xi . \tag{B.8}$$

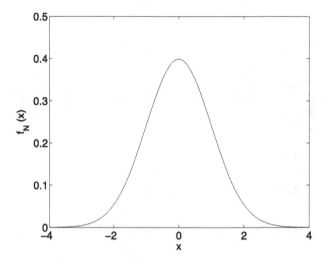

Figure B.2 The PDF of the standard normal random variable.

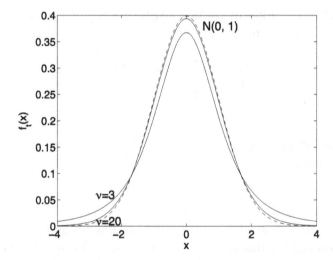

Figure B.3 The Student's t probability density function for $\nu = 3$ and $\nu = 20$. Dashed curve shows the normal distribution $N(0, 1)$ (Fig. B.2)

The Student's t distribution is so named because W. S. Gosset used the pseudonym "Student" in publishing the first paper in which the distribution appeared. In the limit as ν goes to infinity, Student's t distribution approaches a standard normal distribution. However, for small values of ν (B.7) has a greater percentage of extreme values than (B.6).

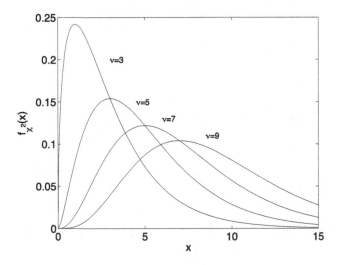

Figure B.4 The χ^2 probability density function for several values of ν.

Definition B.6. The χ^2 random variable has the probability density function (Fig. B.4)

$$f_{\chi^2}(x) = \frac{1}{2^{\nu/2}\Gamma(\nu/2)} x^{\frac{1}{2}\nu-1} e^{-x/2} \tag{B.9}$$

where the parameter ν is the **number of degrees of freedom**.

Definition B.7. The **exponential** random variable (Fig. B.5) has the probability density function

$$f_{exp}(x) = \begin{cases} \lambda e^{-\lambda x} & x \geq 0 \\ 0 & x < 0 . \end{cases} \tag{B.10}$$

Definition B.8. The **double-sided exponential** random variable (Fig. B.6) has the probability density function

$$f_{dexp}(x) = \frac{1}{\sigma\sqrt{2}} e^{-\sqrt{2}|x-\mu|/\sigma} . \tag{B.11}$$

It can be shown that for n independent random variables, X_i, with standard normal distributions, the random variable

$$Z = \sum_{i=1}^{n} X_i^2 \tag{B.12}$$

is a χ^2 random variable with $\nu = n$ degrees of freedom [59].

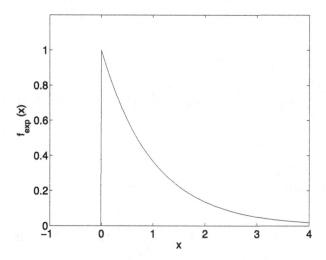

Figure B.5 The exponential probability density function ($\lambda = 1$).

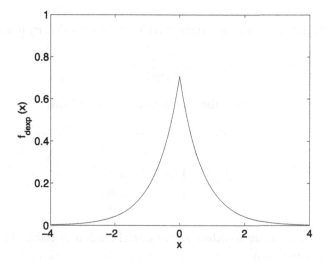

Figure B.6 The double-sided exponential probability density function ($\mu = 0, \sigma = 1$).

The **cumulative distribution function (CDF)** $F_X(a)$ of a one–dimensional random variable X is given by the definite integral of the associated PDF

$$F_X(a) = P(X \leq a) = \int_{-\infty}^{a} f_X(z) \, dz \,. \tag{B.13}$$

Note that $F_X(a)$ must lie in the interval $[0, 1]$ for all a, and is a nondecreasing function of a because of the unit area and nonnegativity of the PDF.

For the uniform PDF on the unit interval, for example, the CDF is a ramp function

$$F_U(a) = \int_{-\infty}^{a} f_u(z) \, dz \tag{B.14}$$

$$F_U(a) = \begin{cases} 0 & a \le 0 \\ a & 0 \le a \le 1 \\ 1 & a > 1 \,. \end{cases} \tag{B.15}$$

The PDF, $f_X(x)$, or CDF, $F_X(a)$, completely determine the properties of a random variable. The probability that a particular realization of X will lie within a general interval $[a, b]$ is

$$P(a \le X \le b) = P(X \le b) - P(X \le a) = F_X(b) - F_X(a) \tag{B.16}$$

$$= \int_{-\infty}^{b} f_X(z) \, dz - \int_{-\infty}^{a} f_X(z) \, dz = \int_{a}^{b} f_X(z) \, dz \,. \tag{B.17}$$

B.2. EXPECTED VALUE AND VARIANCE

Definition B.9. The **expected value** of a random variable X, denoted by $E[X]$ or $\mu(X)$, is

$$E[X] = \int_{-\infty}^{\infty} x f_X(x) \, dx \,. \tag{B.18}$$

In general, if $g(X)$ is some function of a random variable X, then

$$E[g(X)] = \int_{-\infty}^{\infty} g(x) f_X(x) \, dx \,. \tag{B.19}$$

Some authors use the term "mean" for the expected value of a random variable. We will reserve this term for the average of a set of data. Note that the expected value of a random variable is not necessarily identical to the **mode** (the value with the largest value of $f(x)$) nor is it necessarily identical to the **median**, the value of x for which the value of the CDF is $F(x) = 1/2$.

Example B.4

The expected value of an $N(\mu, \sigma)$ random variable X is

$$E[X] = \int_{-\infty}^{\infty} x \frac{1}{\sigma \sqrt{2\pi}} e^{-\frac{(x-\mu)^2}{2\sigma^2}} \, dx \tag{B.20}$$

$$= \int_{-\infty}^{\infty} \frac{1}{\sigma\sqrt{2\pi}} (x + \mu) e^{-\frac{x^2}{2\sigma^2}} \, dx \tag{B.21}$$

$$= \mu \int_{-\infty}^{\infty} \frac{1}{\sigma\sqrt{2\pi}} e^{-\frac{x^2}{2\sigma^2}} \, dx + \int_{-\infty}^{\infty} \frac{1}{\sigma\sqrt{2\pi}} x e^{-\frac{x^2}{2\sigma^2}} \, dx \, . \tag{B.22}$$

The first integral term is μ because the integral of the entire PDF is 1, and the second term is zero because it is an odd function integrated over a symmetric interval. Thus

$$E[X] = \mu \, . \tag{B.23}$$

Definition B.10. The **variance** of a random variable X, denoted by $\text{Var}(X)$ or σ_X^2, is given by

$$\begin{aligned} \text{Var}(X) &= E[(X - E[X])^2] \\ &= E[X^2] - E[X]^2 \\ &= \int_{-\infty}^{\infty} (x - E[X])^2 f_X(x) \, dx \, . \end{aligned} \tag{B.24}$$

The **standard deviation** of X, often denoted σ_X, is

$$\sigma_X = \sqrt{\text{Var}(X)} \, . \tag{B.25}$$

The variance and standard deviation serve as measures of the spread of the random variable about its expected value. Since the units of σ are the same as the units of μ, the standard deviation is generally more practical for reporting a spread measure. However, the variance has properties that make it more useful for certain calculations.

B.3. JOINT DISTRIBUTIONS

Definition B.11. If we have two random variables X and Y, they *may* have a **joint probability density function (JDF)**, $f(x, y)$ with

$$P(X \le a \text{ and } Y \le b) = \int_{-\infty}^{a} \int_{-\infty}^{b} f(x, y) \, dy \, dx \, . \tag{B.26}$$

If X and Y have a joint probability density function, then we can use it to evaluate the expected value of a function of X and Y. The expected value of $g(X, Y)$ is

$$E[g(X, Y)] = \int_{-\infty}^{\infty} \int_{-\infty}^{\infty} g(x, y) f(x, y) \, dy \, dx \, . \tag{B.27}$$

Definition B.12. Two random variables X and Y are **independent** if a JDF exists and is defined by

$$f(x, y) = f_X(x) f_Y(y) \, . \tag{B.28}$$

Definition B.13. If X and Y have a JDF, then the **covariance** of X and Y is

$$\text{Cov}(X, \ Y) = E[(X - E[X])(Y - E[Y])] = E[XY] - E[X]E[Y] \ . \qquad \text{(B.29)}$$

If X and Y are independent, then $E[XY] = E[X]E[Y]$, and $\text{Cov}(X, \ Y) = 0$. However if X and Y are dependent, it is still possible, given some particular distributions, for X and Y to have $\text{Cov}(X, \ Y) = 0$. If $\text{Cov}(X, \ Y) = 0$, X and Y are called **uncorrelated**.

Definition B.14. The **correlation** of X and Y is

$$\rho(X, Y) = \frac{\text{Cov}(X, \ Y)}{\sqrt{\text{Var}(X)\text{Var}(Y)}} \ . \qquad \text{(B.30)}$$

Correlation is thus a scaled covariance.

Theorem B.1. The following properties of Var, Cov, and correlation hold for any random variables X and Y and scalars s and a.
1. $\text{Var}(X) \geq 0$
2. $\text{Var}(X + a) = \text{Var}(X)$
3. $\text{Var}(sX) = s^2\text{Var}(X)$
4. $\text{Var}(X + Y) = \text{Var}(X) + \text{Var}(Y) + 2\text{Cov}(X, \ Y)$
5. $\text{Cov}(X, \ Y) = \text{Cov}(Y, \ X)$
6. $\rho(X, \ Y) = \rho(Y, \ X)$
7. $-1 \leq \rho(X, \ Y) \leq 1$.

The following example demonstrates the use of some of these properties.

●───

Example B.5

Suppose that Z is a standard normal random variable. Let

$$X = \mu + \sigma Z \ . \qquad \text{(B.31)}$$

Then

$$E[X] = E[\mu] + \sigma E[Z] \qquad \text{(B.32)}$$

so

$$E[X] = \mu \ . \qquad \text{(B.33)}$$

Also,

$$\text{Var}(X) = \text{Var}(\mu) + \sigma^2\text{Var}(Z) = \sigma^2 \ . \qquad \text{(B.34)}$$

Thus if we have a program to generate random numbers with the standard normal distribution, we can use it to generate normal random numbers with any desired expected

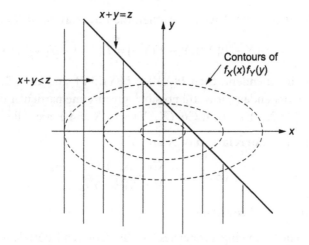

Figure B.7 Integration of a joint probability density function for two independent random variables, X, and Y, to evaluate the CDF of $Z = X + Y$.

value and standard deviation. The MATLAB command **randn** generates independent realizations of an $N(0, 1)$ random variable.

Example B.6

What is the CDF (or PDF) of the sum of two independent random variables $X + Y$? To see this, we write the desired CDF in terms of an appropriate integral over the JDF, $f(x, y)$, which gives (Fig. B.7)

$$F_{X+Y}(z) = P(X + Y \leq z) \tag{B.35}$$

$$= \iint_{x+y \leq z} f(x, y) \, dx \, dy \tag{B.36}$$

$$= \iint_{x+y \leq z} f_X(x) f_Y(y) \, dx \, dy \tag{B.37}$$

$$= \int_{-\infty}^{\infty} \int_{-\infty}^{z-y} f_X(x) f_Y(y) \, dx \, dy \tag{B.38}$$

$$= \int_{-\infty}^{\infty} \int_{-\infty}^{z-y} f_X(x) \, dx \, f_Y(y) \, dy \tag{B.39}$$

$$= \int_{-\infty}^{\infty} F_X(z - y) f_Y(y) \, dy . \tag{B.40}$$

The associated PDF is

$$f_{X+Y}(z) = \frac{d}{dz}\int_{-\infty}^{\infty} F_X(z-y)f_Y(y)\,dy \tag{B.41}$$

$$= \int_{-\infty}^{\infty} \frac{d}{dz}F_X(z-y)f_Y(y)\,dy \tag{B.42}$$

$$= \int_{-\infty}^{\infty} f_X(z-y)f_Y(y)\,dy \tag{B.43}$$

$$= f_X(z)*f_Y(z). \tag{B.44}$$

Adding two independent random variables thus produces a new random variable that has a PDF given by the convolution of the PDF's of the two individual variables.

The JDF can be used to evaluate the CDF or PDF arising from a general function of jointly distributed random variables. The process is identical to the previous example except that the specific form of the integral limits is determined by the specific function.

Example B.7

Consider the product of two independent, identically distributed, standard normal random variables,

$$Z = XY \tag{B.45}$$

with a JDF given by

$$f(x,y) = f_X(x)f_Y(y) = \frac{1}{2\pi\sigma^2}e^{-(x^2+y^2)/2\sigma^2}. \tag{B.46}$$

The CDF of Z is

$$F_Z(z) = P(Z \le z) = P(XY \le z). \tag{B.47}$$

For $z \le 0$, this is the integral of the JDF over the exterior of the hyperbolas defined by $xy \le z \le 0$, whereas for $z \ge 0$, we integrate over the interior of the complementary hyperbolas $xy \le z \ge 0$. At $z = 0$, the integral covers exactly half of the $(x,\ y)$ plane (the 2nd and 4th quadrants) and, because of the symmetry of the JDF, has accumulated half of the probability, or $1/2$.

The integral is thus

$$F_Z(z) = 2\int_{-\infty}^{0}\int_{z/x}^{\infty} \frac{1}{2\pi\sigma^2}e^{-(x^2+y^2)/2\sigma^2}\,dy\,dx \quad (z \le 0) \tag{B.48}$$

and

$$F(z) = 1/2 + 2\int_{-\infty}^{0}\int_{0}^{z/x} \frac{1}{2\pi\sigma^2} e^{-(x^2+y^2)/2\sigma^2}\, dy\, dx \quad (z \geq 0)\,. \tag{B.49}$$

As in the previous example for the sum of two random variables, the PDF may be obtained from the CDF by differentiating with respect to z.

B.4. CONDITIONAL PROBABILITY

In some situations we will be interested in the probability of an event happening given that some other event has also happened.

Definition B.15. The **conditional probability** of A given that B has occurred is given by

$$P(A|B) = \frac{P(A \cap B)}{P(B)}. \tag{B.50}$$

Arguments based on conditional probabilities are often very helpful in computing probabilities. The key to such arguments is the **law of total probability**.

Theorem B.2. Suppose that $B_1, B_2, \ldots, B_n$ are mutually disjoint and exhaustive events. That is, $B_i \cap B_j = \emptyset$ (the empty set) for $i \neq j$, and

$$\cup_{i=1}^{n} B_i = S\,. \tag{B.51}$$

Then

$$P(A) = \sum_{i=1}^{n} P(A|B_i)P(B_i)\,. \tag{B.52}$$

It is often necessary to reverse the order of conditioning in a conditional probability. Bayes' theorem provides a way to do this.

Theorem B.3 (Bayes Theorem).

$$P(B|A) = \frac{P(A|B)P(B)}{P(A)}\,. \tag{B.53}$$

Example B.8

A screening test has been developed for a very serious but rare disease. If a person has the disease, then the test will detect the disease with probability 99%. If a person does not have the disease, then the test will give a false positive detection with probability 1%. The probability that any individual in the population has the disease is 0.01%. Suppose that a randomly selected individual tests positive for the disease. What is the probability that this individual actually has the disease?

Let A be the event "the person tests positive." Let B be the event "the person has the disease." We then want to compute $P(B|A)$. By Bayes theorem,

$$P(B|A) = \frac{P(A|B)P(B)}{P(A)} . \tag{B.54}$$

We have that $P(A|B)$ is 0.99, and that $P(B)$ is 0.0001. To compute $P(A)$, we apply the law of total probability, considering separately the probability of a diseased individual testing positive and the probability of someone without the disease testing positive, and obtain

$$P(A) = 0.99 \cdot 0.0001 + 0.01 \cdot 0.9999 = 0.010098 . \tag{B.55}$$

Thus

$$P(B|A) = \frac{0.99 \cdot 0.0001}{0.010098} = 0.0098 . \tag{B.56}$$

In other words, even after a positive screening test, it is still unlikely that the individual will have the disease. The vast majority of those individuals who test positive will in fact not have the disease.

The concept of conditioning can be extended from simple events to distributions and expected values of random variables. If the distribution of X depends on the value of Y, then we can work with the **conditional PDF** $f_{X|Y}(x)$, the **conditional CDF** $F_{X|Y}(a)$, and the **conditional expected value** $E[X|Y]$. In this notation, we can also specify a particular value of Y by using the notation $f_{X|Y=y}$, $F_{X|Y=y}$, or $E[X|Y=y]$. In working with conditional distributions and expected values, the following versions of the law of total probability can be very useful.

Theorem B.4. Given two random variables X and Y, with the distribution of X depending on Y, we can compute

$$P(X \leq a) = \int_{-\infty}^{\infty} P(X \leq a | Y = y) \, f_Y(y) \, dy \tag{B.57}$$

and

$$E[X] = \int_{-\infty}^{\infty} E[X | Y = y] \, f_Y(y) \, dy . \tag{B.58}$$

●

Example B.9

Let U be a random variable uniformly distributed on $(1,2)$. Let X be an exponential random variable with parameter $\lambda = U$. We will find the expected value of X.

$$E[X] = \int_1^2 E[X|U=u]\, f_U(u)\, du \,. \tag{B.59}$$

Since the expected value of an exponential random variable with parameter λ is $1/\lambda$, and the PDF of a uniform random variable on $(1,2)$ is $f_U(u) = 1$,

$$E[X] = \int_1^2 \frac{1}{u}\, du = \ln 2 \,. \tag{B.60}$$

B.5. THE MULTIVARIATE NORMAL DISTRIBUTION

Definition B.16. If the random variables $X_1, \ldots, X_n$ have a **multivariate normal distribution** (MVN), then the joint probability density function is

$$f(\mathbf{x}) = \frac{1}{(2\pi)^{n/2}} \frac{1}{\sqrt{\det(\mathbf{C})}} e^{-(\mathbf{x}-\boldsymbol{\mu})^T \mathbf{C}^{-1}(\mathbf{x}-\boldsymbol{\mu})/2} \tag{B.61}$$

where $\mathbf{x} = [X_1,\ X_2,\ \ldots,\ X_n]^T$ and $\boldsymbol{\mu} = [\mu_1,\ \mu_2,\ \ldots,\ \mu_n]^T$ is a vector containing the expected values along each of the coordinate directions of $X_1, \ldots, X_n$, and $\mathbf{C}$ contains the covariances between the random variables

$$C_{i,j} = \mathrm{Cov}(X_i,\ X_j) \,. \tag{B.62}$$

Notice that if $\mathbf{C}$ is singular, then the joint probability density function involves a division by zero, and is simply not defined.

The vector $\boldsymbol{\mu}$ and the covariance matrix $\mathbf{C}$ completely characterize the MVN distribution. There are other multivariate distributions that are not completely characterized by the expected values and covariance matrix. The MATLAB statistics toolbox command **mvnrnd** generates random vectors from the MVN distribution.

●

Example B.10

We can also readily generate vectors that are realizations of an MVN distribution with a known mean, $\boldsymbol{\mu}$ and covariance matrix, $\mathbf{C}$ as follows.

1. Find the lower-triangular Cholesky factorization $\mathbf{C} = \mathbf{L}\mathbf{L}^T$.

2. Let $\mathbf{Z}$ be a vector of n independent $N(0, 1)$ random numbers.

3. Let $\mathbf{X} = \boldsymbol{\mu} + \mathbf{LZ}$.

We can easily show that this procedure produces the desired distribution. Because $E[\mathbf{Z}] = \mathbf{0}$, $E[\mathbf{X}] = \boldsymbol{\mu} + \mathbf{L0} = \boldsymbol{\mu}$. Also, since $\text{Cov}(\mathbf{Z}) = \mathbf{I}$ and $\text{Cov}(\boldsymbol{\mu}) = \mathbf{0}$, $\text{Cov}(\mathbf{X}) = \text{Cov}(\boldsymbol{\mu} + \mathbf{LZ}) = \text{Cov}(\mathbf{LZ}) = \mathbf{LIL}^T = \mathbf{C}$ using (B.64).

Theorem B.5. Let $\mathbf{X}$ be a multivariate normal random vector with expected values defined by the vector $\boldsymbol{\mu}$ and covariance matrix $\mathbf{C}$, and let $\mathbf{Y} = \mathbf{AX}$. Then $\mathbf{Y}$ is also multivariate normal, with

$$E[\mathbf{Y}] = \mathbf{A}\boldsymbol{\mu} \tag{B.63}$$

and

$$\text{Cov}(\mathbf{Y}) = \mathbf{ACA}^T . \tag{B.64}$$

Theorem B.6. If we have an n-dimensional MVN distribution $\mathbf{X}$ with covariance matrix $\mathbf{C}$ and expected value $\boldsymbol{\mu}$, and the covariance matrix is of full rank, then the random variable

$$Z = (\mathbf{X} - \boldsymbol{\mu})^T \mathbf{C}^{-1} (\mathbf{X} - \boldsymbol{\mu}) \tag{B.65}$$

has a χ^2 distribution with n degrees of freedom.

B.6. THE CENTRAL LIMIT THEOREM

Theorem B.7. Let $X_1, X_2, \ldots, X_n$ be independent and identically distributed (IID) random variables with a finite expected value μ and variance σ^2. Let

$$Z_n = \frac{X_1 + X_2 + \cdots + X_n - n\mu}{\sqrt{n}\sigma} . \tag{B.66}$$

In the limit as n approaches infinity, the distribution of Z_n approaches the standard normal distribution.

The central limit theorem shows why quasi-normally distributed random variables appear so frequently in nature; the sum of numerous independent random variables produces an approximately normal random variable, regardless of the distribution of the underlying IID variables. In particular, this is one reason that measurement errors are often normally distributed. As we will see in Chapter 2, having normally distributed measurement errors leads us to consider least squares solutions to parameter estimation and inverse problems.

B.7. TESTING FOR NORMALITY

Many of the statistical procedures that we will use assume that data are normally distributed. Fortunately, the statistical techniques that we describe are generally robust in the face of small deviations from normality. Large deviations from the normal distribution, however, can cause problems. Thus it is important to be able to examine a data set to see whether or not the distribution is approximately normal.

Plotting a histogram of the data provides a quick view of the distribution. The histogram should show a roughly "bell-shaped" distribution, symmetrical around a single peak. If the histogram shows that the distribution is obviously skewed, then it would be unwise to assume that the data are normally distributed.

The **Q–Q Plot** provides a more precise graphical test of whether a set of data could have come from a particular distribution. The data points,

$$\mathbf{d} = [d_1, \ d_2, \ \ldots, \ d_n]^T \tag{B.67}$$

are first sorted from smallest to largest into a vector $\mathbf{y}$ (this process sorts the data, when plotted with respect to the $\mathbf{y}$ index, into **quantiles**, or contiguous intervals of equal probability). We then plot $\mathbf{y}$ versus

$$x_i = F^{-1}((i - 0.5)/n) \ \ (i = 1, \ 2, \ \ldots, \ n) \tag{B.68}$$

where $F^{-1}(x)$ is the inverse CDF of the distribution against which we wish to compare our observations.

If we are testing to see if the elements of $\mathbf{d}$ could have come from the normal distribution, then $F(x)$ is the CDF for the standard normal distribution

$$F_N(x) = \frac{1}{\sqrt{2\pi}} \int_{-\infty}^{x} e^{-\frac{1}{2}z^2} \ dz \ . \tag{B.69}$$

If the elements of $\mathbf{d}$ are indeed normally distributed, then the $(y_i, \ x_i)$ will follow a straight line with a slope and intercept determined by the standard deviation and expected value, respectively of the normal distribution that produced the data. The library command **qqplot** produces a basic qqplot with respect to a standard normal distribution.

Example B.11

Fig. B.8 shows the histogram from a set of 1000 data points. The characteristic bell-shaped curve in the histogram might make it appear at first that these data are normally distributed. The sample mean is -0.01 and the sample standard deviation is 1.41.

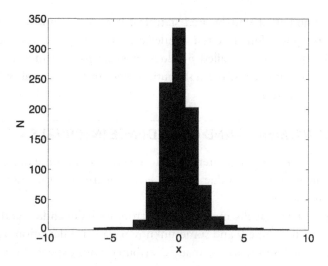

Figure B.8 Histogram of a sample data set.

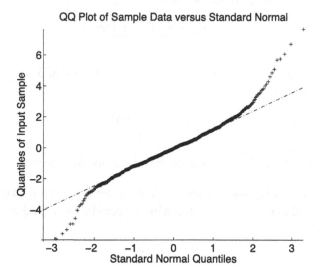

Figure B.9 Q–Q Plot for the sample data set.

Fig. B.9 shows the Q–Q plot for these data compared to a normal distribution with the same mean and standard deviation. The plot makes it apparent that the data set contains substantially more extreme values than the normal distribution would predict. In fact, these data were generated according to a Student's t distribution (B.7) with 5 degrees of freedom. This distribution has appreciably broader tails than the normal distribution (Fig. B.3).

There are a number of widely-used statistical tests for normality. These tests, including the Kolmogorov–Smirnov test, Anderson–Darling test, and Lilliefors test each produce probabilistic measures called p-values. A small p-value indicates that the observed data would be unlikely if the distribution were in fact normal, whereas a larger p-value is consistent with normality.

B.8. ESTIMATING MEANS AND CONFIDENCE INTERVALS

Given a collection of n noisy measurements $m_1, m_2, \ldots, m_n$ of some quantity of interest, how can we estimate the true value m, and how uncertain is this estimate? This is a classic problem in statistics.

We will assume first that the measurement errors are independent and normally distributed with expected value 0 and some unknown standard deviation σ. Equivalently, the measurements themselves are normally distributed with expected value m and standard deviation σ.

We begin by computing the measurement average

$$\bar{m} = \frac{m_1 + m_2 + \cdots + m_n}{n} . \tag{B.70}$$

This **sample mean** $\bar{m}$ will serve as our estimate of m. We will also compute an estimate s of the standard deviation

$$s = \sqrt{\frac{\sum_{i=1}^{n}(m_i - \bar{m})^2}{n-1}} . \tag{B.71}$$

The key to our approach to estimating m is the following theorem.

Theorem B.8 (The Sampling Theorem). Under the assumption that measurements are independent and normally distributed with expected value m and standard deviation σ, the random quantity

$$t = \frac{m - \bar{m}}{s/\sqrt{n}} \tag{B.72}$$

has a **Student's t distribution** with $n - 1$ degrees of freedom.

If we had the true standard deviation σ instead of the estimate s, then t would in fact be normally distributed with expected value 0 and standard deviation 1. This does not quite work out because we have used an estimate s of the standard deviation. For smaller values of n, the estimate s is less accurate, and the t distribution therefore has fatter tails than the standard normal distribution. As n becomes large, s becomes a better estimate of σ and it can be shown that the t distribution converges to a standard normal distribution [59].

Let $t_{n-1,0.975}$ be the 97.5%-tile of the t distribution and let $t_{n-1,0.025}$ be the 2.5%-tile of the t distribution. Then

$$P\left(t_{n-1,0.025} \le \frac{m - \bar{m}}{s/\sqrt{n}} \le t_{n-1,0.975}\right) = 0.95 . \tag{B.73}$$

This can be rewritten as

$$P\left(\left(t_{n-1,0.025}\ s/\sqrt{n}\right) \le (m - \bar{m}) \le \left(t_{n-1,0.975}\ s/\sqrt{n}\right)\right) = 0.95 . \tag{B.74}$$

We can construct the 95% **confidence interval** for m as the interval from $\bar{m} + t_{n-1,0.025}\ s/\sqrt{n}$ to $\bar{m} + t_{n-1,0.975}\ s/\sqrt{n}$. Because the t distribution is symmetric, this can also be written as $\bar{m} - t_{n-1,0.975}\ s/\sqrt{n}$ to $\bar{m} + t_{n-1,0.975}\ s/\sqrt{n}$.

As we have seen, there is a 95% probability that when we construct the confidence interval, that interval will contain the true mean, m. Note that we have not said that, given a particular set of data and the resulting confidence interval, there is a 95% probability that m is in the confidence interval. The semantic difficulty here is that m is not a random variable, but is rather some true fixed quantity that we are estimating; the measurements $m_1, m_2, \ldots, m_n$, and the calculated $\bar{m}$, s and confidence interval are the random quantities.

●———————————————————————————————

Example B.12

Suppose that we want to estimate the mass of an object and obtain the following ten measurements of the mass (in grams, g):

$$\begin{array}{ccccc} 9.98 & 10.07 & 9.94 & 10.22 & 9.98 \\ 10.01 & 10.11 & 10.01 & 9.99 & 9.92 \end{array} \tag{B.75}$$

The sample mean is $\bar{m} = 10.02$ g. The sample standard deviation is $s = 0.0883$ g. The 97.5%-tile of the t distribution with $n - 1 = 9$ degrees of freedom is (from a t-table or function) 2.262. Thus our 95% confidence interval for the mean is

$$\left[\bar{m} - 2.262 s/\sqrt{n},\ \bar{m} + 2.262 s/\sqrt{n}\right] \text{ g} . \tag{B.76}$$

Substituting the values for $\bar{m}$, s, and n, we get an interval of

$$\left[10.02 - 2.262 \cdot 0.0883/\sqrt{10},\ 10.02 + 2.262 \cdot 0.0883/\sqrt{10}\right] \text{g} \tag{B.77}$$

or

$$[9.96, 10.08] \text{ g} . \tag{B.78}$$

The above procedure for constructing a confidence interval for the mean using the t distribution was based on the assumption that the measurements were normally distributed. In situations where the data are not normally distributed this procedure can fail in a very dramatic fashion (Exercise B.10). However, it may be safe to generate an approximate confidence interval using this procedure if (1) the number n of data is large (50 or more) or (2) the distribution of the data is not strongly skewed and n is at least 15. Furthermore, the confidence interval also depends on the assumption that the measurements were independent. In situations where the data are correlated the procedure can also fail (Exercise B.9).

B.9. EXERCISES

1. Compute the expected value and variance of a uniform random variable in terms of the parameters a and b.
2. Compute the CDF of an exponential random variable with parameter λ. You may find the library function **exprand** to be useful here.
3. Show that

$$\text{Cov}(aX,\ Y) = a\text{Cov}(X,\ Y) \tag{B.79}$$

and that

$$\text{Cov}(X + Y, Z) = \text{Cov}(X, Z) + \text{Cov}(Y, Z). \tag{B.80}$$

4. Show that the PDF for the sum of two independent uniform random variables on $[a,\ b] = [0,\ 1]$ is

$$f(x) = \begin{cases} 0 & (x \leq 0) \\ x & (0 \leq x \leq 1) \\ 2 - x & (1 \leq x \leq 2) \\ 0 & (x \geq 2). \end{cases} \tag{B.81}$$

5. Suppose that X and Y are independent random variables. Use conditioning to find a formula for the CDF of $X + Y$ in terms of the PDF's and CDF's of X and Y.
6. Suppose that $\mathbf{x}$ is a two-dimensional multivariate normal distributed random variable with expected value $\boldsymbol{\mu}$ and covariance matrix $\mathbf{C}$. For a 2 by 2 matrix $\mathbf{A}$, use properties of expected value and covariance to show that $\mathbf{y} = \mathbf{A}\mathbf{x}$ has expected value $\mathbf{A}\boldsymbol{\mu}$ and covariance $\mathbf{A}\mathbf{C}\mathbf{A}^T$.
7. Suppose that $\mathbf{x}$ is a multivariate normal random variable with expected value $\boldsymbol{\mu}$ and covariance matrix $\mathbf{C}$. Use Theorem B.5 to show that the ith element of $\mathbf{x}$, x_i, is normally distributed with mean μ_i and variance $\mathbf{C}_{i,i}$. Hint: Let $\mathbf{A} = \mathbf{e}_i^T$, where $\mathbf{e}_i$ is the ith column of the identity matrix.

8. Consider the following data, which we will assume are drawn from a normal distribution.

$$
\begin{array}{ccccc}
-0.4326 & -1.6656 & 0.1253 & 0.2877 & -1.1465 \\
1.1909 & 1.1892 & -0.0376 & 0.3273 & 0.1746
\end{array}
$$

Find the sample mean and standard deviation. Use these to construct a 95% confidence interval for the mean.

9. Our formula for the 95% confidence interval for the mean is based on the assumption that the data are independent and identically distributed. In this exercise we'll consider what can go wrong if the data are not independent. Using MATLAB, repeat the following experiment 1000 times. Use the library function **simmvn** to generate $N = 100$ multivariate normal random numbers with mean $E[X] = 10$ and covariance matrix C, with

$$
C_{i,j} = e^{-|i-j|/N}. \tag{B.82}
$$

Use (B.74) to calculate a 95% confidence interval for the mean. How many times out of the 1000 experiments did the 95% confidence interval include the expected value of 10? Examine the sample means $\hat{\mu}$ and sample standard deviations s. Explain why the actual coverage of the confidence interval is far less than 95%.

10. Using MATLAB, repeat the following experiment 1000 times. Use the library function **exprand** to generate 5 exponentially distributed random numbers from the exponential probability density function (B.10) with means $\mu = 1/\lambda = 10$. Use (B.74) to calculate a 95% confidence interval for the mean. How many times out of the 1000 experiments did the 95% confidence interval include the expected value of 10? What happens if you instead generate 50 exponentially distributed random numbers at a time? Discuss your results.

11. Using MATLAB, repeat the following experiment 1000 times. Using the library function **exprand**, generate 5 exponentially distributed random numbers with expected value 10. Take the average of the 5 random numbers. Plot a histogram and make a Q–Q plot of the 1000 averages that you computed. Are the averages approximately normally distributed? Explain why or why not. What would you expect to happen if you took averages of 50 exponentially distributed random numbers at a time? Try it and discuss the results.

B.10. NOTES AND FURTHER READING

Most of the material in this Appendix can be found in introductory textbooks in probability and statistics. Some recent textbooks include [4,38]. The early history of the normal distribution and Laplace's proposal to use the double sided exponential distribution for measurement errors are discussed in [189]. The multivariate normal distribution is a somewhat more advanced topic that is often ignored in introductory courses. [178] has a good discussion of the multivariate normal distribution and its properties. Numerical methods for probability and statistics are a specialized topic. Two standard references include [111,202].

Review of Vector Calculus

Synopsis

A review is given of key vector calculus topics, including the gradient, Hessian, Jacobian, Taylor's theorem, and Lagrange multipliers.

C.1. THE GRADIENT, HESSIAN, AND JACOBIAN

In vector calculus, the familiar first and second derivatives of a single-variable function are generalized to operate on vectors.

Definition C.1. Given a scalar-valued function with a vector argument, $f(\mathbf{x})$, the **gradient** of f is

$$\nabla f(\mathbf{x}) = \begin{bmatrix} \frac{\partial f}{\partial x_1} \\ \frac{\partial f}{\partial x_2} \\ \vdots \\ \frac{\partial f}{\partial x_n} \end{bmatrix}. \tag{C.1}$$

The vector $\nabla f(\mathbf{x})$ has an important geometric interpretation in that it points in the direction in which $f(\mathbf{x})$ increases most rapidly at the point $\mathbf{x}$.

Recall from single-variable calculus that if a function f is continuously differentiable, then a point x^* can only be a minimum or maximum point of f if $f'(x)|_{x=x^*} = 0$. Similarly in vector calculus, if $f(\mathbf{x})$ is continuously differentiable, then a point $\mathbf{x}^*$ can only be a minimum or maximum point if $\nabla f(\mathbf{x}^*) = \mathbf{0}$. In more than one dimension, a point $\mathbf{x}^*$ where $\nabla f(\mathbf{x}^*) = \mathbf{0}$ can also be a **saddle point**. Any point where $\nabla f(\mathbf{x}^*) = \mathbf{0}$ is called a **critical point**.

Definition C.2. Given a scalar-valued function of a vector, $f(\mathbf{x})$, the **Hessian** of f is a square matrix of partial derivatives given by

$$\mathbf{H}(f(\mathbf{x})) = \begin{bmatrix} \frac{\partial^2 f}{\partial x_1^2} & \frac{\partial^2 f}{\partial x_1 \partial x_2} & \cdots & \frac{\partial^2 f}{\partial x_1 \partial x_n} \\ \frac{\partial^2 f}{\partial x_2 \partial x_1} & \frac{\partial^2 f}{\partial x_2^2} & \cdots & \frac{\partial^2 f}{\partial x_2 \partial x_n} \\ \vdots & \vdots & \ddots & \vdots \\ \frac{\partial^2 f}{\partial x_n \partial x_1} & \frac{\partial^2 f}{\partial x_n \partial x_2} & \cdots & \frac{\partial^2 f}{\partial x_n^2} \end{bmatrix}. \tag{C.2}$$

If f is twice continuously differentiable, the Hessian is symmetric. It is common in mathematics to write the Hessian using the operator ∇^2, but this sometimes leads to confusion with another vector calculus operator, the Laplacian.

Theorem C.1. If $f(\mathbf{x})$ is a twice continuously differentiable function, and $\mathbf{H}(f(\mathbf{x}_0))$ is a positive semidefinite matrix, then $f(\mathbf{x})$ is a **convex function** at $\mathbf{x}_0$. If $\mathbf{H}(f(\mathbf{x}_0))$ is positive definite, then $f(\mathbf{x})$ is **strictly convex** at $\mathbf{x}_0$.

This theorem can be used to check whether a critical point is a minimum of f. If $\mathbf{x}^*$ is a critical point of f and $\mathbf{H}(f(\mathbf{x}^*))$ is positive definite, then f is convex at $\mathbf{x}^*$, and $\mathbf{x}^*$ is thus a local minimum of f.

It will be necessary to compute derivatives of quadratic forms.

Theorem C.2. Let $f(\mathbf{x}) = \mathbf{x}^T \mathbf{A} \mathbf{x}$ where $\mathbf{A}$ is an n by n symmetric matrix. Then

$$\nabla f(\mathbf{x}) = 2\mathbf{A}\mathbf{x} \tag{C.3}$$

and

$$\mathbf{H}(f(\mathbf{x})) = 2\mathbf{A} . \tag{C.4}$$

Definition C.3. Given a vector-valued function of a vector, $\mathbf{F}(\mathbf{x})$, where

$$\mathbf{F}(\mathbf{x}) = \begin{bmatrix} f_1(\mathbf{x}) \\ f_2(\mathbf{x}) \\ \vdots \\ f_m(\mathbf{x}) \end{bmatrix} \tag{C.5}$$

the **Jacobian** of $\mathbf{F}$ is

$$\mathbf{J}(\mathbf{x}) = \begin{bmatrix} \frac{\partial f_1}{\partial x_1} & \frac{\partial f_1}{\partial x_2} & \cdots & \frac{\partial f_1}{\partial x_n} \\ \frac{\partial f_2}{\partial x_1} & \frac{\partial f_2}{\partial x_2} & \cdots & \frac{\partial f_2}{\partial x_n} \\ \vdots & \vdots & \ddots & \vdots \\ \frac{\partial f_m}{\partial x_1} & \frac{\partial f_m}{\partial x_2} & \cdots & \frac{\partial f_m}{\partial x_n} \end{bmatrix} . \tag{C.6}$$

Some authors use the notation $\nabla \mathbf{F}(\mathbf{x})$ for the Jacobian. Notice that the rows of $\mathbf{J}(\mathbf{x})$ are the gradients (C.1) of the functions $f_1(\mathbf{x}), f_2(\mathbf{x}), \ldots, f_m(\mathbf{x})$.

C.2. TAYLOR'S THEOREM

In the calculus of single-variable functions, Taylor's theorem produces an infinite series for $f(x + \Delta x)$ in terms of $f(x)$ and its derivatives. Taylor's theorem can be extended to a function of a vector $f(\mathbf{x})$, but in practice, derivatives of order higher than two are extremely inconvenient. The following form of Taylor's theorem is often used in optimization theory.

Theorem C.3. Suppose that $f(\mathbf{x})$ and its first and second partial derivatives are continuous. For any vectors $\mathbf{x}$ and $\Delta\mathbf{x}$, there is a vector $\mathbf{c}$, with $\mathbf{c}$ on the line between $\mathbf{x}$ and $\mathbf{x} + \Delta\mathbf{x}$, such that

$$f(\mathbf{x} + \Delta\mathbf{x}) = f(\mathbf{x}) + \nabla f(\mathbf{x})^T \Delta\mathbf{x} + \frac{1}{2}\Delta\mathbf{x}^T \mathbf{H}(f(\mathbf{c}))\Delta\mathbf{x} . \tag{C.7}$$

This form of **Taylor's theorem with remainder term** is useful in many proofs. However, in computational work there is no way to determine $\mathbf{c}$. For that reason, when $\Delta\mathbf{x}$ is a small perturbation, we often make use of the approximation

$$f(\mathbf{x} + \Delta\mathbf{x}) \approx f(\mathbf{x}) + \nabla f(\mathbf{x})^T \Delta\mathbf{x} + \frac{1}{2}\Delta\mathbf{x}^T \mathbf{H}(f(\mathbf{x}))\Delta\mathbf{x} . \tag{C.8}$$

An even simpler version of Taylor's theorem, the **mean value theorem** uses only the first derivative.

Theorem C.4. Suppose that $f(\mathbf{x})$ and its first partial derivatives are continuous. For any vectors $\mathbf{x}$ and $\Delta\mathbf{x}$ there is a vector $\mathbf{c}$, with $\mathbf{c}$ on the line between $\mathbf{x}$ and $\mathbf{x} + \Delta\mathbf{x}$ such that

$$f(\mathbf{x} + \Delta\mathbf{x}) = f(\mathbf{x}) + \nabla f(\mathbf{c})^T \Delta\mathbf{x} . \tag{C.9}$$

We will make use of a truncated version of (C.8)

$$f(\mathbf{x} + \Delta\mathbf{x}) \approx f(\mathbf{x}) + \nabla f(\mathbf{x})^T \Delta\mathbf{x} . \tag{C.10}$$

By applying (C.10) to each of the functions $f_1(\mathbf{x}), f_2(\mathbf{x}), \ldots, f_m(\mathbf{x})$, we obtain the approximation

$$\mathbf{F}(\mathbf{x} + \Delta\mathbf{x}) \approx \mathbf{F}(\mathbf{x}) + \mathbf{J}(\mathbf{x})\Delta\mathbf{x} . \tag{C.11}$$

C.3. LAGRANGE MULTIPLIERS

The method of **Lagrange multipliers** is an important technique for solving optimization problems of the form

$$\begin{aligned} \min &f(\mathbf{x}) \\ &g(\mathbf{x}) = 0 \end{aligned} \tag{C.12}$$

where the scalar-valued function of a vector argument, $f(\mathbf{x})$, is called the **objective function**.

Fig. C.1 shows a general situation. The solid contour represents the set of points where the (non-constant) function $g(\mathbf{x}) = 0$, and the dashed contours are those of another function $f(\mathbf{x})$ that has a minimum as indicated. Moving along the $g(\mathbf{x}) = 0$ contour, we can trace out the curve $\mathbf{x}(t)$, parameterized by the variable $t \geq 0$, where

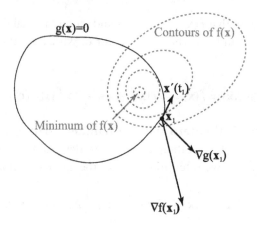

Figure C.1 The situation at a point $\mathbf{x}_1 = \mathbf{x}(t_1)$ along the contour $g(\mathbf{x}) = 0$ that is not a minimum of $f(\mathbf{x})$ and thus does not satisfy (C.12).

$g(\mathbf{x}(t)) = 0$ and t increases as we progress counter-clockwise. At any point $\mathbf{x}(t)$ on the contour, the gradient of $g(\mathbf{x}(t))$ must be perpendicular to the contour because the function is constant along this curve. Note that in Fig. C.1, $g(\mathbf{x})$ increases in the outward direction relative to the contour, so the gradient of $g(\mathbf{x})$ will be outward.

By the chain rule,

$$f'(\mathbf{x}(t)) = \mathbf{x}'(t)^T \nabla f(\mathbf{x}(t)) \qquad (C.13)$$

where $\mathbf{x}'(t)$ is the counter-clockwise tangent to the contour $g(x) = 0$. For the point $\mathbf{x}_1 = \mathbf{x}(t_1)$ in Fig. C.1, $\nabla f(\mathbf{x}_1)$ and $\mathbf{x}'(t_1)$ are at an obtuse angle, and their dot product $f'(\mathbf{x}_1)$ (C.13) will therefore be negative. Thus, $f(\mathbf{x})$ is decreasing as we move counter-clockwise around the contour $g(\mathbf{x}) = 0$ from $\mathbf{x}_1$, and $\mathbf{x}_1$ cannot satisfy (C.12).

In Fig. C.2, for the point $\mathbf{x}_0 = \mathbf{x}(t_0)$, $\nabla f(\mathbf{x}_0)$ is perpendicular to the curve $g(\mathbf{x}) = 0$. In this case, by (C.13), $f'(\mathbf{x}_0) = 0$, and the point $\mathbf{x}_0$ may or may not be a minimum for $f(\mathbf{x})$ along the contour. Fig. C.2 shows that a point $\mathbf{x}_0$ on the curve $g(\mathbf{x}) = 0$ can only be a possible minimum point for $f(\mathbf{x})$ if $\nabla g(\mathbf{x}_0)$ and $\nabla f(\mathbf{x}_0)$ are parallel or antiparallel. A point where this occurs is called a **stationary point**.

Finding a stationary point is necessary, but not sufficient, for finding a minimum of $f(\mathbf{x})$ along the contour $g(\mathbf{x}) = 0$, because such a point may be a minimum, maximum, or saddle point. Furthermore, a problem may have several local minima. Thus it is necessary to examine the behavior of $f(\mathbf{x})$ at all stationary points to find a global minimum.

Theorem C.5. (C.12) can only be satisfied at a point $\mathbf{x}_0$ where

$$\nabla f(\mathbf{x}_0) + \lambda \nabla g(\mathbf{x}_0) = \mathbf{0} \qquad (C.14)$$

for some λ. λ is called a **Lagrange multiplier**.

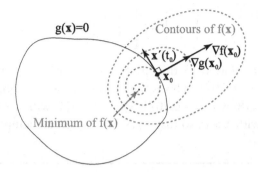

Figure C.2 The situation at a point $\mathbf{x}_0 = \mathbf{x}(t_0)$ along the contour $g(\mathbf{x}) = 0$ that is a minimum of $f(\mathbf{x})$ and thus satisfies (C.12). Note that $\nabla g(\mathbf{x}_0)$ and $\nabla f(\mathbf{x}_0)$ are parallel, and that (C.14) is thus satisfied, for some Lagrange multiplier λ.

The Lagrange multiplier condition can be extended to problems of the form

$$\min f(\mathbf{x})$$
$$g(\mathbf{x}) \leq 0 . \tag{C.15}$$

Since points along the curve $g(\mathbf{x}) = 0$ are still feasible in (C.15), (C.14) must still hold true. However, there is an additional restriction. Suppose that $\nabla g(\mathbf{x}_0)$ and $\nabla f(\mathbf{x}_0)$ both point in the outward direction, as in Fig. C.2. In this case, we can move in the opposite direction, into the feasible region to decrease $f(\mathbf{x})$ (e.g., in the situation depicted in Fig. C.2, the solution to (C.15) is simply the indicated minimum of $f(\mathbf{x})$). Thus, a point $\mathbf{x}_0$ satisfying (C.14) cannot satisfy (C.15) unless the gradients of $g(\mathbf{x}_0)$ and $f(\mathbf{x}_0)$ point in opposite directions.

Theorem C.6. (C.15) can only be satisfied at a point $\mathbf{x}_0$ where

$$\nabla f(\mathbf{x}_0) + \lambda \nabla g(\mathbf{x}_0) = \mathbf{0} \tag{C.16}$$

for some Lagrange multiplier $\lambda > 0$.

Example C.1

Consider a simple example in two variables where $f(\mathbf{x})$ defines linear contours and $g(\mathbf{x}) = 0$ defines a unit circle

$$\min x_1 + x_2$$
$$x_1^2 + x_2^2 - 1 \leq 0 . \tag{C.17}$$

The Lagrange multiplier condition is

$$\begin{bmatrix} 1 \\ 1 \end{bmatrix} + \lambda \begin{bmatrix} 2x_1 \\ 2x_2 \end{bmatrix} = \mathbf{0} \ . \tag{C.18}$$

One stationary point solution to this nonlinear system of equations is $x_1 = 0.7071$, $x_2 = 0.7071$, with $\lambda = -0.7071$. This is the maximum of $f(\mathbf{x})$ subject to $g(\mathbf{x}) \leq 0$. The second solution to (C.18) is $x_1 = -0.7071$, $x_2 = -0.7071$, with $\lambda = 0.7071$. Because this is the only solution with $\lambda > 0$, so that $\nabla f(\mathbf{x})$ and $\nabla g(\mathbf{x})$ are antiparallel, this solves the minimization problem.

Note that (C.16) is (except for the nonnegativity constraint on λ), the necessary condition for a minimum point of the unconstrained minimization problem

$$\min f(\mathbf{x}) + \lambda g(\mathbf{x}) \ . \tag{C.19}$$

Here the parameter λ can be adjusted so that, for the optimal solution, $\mathbf{x}^*$, $g(\mathbf{x}^*) \leq 0$. We will make frequent use of this technique to convert constrained optimization problems into unconstrained optimization problems.

C.4. EXERCISES

1. Let

$$f(\mathbf{x}) = x_1^2 x_2^2 - 2x_1 x_2^2 + x_2^2 - 3x_1^2 x_2 + 12x_1 x_2 - 12x_2 + 6 \ . \tag{C.20}$$

   Find the gradient, $\nabla f(\mathbf{x})$, and Hessian, $\mathbf{H}(f(\mathbf{x}))$. What are the critical points of f? Which of these are minima and maxima of f?

2. Find a Taylor's series approximation for $f(\mathbf{x} + \Delta\mathbf{x})$, where

$$f(\mathbf{x}) = e^{-(x_1 + x_2)^2} \tag{C.21}$$

   is near the point

$$\mathbf{x} = \begin{bmatrix} 2 \\ 3 \end{bmatrix} \ . \tag{C.22}$$

3. Use the method of Lagrange multipliers to solve the problem

$$\min 2x_1 + x_2 \\ 4x_1^2 + 3x_2^2 - 5 \leq 0 \ . \tag{C.23}$$

4. Derive the formula (A.89) for the 2-norm of a matrix. Begin with the maximization problem

$$\max_{\|\mathbf{x}\|_2 = 1} \|\mathbf{A}\mathbf{x}\|_2^2 \ . \tag{C.24}$$

Note that we have squared $\|\mathbf{Ax}\|_2$. We will take the square root at the end of the problem.

a. Using the formula $\|\mathbf{x}\|_2 = \sqrt{\mathbf{x}^T\mathbf{x}}$, rewrite the above maximization problem without norms.

b. Use the Lagrange multiplier method to find a system of equations that must be satisfied by any stationary point of the maximization problem.

c. Explain how the eigenvalues and eigenvectors of $\mathbf{A}^T\mathbf{A}$ are related to this system of equations. Express the solution to the maximization problem in terms of the eigenvalues and eigenvectors of $\mathbf{A}^T\mathbf{A}$.

d. Use this solution to get $\|\mathbf{A}\|_2$.

5. Derive the normal equations (2.3) using vector calculus, by letting

$$f(\mathbf{m}) = \|\mathbf{Gm} - \mathbf{d}\|_2^2 \tag{C.25}$$

and minimizing $f(\mathbf{m})$. Note that in problems with many least squares solutions, all least squares solutions will satisfy the normal equations.

a. Rewrite $f(\mathbf{m})$ as a dot product and then expand the expression.

b. Find $\nabla f(\mathbf{m})$.

c. Set $\nabla f(\mathbf{m}) = \mathbf{0}$, and obtain the normal equations.

C.5. NOTES AND FURTHER READING

Basic material on vector calculus can be found in calculus textbooks. However, more advanced topics, such as Lagrange multipliers and Taylor's theorem for functions of a vector are often not covered in basic texts. The material in this chapter is particularly important in optimization, and can often be found in associated references [76,134,150].

Glossary of Notation

- α, β, γ, ...: Scalars.
- a, b, c, ...: Scalar–valued functions or scalars.
- $\mathbf{a}$, $\mathbf{b}$, $\mathbf{c}$, ...: Column vectors.
- a_i: ith element of vector $\mathbf{a}$.
- A, B, C, ...: Scalar-valued functions or random variables.
- $\mathcal{A}$, $\mathcal{B}$, $\mathcal{C}$, ...: Fourier transforms.
- $\mathbf{A}$, $\mathbf{B}$, $\mathbf{C}$, ...: Vector-valued functions or matrices.
- $\mathbf{A}_{i,.}$: ith row of matrix $\mathbf{A}$.
- $\mathbf{A}_{.,j}$: ith column of matrix $\mathbf{A}$.
- $A_{i,j}$: (i,j)th element of matrix $\mathbf{A}$.
- $\mathbf{m}^{(k)}$: kth iterate of vector $\mathbf{m}$.
- $\mathbf{A}^{-1}$: Inverse of the matrix $\mathbf{A}$.
- $\mathbf{A}^{T}$: Transpose of the matrix $\mathbf{A}$.
- $\mathbf{A}^{*}$: Complex conjugate of the matrix $\mathbf{A}$.
- diag($\mathbf{A}$): Vector constructed by extracting the diagonal elements of the matrix $\mathbf{A}$.
- $\mathbf{a} \odot \mathbf{b}$: Vector constructed by element-by-element multiplication of the vectors $\mathbf{a}$ and $\mathbf{b}$.
- $\mathbf{a} \oslash \mathbf{b}$: Vector constructed by element-by-element division of the vector $\mathbf{a}$ by $\mathbf{b}$.
- $\mathbf{A}^{-T}$: Transpose of the matrix $\mathbf{A}^{-1}$.
- R^{n}: Space of n–dimensional real vectors.
- $N(\mathbf{A})$: Null space of the matrix $\mathbf{A}$.
- $R(\mathbf{A})$: Range of the matrix $\mathbf{A}$.
- rank($\mathbf{A}$): Rank of the matrix $\mathbf{A}$.
- cond($\mathbf{A}$) $= \kappa(\mathbf{A})$: Condition number of the matrix $\mathbf{A}$.
- Tr($\mathbf{A}$): Trace of the matrix $\mathbf{A}$.
- $\|\mathbf{x}\|$: Norm of a vector $\mathbf{x}$. A subscript is used to specify the one norm, two norm, or infinity norm.
- $\|\mathbf{A}\|$: Norm of a matrix $\mathbf{A}$. A subscript is used to specify the one norm, two norm, or infinity norm.
- $\mathbf{G}^{\dagger}$: Generalized inverse of the matrix $\mathbf{G}$.
- $\mathbf{m}_{\dagger}$: Generalized inverse solution $\mathbf{m}_{\dagger} = \mathbf{G}^{\dagger}\mathbf{d}$.
- $\mathbf{G}^{\sharp}$: A regularized generalized inverse of the matrix $\mathbf{G}$.
- $E[X]$: Expected value of the random variable X.
- $\bar{\mathbf{a}}$: Mean value of the elements in vector $\mathbf{a}$.
- $N(\mu, \sigma^2)$: Normal probability distribution with expected value μ and variance σ^2.
- Cov(X, Y): Covariance of the random variables X and Y.

- Cov($\mathbf{x}$): Matrix of covariances of elements of the vector $\mathbf{x}$.
- $\rho(X, Y)$: Correlation between the random variables X and Y.
- Var(X): Variance of the random variable X.
- $f(\mathbf{d}|\mathbf{m})$: Conditional probability density for $\mathbf{d}$, conditioned on a particular model $\mathbf{m}$.
- $L(\mathbf{m}|\mathbf{d})$: Likelihood function for a model $\mathbf{m}$, given a particular data vector $\mathbf{d}$.
- σ: Standard deviation.
- σ^2: Variance.
- $\nabla f(\mathbf{x})$: Gradient of the function $f(\mathbf{x})$.
- $\mathbf{J}(\mathbf{x})$: Jacobian of the vector-valued function, $\mathbf{F}(\mathbf{x})$.
- $\mathbf{H}(f(\mathbf{x}))$: Hessian of the scalar-valued function $f(\mathbf{x})$.

BIBLIOGRAPHY

[1] R.C. Aster, On projecting error ellipsoids, Bulletin of the Seismological Society of America 78 (3) (1988) 1373–1374.

[2] G. Backus, F. Gilbert, Uniqueness in the inversion of inaccurate gross earth data, Philosophical Transactions of the Royal Society A 266 (1970) 123–192.

[3] Z. Bai, J.W. Demmel, Computing the generalized singular value decomposition, SIAM Journal on Scientific Computing 14 (1993) 1464–1486.

[4] L.J. Bain, M. Englehardt, Introduction to Probability and Mathematical Statistics, Brooks/Cole, Pacific Grove, CA, 2000.

[5] R. Barrett, M. Berry, T.F. Chan, J. Demmel, J. Donato, V. Eijkhout, R. Pozo, C. Romine, H. van der Vorst, Templates for the Solution of Linear Systems: Building Blocks for Iterative Methods, 2nd ed., SIAM, Philadelphia, 1994.

[6] I. Barrowdale, F.D.K. Roberts, Solution of an overdetermined system of equations in the l_1 norm, Communications of the ACM 17 (6) (1974) 319–326.

[7] D.M. Bates, D.M. Watts, Nonlinear Regression Analysis and Its Applications, Wiley, 2007.

[8] J. Baumeister, Stable Solution of Inverse Problems, Vieweg, Braunschweig, 1987.

[9] A. Beck, First-Order Methods in Optimization, SIAM, 2017.

[10] A. Beck, M. Teboulle, A fast iterative shrinkage-thresholding algorithm for linear inverse problems, SIAM Journal on Imaging Sciences 2 (1) (2009) 183–202.

[11] S.R. Becker, E.J. Candes, M.C. Grant, Templates for convex cone problems with applications to sparse signal recovery, Mathematical Programming Computation 3 (3) (August 2011) 165–218.

[12] C. Bekas, E. Kokiopoulou, Y. Saad, An estimator for the diagonal of a matrix, Applied Numerical Mathematics 57 (11–12) (2007) 1214–1229.

[13] A. Ben-Israel, T.N.E. Greville, Generalized Inverses: Theory and Applications, 2nd ed., Springer, 2004.

[14] A. Ben-Tal, A. Nemirovski, Lectures on Modern Convex Optimization: Analysis, Algorithms, and Engineering Applications, SIAM, Philadelphia, 2001.

[15] J.G. Berryman, Analysis of approximate inverses in tomography I. Resolution analysis, Optimization and Engineering 1 (1) (2000) 87–115.

[16] J.G. Berryman, Analysis of approximate inverses in tomography II. Iterative inverses, Optimization and Engineering 1 (4) (2000) 437–473.

[17] M. Bertero, P. Boccacci, Introduction to Inverse Problems in Imaging, Institute of Physics, London, 1998.

[18] D.P. Bertsekas, Convex Optimization Algorithms, Athena Scientific, Nashua, NH, 2015.

[19] Å. Björck, Numerical Methods for Least Squares Problems, SIAM, Philadelphia, 1996.

[20] R.J. Blakely, Potential Theory in Gravity and Magnetic Applications, Stanford–Cambridge Program, Cambridge University Press, 1996.

[21] C.G.E. Boender, H.E. Romeijn, Stochastic methods, in: R. Horst, P.M. Pardalos (Eds.), Handbook of Global Optimization, Kluwer Academic Publishers, Dordrecht, 1995, pp. 829–869.

[22] P.T. Boggs, J.R. Donaldson, R.H. Byrd, R.B. Schnabel, ODRPACK software for weighted orthogonal distance regression, ACM Transactions on Mathematical Software 15 (4) (1989) 348–364. Available at http://www.netlib.org/odrpack/.

[23] A. Borzì, V. Schulz, Computational Optimization of Systems Governed by Partial Differential Equations, vol. 8, SIAM, 2011.

[24] L. Boschi, Measures of resolution in global body wave tomography, Geophysical Research Letters 30 (2003) 4.

[25] S. Boyd, Distributed optimization and statistical learning via the alternating direction method of multipliers, Foundations and Trends® in Machine Learning 3 (1) (2010) 1–122.

[26] R. Bracewell, The Fourier Transform and Its Applications, McGraw-Hill, Boston, 2005.

[27] A.M. Bruckstein, D.L. Donoho, M. Elad, From sparse solutions of systems of equations to sparse modeling of signals and images, SIAM Review 51 (1) (2009) 34–81.

[28] J.N. Brune, Tectonic stress and the spectra of seismic shear waves from earthquakes, Journal of Geophysical Research 75 (26) (1970) 4997–5009.

[29] D. Calvetti, Preconditioned iterative methods for linear discrete ill-posed problems from a Bayesian inversion perspective, Journal of Computational and Applied Mathematics 198 (2) (January 2007) 378–395.

[30] D. Calvetti, E. Somersalo, Introduction to Bayesian Scientific Computing: Ten Lectures on Subjective Computing, Springer, 2007.

[31] S.L. Campbell, C.D. Meyer Jr., Generalized Inverses of Linear Transformations, Dover, Mineola, NY, 1991.

[32] E.J. Candes, Compressive sampling, in: Proceedings of the International Congress of Mathematicians, vol. 3, 2006, pp. 1433–1452.

[33] E.J. Candes, J.K. Romberg, T. Tao, Robust uncertainty principles: Exact signal reconstruction from highly incomplete frequency information, IEEE Transactions on Information Theory 52 (2) (2006) 489–509.

[34] E.J. Candes, J.K. Romberg, T. Tao, Stable signal recovery from incomplete and inaccurate measurements, Communications on Pure and Applied Mathematics 59 (8) (2006) 1207–1223.

[35] E.J. Candes, T. Tao, Near-optimal signal recovery from random projections: Universal encoding strategies?, IEEE Transactions on Information Theory 52 (12) (2006) 5406–5425.

[36] E.J. Candes, M.B. Wakin, An introduction to compressive sampling, IEEE Signal Processing Magazine 25 (2) (2008) 21–30.

[37] P. Carrion, Inverse Problems and Tomography in Acoustics and Seismology, Penn, Atlanta, 1987.

[38] G. Casella, R.L. Berger, Statistical Inference, 2nd ed., Cengage Learning, 2001.

[39] Y. Censor, P.P.B. Eggermont, D. Gordon, Strong underrelaxation in Kaczmarz's method for inconsistent systems, Numerische Mathematik 41 (1) (February 1983) 83–92.

[40] Y. Censor, S.A. Zenios, Parallel Optimization: Theory, Algorithms, and Applications, Oxford University Press, NY, 1997.

[41] J. Christensen-Dalsgaard, J. Schou, M.J. Thompson, A comparison of methods for inverting helioseismic data, Monthly Notices of the Royal Astronomical Society 242 (3) (June 1990) 353–369.

[42] J.F. Claerbout, F. Muir, Robust modeling with erratic data, Geophysics 38 (5) (1973) 826–844.

[43] T.F. Coleman, Y. Li, A globally and quadratically convergent method for linear l_1 problems, Mathematical Programming 56 (1992) 189–222.

[44] P.L. Combettes, J.-C. Pesquet, Proximal splitting methods in signal processing, in: Fixed-Point Algorithms for Inverse Problems in Science and Engineering, Springer, 2011, pp. 185–212.

[45] S.C. Constable, R.L. Parker, C.G. Constable, Occam's inversion: A practical algorithm for generating smooth models from electromagnetic sounding data, Geophysics 52 (3) (1987) 289–300.

[46] G. Corliss, C. Faure, A. Griewank, L. Hascoet, U. Naumann, Automatic Differentiation of Algorithms, Springer-Verlag, Berlin, 2002.

[47] R.T. Cox, Algebra of Probable Inference, The Johns Hopkins University Press, Baltimore, 2002.

[48] P. Craven, G. Wahba, Smoothing noisy data with spline functions: Estimating the correct degree of smoothing by the method of generalized cross-validation, Numerische Mathematik 31 (1979) 377–403.

[49] M.M. Deal, G. Nolet, Comment on 'Estimation of resolution and covariance for large matrix inversions' by J. Zhang and G. A. McMechan, Geophysical Journal International 127 (1) (1996) 245–250.

[50] J.W. Demmel, Applied Numerical Linear Algebra, SIAM, Philadelphia, 1997.

[51] J.E. Dennis Jr., R.B. Schnabel, Numerical Methods for Unconstrained Optimization and Nonlinear Equations, SIAM, Philadelphia, 1996.

[52] N.R. Draper, H. Smith, Applied Regression Analysis, Wiley, NY, 1998.

[53] C. Eckart, G. Young, A principal axis transformation for non-hermitian matrices, Bulletin of the American Mathematical Society 45 (1939) 118–121.

[54] A.W.F. Edwards, Likelihood, The Johns Hopkins University Press, Baltimore, 1992.

[55] L. ElGhaoui, H. Lebret, Robust solutions to least-squares problems with uncertain data, SIAM Journal on Matrix Analysis and Applications 18 (4) (1997) 1035–1064.

[56] H.W. Engl, Regularization methods for the stable solution of inverse problems, Surveys on Mathematics for Industry 3 (1993) 71–143.

[57] H.W. Engl, M. Hanke, A. Neubauer, Regularization of Inverse Problems, Kluwer Academic Publishers, Boston, 1996.

[58] R.M. Errico, What is an adjoint model?, Bulletin of the American Meteorological Society 78 (11) (1997) 2577–2591.

[59] M. Evans, N. Hasting, B. Peacock, Statistical Distributions, John Wiley & Sons, NY, 1993.

[60] S.N. Evans, P.B. Stark, Inverse problems as statistics, Inverse Problems 18 (2002) R1–R43.

[61] J.G. Ferris, D.B. Knowles, The Slug-Injection Test for Estimating the Coefficient of Transmissibility of an Aquifer, U.S. Geological Survey, Washington, DC, 1963, pp. 299–304.

[62] M.A.T. Figueiredo, R.D. Nowak, S.J. Wright, Gradient projection for sparse reconstruction: Application to compressed sensing and other inverse problems, IEEE Journal of Selected Topics in Signal Processing 1 (4) (December 2007) 586–597.

[63] R. Fletcher, C.M. Reeves, Function minimization by conjugate gradients, The Computer Journal 7 (2) (1964) 149–154.

[64] D. Fong, M. Saunders, LSMR: An iterative algorithm for sparse least-squares problems, SIAM Journal on Scientific Computing 33 (8) (2011) 2950–2971.

[65] A. Frommer, P. Maass, Fast CG-based methods for Tikhonov–Phillips regularization, SIAM Journal on Scientific Computing 20 (5) (1999) 1831–1850.

[66] A. Gelman, J.B. Carlin, H.S. Stern, D.B. Dunson, A. Vehtari, D.B. Rubin, Bayesian Data Analysis, 3rd ed., Chapman and Hall/CRC, 2013.

[67] M.Th. van Genuchten, A closed-form equation for predicting the hydraulic conductivity of unsaturated soils, Soil Science Society of America Journal 44 (1980) 892–898.

[68] A. Gholami, H.R. Siahkoohi, Regularization of linear and non-linear geophysical ill-posed problems with joint sparsity constraints, Geophysical Journal International 180 (2) (February 2010) 871–882.

[69] M.B. Giles, N.A. Pierce, An introduction to the adjoint approach to design, Flow, Turbulence and Combustion 65 (3–4) (2000) 393–415.

[70] R. Glowinski, On alternating direction methods of multipliers: A historical perspective, in: W. Fitzgibbon, Y.A. Kuznetsov, P. Neittaanmäki, O. Pironneau (Eds.), Modeling, Simulation and Optimization for Science and Technology, vol. 34, Springer, Netherlands, Dordrecht, 2014, pp. 59–82.

[71] G.H. Golub, D.P. O'Leary, Some history of the conjugate gradient and Lanczos methods, SIAM Review 31 (1) (1989) 50–102.

[72] G.H. Golub, C.F. Van Loan, Matrix Computations, 4th ed., Johns Hopkins University Press, 2012.

[73] G.H. Golub, U. von Matt, Generalized cross-validation for large-scale problems, Journal of Computational and Graphical Statistics 6 (1) (1997) 1–34.

[74] W.P. Gouveia, J.A. Scales, Resolution of seismic waveform inversion: Bayes versus Occam, Inverse Problems 13 (2) (1997) 323–349.

[75] A. Griewank, Evaluating Derivatives: Principles and Techniques of Algorithmic Differentiation, SIAM, Philadelphia, 2000.

[76] I. Griva, S.G. Nash, A. Sofer, Linear and Nonlinear Programming, SIAM, Philadelphia, 2008.

[77] C.W. Groetsch, Inverse Problems in the Mathematical Sciences, Vieweg, Braunschweig, 1993.

[78] D. Gubbins, Time Series Analysis and Inverse Theory for Geophysicists, Cambridge University Press, Cambridge, U.K., 2004.

[79] M. Hanke, Conjugate Gradient Type Methods for Ill-Posed Problems, CRC Press, 1995.

[80] M. Hanke, W. Niethammer, On the acceleration of Kaczmarz's method for inconsistent linear systems, Linear Algebra and Its Applications 130 (Supplement C) (March 1990) 83–98.

[81] P.C. Hansen, Relations between SVD and GSVD of discrete regularization problems in standard and general form, Linear Algebra and Its Applications 141 (1990) 165–176.

[82] P.C. Hansen, Analysis of discrete ill-posed problems by means of the L-curve, SIAM Review 34 (4) (1992) 561–580.

[83] P.C. Hansen, Regularization tools: A MATLAB package for analysis and solution of discrete ill-posed problems, Numerical Algorithms 6 (I–II) (1994) 1–35, https://www.mathworks.com/matlabcentral/fileexchange/52-regtools.

[84] P.C. Hansen, Rank-Deficient and Discrete Ill-Posed Problems: Numerical Aspects of Linear Inversion, SIAM, Philadelphia, 1998.

[85] P.C. Hansen, Deconvolution and regularization with Toeplitz matrices, Numerical Algorithms 29 (2002) 323–378.

[86] P.C. Hansen, Discrete Inverse Problems: Insight and Algorithms, Society for Industrial and Applied Mathematics, Philadelphia, 2010.

[87] J.M.H. Hendrickx, Bosque del Apache soil data, 2003, Personal communication.

[88] J.M.H. Hendrickx, B. Borchers, J.D. Rhoades, D.L. Corwin, S.M. Lesch, A.C. Hilgendorf, J. Schlue, Inversion of soil conductivity profiles from electromagnetic induction measurements; theory and experimental verification, Soil Science Society of America Journal 66 (3) (2002) 673–685.

[89] G.T. Herman, Image Reconstruction from Projections, Academic Press, San Francisco, 1980.

[90] M.R. Hestenes, Conjugacy and gradients, in: S.G. Nash (Ed.), A History of Scientific Computing, ACM Press, NY, 1990, pp. 167–179.

[91] M.R. Hestenes, E. Stiefel, Methods of conjugate gradients for solving linear systems, Journal of Research, National Bureau of Standards 49 (1952) 409–436.

[92] J.A. Hildebrand, J.M. Stevenson, P.T.C. Hammer, M.A. Zumberge, R.L. Parker, C.J. Fox, P.J. Meis, A sea-floor and sea-surface gravity survey of Axial Volcano, Journal of Geophysical Research 95 (B8) (1990) 12751–12763.

[93] M. Hinze (Ed.), Optimization with PDE Constraints, Mathematical Modelling, vol. 23, Springer, Dordrecht, 2009.

[94] R.A. Horn, C.R. Johnson, Matrix Analysis, 2nd ed., Cambridge University Press, 2012.

[95] R. Horst, P.M. Pardalos, Handbook of Global Optimization, Kluwer Academic Publishers, Dordrecht, 1995.

[96] R. Horst, P.M. Pardalos, N.V. Thoai, Introduction to Global Optimization, Kluwer Academic Publishers, Dordrecht, 1995.

[97] P.J. Huber, Robust Statistical Procedures, SIAM, Philadelphia, 1996.

[98] S. Von Huffel, J. Vandewalle, The Total Least Squares Problem: Computational Aspects and Analysis, SIAM, Philadelphia, 1991.

[99] H. Iyer, K. Hirahara (Eds.), Seismic Tomography, Chapman and Hall, NY, 1993.

[100] E.T. Jaynes, Probability Theory: The Logic of Science, Cambridge University Press, Cambridge, 2003.

[101] H. Jeffreys, Theory of Probability, Oxford University Press, NY, 1998.

[102] W.H. Jeffreys, M.J. Fitzpatrick, B.E. McArthur, Gaussfit – a system for least squares and robust estimation, Celestial Mechanics 41 (1–4) (1987) 39–49. Available at http://clyde.as.utexas.edu/Gaussfit.html.

[103] S. Kaczmarz, Angenaherte Auflizisung von Systemen linearer Gleichungen, Bulletin International de l'Académie Polonaise des Sciences et des Lettres 35 (1937) 355–357.

[104] S. Kaczmarz, Approximate solution of systems of linear equations†, International Journal of Control 57 (6) (June 1993) 1269–1271.

[105] J. Kaipio, E. Somersalo, Statistical and Computational Inverse Problems, Applied Mathematical Sciences, Springer, 2004.

[106] A.C. Kak, M. Slaney, Principles of Computerized Tomographic Imaging, SIAM, Philadelphia, 2001.

[107] L. Kaufman, A. Neumaier, PET regularization by envelope guided conjugate gradients, IEEE Transactions on Medical Imaging 15 (3) (1996) 385–389.

[108] L. Kaufman, A. Neumaier, Regularization of ill-posed problems by envelope guided conjugate gradients, Journal of Computational and Graphical Statistics 6 (4) (1997) 451–463.

[109] C.T. Kelley, Iterative Methods for Solving Linear and Nonlinear Equations, SIAM, Philadelphia, 1995.

[110] C.T. Kelley, Solving Nonlinear Equations with Newton's Method, SIAM, Philadelphia, 2003.

[111] W.J. Kennedy Jr., J.E. Gentle, Statistical Computing, Marcel Dekker, NY, 1980.

[112] D.R. Kincaid, E.W. Cheney, Numerical Analysis: Mathematics of Scientific Computing, vol. 2, American Mathematical Society, 2002.

[113] A. Kirsch, An Introduction to the Mathematical Theory of Inverse Problems, Springer-Verlag, NY, 1996.

[114] K. Kiwiel, Convergence of approximate and incremental subgradient methods for convex optimization, SIAM Journal on Optimization 14 (3) (2004) 807–840.

[115] F.J. Klopping, G. Peter, D.S. Robertson, K.A. Berstis, R.E. Moose, W.E. Carter, Improvements in absolute gravity observations, Journal of Geophysical Research 96 (B5) (1991) 8295–8303.

[116] C. Lanczos, Solutions of systems of linear equations by minimized iterations, Journal of Research, National Bureau of Standards 49 (1952) 33–53.

[117] C. Lanczos, Linear Differential Operators, Dover, Mineola, NY, 1997.

[118] K. Lange, MM Optimization Algorithms, SIAM, Philadelphia, July 2016.

[119] R.M. Larsen, P.C. Hansen, Efficient implementations of the SOLA mollifier method, Astronomy and Astrophysics Supplement Series 121 (3) (March 1997) 587–598.

[120] C.L. Lawson, R.J. Hanson, Solving Least Squares Problems, SIAM, Philadelphia, 1995.

[121] D.C. Lay, Linear Algebra and Its Applications, Addison-Wesley, Boston, 2003.

[122] T. Lay, T. Wallace, Modern Global Seismology, Academic Press, San Diego, 1995.

[123] J.-J. Leveque, L. Rivera, G. Wittlinger, On the use of the checker-board test to assess the resolution of tomographic inversions, Geophysical Journal International 115 (1) (1993) 313–318.

[124] Z.-P. Liang, P.C. Lauterbur, Principles of Magnetic Resonance Imaging: A Signal Processing Perspective, IEEE Press, NY, 2000.

[125] L.R. Lines (Ed.), Inversion of Geophysical Data, Society of Exploration Geophysicists, Tulsa, OK, 1988.

[126] T.-W. Lo, P. Inderwiesen, Fundamentals of Seismic Tomography, Society of Exploration Geophysicists, Tulsa, OK, 1994.

[127] C.F. Van Loan, Generalizing the singular value decomposition, SIAM Journal on Numerical Analysis 13 (1976) 76–83.

[128] M.H. Loke, T. Dahlin, A comparison of the Gauss–Newton and quasi-Newton methods in resistivity imaging inversion, Journal of Applied Geophysics 49 (3) (2002) 149–162.

[129] A.K. Louis, P. Maass, A mollifier method for linear operator equations of the first kind, Inverse Problems 6 (3) (1990) 427.

[130] D.G. Luenberger, Optimization by Vector Space Methods, John Wiley and Sons, NY, 1969.

[131] J.K. MacCarthy, B. Borchers, R.C. Aster, Efficient stochastic estimation of the model resolution matrix diagonal and generalized cross-validation for large geophysical inverse problems, Journal of Geophysical Research: Solid Earth 116 (2011) B10304.

[132] D. Mackenzie, Compressed sensing makes every pixel count, in: What's Happening in the Mathematical Sciences, American Mathematical Society, 2009.

[133] G.I. Marcuk, Adjoint Equations and Analysis of Complex Systems, Springer, Dordrecht, London, 2011, OCLC: 945924765.

[134] W.H. Marlow, Mathematics for Operations Research, Dover, Mineola, NY, 1993.

[135] P.J. McCarthy, Direct analytic model of the L-curve for Tikhonov regularization parameter selection, Inverse Problems 19 (2003) 643–663.

[136] P.R. McGillivray, D. Oldenburg, Methods for calculating Frechet derivatives and sensitivities for the non-linear inverse problem: A comparative study, Geophysical Prospecting 38 (1990) 499–524.

[137] W. Menke, Geophysical Data Analysis: Discrete Inverse Theory, 4th ed., Academic Press, 2018.

[138] C.D. Meyer, Matrix Analysis and Applied Linear Algebra, SIAM, Philadelphia, 2000.

[139] E.H. Moore, On the reciprocal of the general algebraic matrix, Bulletin of the American Mathematical Society 26 (1920) 394–395.

[140] J.J. More, B.S. Garbow, K.E. Hillstrom, User Guide for MINPACK-1, Technical Report ANL–80–74, Argonne National Laboratory, 1980.

[141] V.A. Morozov, Methods for Solving Incorrectly Posed Problems, Springer-Verlag, NY, 1984.

[142] R.H. Myers, Classical and Modern Regression with Applications, PWS Kent, Boston, 1990.

[143] F. Natterer, The Mathematics of Computerized Tomography, SIAM, Philadelphia, 2001.

[144] F. Natterer, F. Wübbeling, Mathematical Methods in Image Reconstruction, SIAM, Philadelphia, 2001.

[145] D. Needell, R. Ward, N. Srebro, Stochastic gradient descent, weighted sampling, and the randomized Kaczmarz algorithm, in: Advances in Neural Information Processing Systems, 2014, pp. 1017–1025.

[146] Y. Nesterov, A method of solving a convex programming problem with convergence rate O (1/k2), Soviet Mathematics Doklady 27 (1983) 372–376.

[147] A. Neumaier, Solving ill-conditioned and singular linear systems: A tutorial on regularization, SIAM Review 40 (3) (1998) 636–666.

[148] R. Neupauer, B. Borchers, J.L. Wilson, Comparison of inverse methods for reconstructing the release history of a groundwater contamination source, Water Resources Research 36 (9) (2000) 2469–2475.

[149] I. Newton, The Principia, Mathematical Principles of Natural Philosophy, University of California Press, Berkeley, 1999.

[150] J. Nocedal, S.J. Wright, Numerical Optimization, 2nd ed., Springer-Verlag, NY, 2006.

[151] G. Nolet, Solving or resolving inadequate and noisy tomographic systems, Journal of Computational Physics 61 (3) (1985) 463–482.

[152] G. Nolet (Ed.), Seismic Tomography with Applications in Global Seismology and Exploration Geophysics, D. Reidel, Boston, 1987.

[153] S.J. Osher, R.P. Fedkiw, Level Set Methods and Dynamic Implicit Surfaces, Springer-Verlag, NY, 2002.

[154] C.C. Paige, M.A. Saunders, Algorithm 583 LSQR: Sparse linear equations and least-squares problems, ACM Transactions on Mathematical Software 8 (2) (1982) 195–209.

[155] C.C. Paige, M.A. Saunders LSQR, An algorithm for sparse linear equations and sparse least squares, ACM Transactions on Mathematical Software 8 (1) (1982) 43–71.

[156] N. Parikh, S.P. Boyd, Proximal algorithms, Foundations and Trends in Optimization 1 (3) (2014) 127–239.

[157] R.L. Parker, Understanding inverse theory, Annual Review of Earth and Planetary Sciences 5 (1) (1977) 35–64.

[158] R.L. Parker, A theory of ideal bodies for seamount magnetism, Journal of Geophysical Research 96 (B10) (1991) 16101–16112.

[159] R.L. Parker, Geophysical Inverse Theory, Princeton University Press, Princeton, NJ, 1994.

[160] R.L. Parker, M.K. McNutt, Statistics for the one-norm misfit measure, Journal of Geophysical Research 85 (1980) 4429–4430.

[161] R.L. Parker, M.A. Zumberge, An analysis of geophysical experiments to test Newton's law of gravity, Nature 342 (1989) 29–32.

[162] R. Penrose, A generalized inverse for matrices, Mathematical Proceedings of the Cambridge Philosophical Society 51 (1955) 406–413.

[163] F.M. Phillips, D.C. Argento, G. Balco, M.W. Caffee, J. Clem, T.J. Dunai, R. Finkel, B. Goehring, J.C. Gosse, A.M. Hudson, A.J.T. Jull, M.A. Kelly, M. Kurz, D. Lal, N. Lifton, S.M. Marrero, K. Nishiizumi, R.C. Reedy, J. Schaefer, J.O.H. Stone, T. Swanson, M.G. Zreda, The CRONUS-Earth project: A synthesis, Quaternary Geochronology 31 (2016) 119–154.

[164] F.P. Pijpers, M.J. Thompson, Faster formulations of the optimally localized averages method for helioseismic inversions, Astronomy and Astrophysics 262 (September 1992) L33–L36.

[165] F.P. Pijpers, M.J. Thompson, The SOLA method for helioseismic inversion, Astronomy and Astrophysics 281 (January 1994) 231–240.

[166] B.T. Polyak, Some methods of speeding up the convergence of iteration methods, USSR Computational Mathematics and Mathematical Physics 4 (5) (1964) 1–17.

[167] S. Portnoy, R. Koenker, The Gaussian hare and the Laplacian tortoise: Computability of squared-error versus absolute-error estimators, Statistical Science 12 (1997) 279–296.

[168] M.B. Priestley, Spectral Analysis and Time Series, Academic Press, London, 1983.

[169] J. Pujol, The Backus–Gilbert method and their minimum-norm solution, Geophysics 78 (3) (2013) W9–W30.

[170] W. Rison, R.J. Thomas, P.R. Krehbiel, T. Hamlin, J. Harlin, A GPS-based three-dimensional lightning mapping system: Initial observations in central New Mexico, Geophysical Research Letters 26 (23) (1999) 3573–3576.

[171] C.D. Rodgers, Inverse Methods for Atmospheric Sounding: Theory and Practice, Word Scientific Publishing, Singapore, 2000.

[172] L.I. Rudin, S. Osher, E. Fatemi, Nonlinear total variation based noise removal algorithms, Physics D 60 (1992) 259–268.

[173] W. Rudin, Real and Complex Analysis, McGraw-Hill, NY, 1987.

[174] Y. Saad, Iterative Methods for Sparse Linear Systems, 2nd ed., SIAM, Philadelphia, 2003.

[175] L.J. Savage, The Foundation of Statistics, Dover, Mineola, NY, 1972.

[176] J. Scales, M. Smith, Introductory Geophysical Inverse Theory (draft), http://cdn.preterhuman.net/texts/science_and_technology/physics, 1997.

[177] J.A. Scales, A. Gersztenkorn, S. Treitel, Fast lp solution of large, sparse, linear systems: Application to seismic travel time tomography, Journal of Computational Physics 75 (2) (1988) 314–333.

[178] S.R. Searle, Matrix Algebra Useful for Statistics, Wiley-Interscience, NY, 2006.

[179] M.K. Sen, P.L. Stoffa, Global Optimization Methods in Geophysical Inversion, 2nd ed., Cambridge University Press, NY, 2013.

[180] C.B. Shaw Jr., Improvement of the resolution of an instrument by numerical solution of an integral equation, Journal of Mathematical Analysis and Applications 37 (1972) 83–112.

[181] J.R. Shewchuk, An Introduction to the Conjugate Gradient Method without the Agonizing Pain, Technical Report, School of Computer Science, Carnegie Mellon University, 1994, https://www.cs.cmu.edu/~quake-papers/painless-conjugate-gradient.pdf.

[182] Z. Sirkes, E. Tziperman, Finite difference of adjoint or adjoint of finite difference? Monthly Weather Review 125 (12) (1997) 3373–3378.

[183] D.S. Sivia, J. Skilling, Data Analysis, A Bayesian Tutorial, Oxford University Press, NY, 2006.

[184] T.H. Skaggs, Z.J. Kabala, Recovering the release history of a groundwater contaminant, Water Resources Research 30 (1) (1994) 71–79.

[185] T.H. Skaggs, Z.J. Kabala, Recovering the history of a groundwater contaminant plume: Method of quasi-reversibility, Water Resources Research 31 (1995) 2669–2673.

[186] N. Sleep, K. Fujita, Principles of Geophysics, Wiley, 1997.

[187] A. van der Sluis, H.A. van der Vorst, Numerical solution of large, sparse linear algebraic systems arising from tomographic problems, in: G. Nolet (Ed.), Seismic Tomography, D. Reidel, 1987, pp. 49–83.

[188] W. Spakman, G. Nolet, Imaging algorithms, accuracy and resolution in delay time tomography, in: M.J.R. Wortel, N.J. Vlaar, G. Nolet, S.A.P.L. Cloetingh (Eds.), Mathematical Geophysics: A Survey of Recent Developments In Seismology and Geodynamics, D. Reidel, Dordrecht, 1988, pp. 155–187.

[189] S. Stahl, The evolution of the normal distribution, Mathematics Magazine 79 (2) (2006) 96–113.

[190] P.B. Stark, R.L. Parker, Velocity bounds from statistical estimates of $\tau(p)$ and $x(p)$, Journal of Geophysical Research 92 (B3) (1987) 2713–2719.

[191] P.B. Stark, R.L. Parker, Correction to 'Velocity bounds from statistical estimates of $\tau(p)$ and $x(p)$', Journal of Geophysical Research 93 (1988) 13821–13822.

[192] P.B. Stark, R.L. Parker, Bounded-variable least-squares: An algorithm and applications, Computational Statistics 10 (2) (1995) 129–141.

[193] G.W. Stewart, On the early history of the singular value decomposition, SIAM Review 35 (1993) 551–566.

[194] G. Strang, Linear Algebra and Its Applications, 4th ed., Cengage Learning, 2006.

[195] G. Strang, Computational Science and Engineering, vol. 791, Wellesley-Cambridge Press, Wellesley, 2007.

[196] G. Strang, K. Borre, Linear Algebra, Geodesy, and GPS, Wellesley-Cambridge Press, Wellesley, MA, 1997.

[197] T. Strohmer, R. Vershynin, A randomized Kaczmarz algorithm with exponential convergence, Journal of Fourier Analysis and Applications 15 (2) (April 2009) 262.

[198] N.Z. Sun, Inverse Problems in Groundwater Modeling, Kluwer Academic Publishers, Boston, 1984.

[199] A. Tarantola, Inverse Problem Theory and Methods for Model Parameter Estimation, SIAM, Philadelphia, 2004.

[200] A. Tarantola, B. Valette, Inverse problems = quest for information, Journal of Geophysics 50 (3) (1982) 159–170.

[201] W.M. Telford, W.M. Telford, L.P. Geldart, R.E. Sheriff, Applied Geophysics, Monograph Series, Cambridge University Press, 1990.

[202] R.A. Thisted, Elements of Statistical Computing, Chapman and Hall, NY, 1988.

[203] C. Thurber, Hypocenter velocity structure coupling in local earthquake tomography, Physics of the Earth and Planetary Interiors 75 (1–3) (1992) 55–62.

[204] C. Thurber, K. Aki, Three-dimensional seismic imaging, Annual Review of Earth and Planetary Sciences 15 (1987) 115–139.

[205] C. Thurber, J. Ritsema, Theory and observations – seismic tomography and inverse methods, in: Treatise on Geophysics, Elsevier, Amsterdam, 2007, pp. 323–360.

[206] A.N. Tikhonov, V.Y. Arsenin, Solutions of Ill-Posed Problems, Halsted Press, NY, 1977.

[207] A.N. Tikhonov, A.V. Goncharsky (Eds.), Ill-Posed Problems in the Natural Sciences, MIR Publishers, Moscow, 1987.

[208] J. Trampert, J.-J. Leveque, Simultaneous iterative reconstruction technique: Physical interpretation based on the generalized least squares solution, Journal of Geophysical Research 95 (B8) (1990) 12553–12559.

[209] L.N. Trefethen, D. Bau, Numerical Linear Algebra, SIAM, Philadelphia, 1997.

[210] J. Tromp, C. Tape, Q. Liu, Seismic tomography, adjoint methods, time reversal, and banana–doughnut kernels, Geophysical Journal International 160 (2005) 195–216.

[211] S. Twomey, Introduction to the Mathematics of Inversion in Remote Sensing and Indirect Measurements, Dover, Mineola, NY, 1996.

[212] J. Um, C. Thurber, A fast algorithm for 2-point seismic ray tracing, Bulletin of the Seismological Society of America 77 (3) (1987) 972–986.

[213] A. Valentine, M. Sambridge, Optimal regularisation for a class of linear inverse problem, Geophysical Journal International ggy303 (2018).

[214] E. van den Berg, M.P. Friedlander, Probing the pareto frontier for basis pursuit solutions, SIAM Journal on Scientific Computing 31 (2) (2008) 890–912.

[215] C.R. Vogel, Non-convergence of the L-curve regularization parameter selection method, Inverse Problems 12 (1996) 535–547.

[216] C.R. Vogel, Computational Methods for Inverse Problems, SIAM, Philadelphia, 2002.

[217] G. Wahba, Spline Models for Observational Data, SIAM, Philadelphia, 1990.

[218] D.S. Watkins, Fundamentals of Matrix Computations, 3rd ed., Wiley, 2010.

[219] G.A. Watson, Approximation in normed linear spaces, Journal of Computational and Applied Mathematics 121 (1–2) (2000) 1–36.

[220] G.M. Wing, A Primer on Integral Equations of the First Kind: The Problem of Deconvolution and Unfolding, SIAM, Philadelphia, 1991.

[221] C. Zaroli, Global seismic tomography using Backus–Gilbert inversion, Geophysical Journal International 207 (2) (2016) 876–888.

[222] C. Zaroli, P. Koelemeijer, S. Lambotte, Toward seeing the Earth's interior through unbiased tomographic lenses, Geophysical Research Letters 44 (22) (2017) 11399–11408.

[223] H. Zhang, C.H. Thurber, Estimating the model resolution matrix for large seismic tomography problems based on Lanczos bidiagonalization with partial reorthogonalization, Geophysical Journal International 170 (1) (2007) 337–345.

[224] J. Zhang, G.A. McMechan, Estimation of resolution and covariance for large matrix inversions, Geophysical Journal International 121 (2) (1995) 409–426.

[225] Y.-Q. Zhang, W.-K. Lu, 2d and 3d prestack seismic data regularization using an accelerated sparse time-invariant Radon transform, Geophysics 79 (5) (2014) V165–V177.

INDEX